216
Topics in Current Chemistry

Springer

Berlin
Heidelberg
New York
Barcelona
Hong Kong
London
Milan
Paris
Singapore
Tokyo

Stereoselective Heterocyclic Synthesis III

Volume Editor: Peter Metz

With contributions by
A. Hassner, I. N. N. Namboothiri,
U. Nubbemeyer, S. D. Rychnovsky, C. J. Sinz,
P. ten Holte, B. Zwanenburg

Springer

The series *Topics in Current Chemistry* presents critical reviews of the present and future trends in modern chemical research. The scope of coverage includes all areas of chemical science including the interfaces with related disciplines such as biology, medicine and materials science. The goal of each thematic volume is to give the nonspecialist reader, whether at the university or in industry, a comprehensive overview of an area where new insights are emerging that are of interest to a larger scientific audience.

As a rule, contributions are specially commissioned. The editors and publishers will, however, always be pleased to receive suggestions and supplementary information. Papers are accepted for *Topics in Current Chemistry* in English.

In references *Topics in Current Chemistry* is abbreviated Top. Curr. Chem. and is cited as a journal.

Springer WWW home page: http://www.springer.de
Visit the TCC home page at http:/link.springer.de/series/tcc/
or http://link.springer-ny.com/series/tcc/

ISSN 0340-1022
ISBN 3-540-41763-X
Springer-Verlag Berlin Heidelberg New York

Library of Congress Catalog Card Number 74-644622

Springer-Verlag Berlin Heidelberg New York
a member of BertelsmannSpringer Science+Business Media GmbH

Printed in Germany

Cover design: Friedhelm Steinen-Broo, Barcelona; MEDIO, Berlin
Typesetting: Fotosatz-Service Köhler GmbH, 97084 Würzburg

SPIN: 10779693 02/3020 ra – 5 4 3 2 1 0 – Printed on acid-free paper

Volume Editor

Prof. Dr. Peter Metz
Institut für Organische Chemie
Technische Universität Dresden
Mommsenstraße 13
01062 Dresden, Germany
E-mail: metz@coch01.chm.tu-dresden.de

Editorial Board

Topics in Current Chemistry
Now Also Available Electronically

For all customers with a standing order for Topics in Current Chemistry we offer the electronic form via LINK free of charge. Please contact your librarian who can receive a password for free access to the full articles by registration at:

http://link.springer.de/orders/index.htm

If you do not have a standing order you can nevertheless browse through the table of contents of the volumes and the abstracts of each article at:

http://link.springer.de/series/tcc
http://link.springer-ny.com/series/tcc

There you will also find information about the

- Editorial Board
- Aims and Scope
- Instructions for Authors

Preface

Keeping up with the advances in modern heterocyclic chemistry is essential for many of our colleagues in academia and industry. It is the aim of this series on "Stereoselective Heterocyclic Synthesis" to assist the chemical community in this respect by presenting a selection of exciting recent developments. As it was for the first two volumes (1997), the stereoselective synthesis of – or with the aid of – heterocycles is the common motif for all the chapters in this third volume.

I am very glad that again leading researchers in this area have contributed highly stimulating accounts with up-to-date coverage. "Stereoselective Heterocyclic Synthesis III" features chapters on "*Stereoselective Intramolecular 1,3-Dipolar Cycloadditions*" by *I.N.N. Nambothiiri* and *A. Hassner* giving an in depth survey of the generation and synthetic application of valuable 1,3-dipoles, "*4-Acetoxy- and 4-Cyano-1,3-dioxanes in Synthesis*" by *C.J. Sinz* and *S.D. Rychnovsky* presenting a comprehensive summary of the utility of the versatile title compounds in natural products synthesis, "*The Synthetic Potential of Three-Membered Ring Aza-Heterocycles*" by *B. Zwanenburg* and *P. ten Holte* highlighting the fascinating chemistry of aziridine and azirine carboxylic esters, and "*Synthesis of Medium-Sized Ring Lactams*" by *U. Nubbemeyer* discussing a wide range of modern strategies for the stereoselective preparation of these important heterocycles.

I wish to express my thanks to all contributors for their dedicated effort and to Mrs. Kollmar-Thoni, Springer-Verlag, for her continuous support.

Dresden, March 2001 — Peter Metz

Contents

Stereoselective Intramolecular 1,3-Dipolar Cycloadditions
I.N.N. Namboothiri, A. Hassner 1

4-Acetoxy- and 4-Cyano-1,3-dioxanes in Synthesis
C.J. Sinz, S.D. Rychnovsky 51

The Synthetic Potential of Three-Membered Ring Aza-Heterocycles
B. Zwanenburg, P. ten Holte 93

Synthesis of Medium-Sized Ring Lactams
U. Nubbemeyer 125

Author Index Volumes 201–216 197

Contents of Volume 189

Stereoselective Heterocyclic Synthesis I

Volume Editors: P. Metz

ISBN 3-540-62678-6

Hetero Diels-Alder Reactions in Organic Chemistry
L. F. Tietze, G. Kettschau

Tandem Processes of Metallo Carbenoids for the Synthesis of Azapolycycles
A. Pawda

Contents of Volume 190

Stereoselective Heterocyclic Synthesis I I

Volume Editors: P. Metz

ISBN 3-540-62700-6

Using Ring-Opening Reactions of Oxabicyclic Compounds as a Strategy in Organic Synthesis
P. Chiu, M. Lautens

The Nucleophilic Addition/Ring Closure (NARC) Sequence for the Stereocontrolled Synthesis of Heterocycles
P. Perlmutter

Chiral Acetylenic Sulfoxides and Related Compounds in Organic Synthesis
A.W.M. Lee, W.H. Chan

***N*-Sulfonyl Imines – Useful Synthons in Stereoselective Organic Synthesis**
S.M. Weinreb

Stereoselective Intramolecular 1,3-Dipolar Cycloadditions

Irishi N. N. Namboothiri[1] · Alfred Hassner[2]

[1] Department of Chemistry, Indian Institute of Technology, Bombay, Mumbai 400 076, India
E-mail: irishi@chem.iitb.ac.in

[2] Department of Chemistry, Bar-Ilan University, Ramat Gan 52900, Israel
E-mail: hassna@mail.biu.ac.il

An in depth account of intramolecular 1,3-dipolar cycloadditions involving dipoles such as nitrile oxides, silyl nitronates, *H*-nitrones, azides, and nitrilimines is presented with particular emphasis on the stereochemistry during the cycloaddition. Various methods employed for the generation of the dipoles and their applications to stereoselective synthesis are also discussed.

Keywords: Intramolecular 1,3-dipolar cycloadditions, Stereoselectivity, Nitrile oxides, Silyl nitronates, Oximes, *H*-Nitrones, Azides, Nitrilimines

1 Introduction 2

2 Intramolecular Nitrile Oxide Cycloaddition (INOC) 2

2.1 Some General Factors Influencing the Cycloaddition 3
2.2 Nitrile Oxides from Oximes 4
2.2.1 α-Bromoaldoximes 8
2.3 Nitrile Oxides from Nitro Alkanes 13

3 Intramolecular Silyl Nitronate-Olefin Cycloaddition (ISOC) 21

3.1 INOC vs ISOC 22
3.2 One-Pot Reactions Involving Michael Addition and ISOC 23
3.3 Sequential 1,3-Dipolar Cycloadditions 27
3.4 Miscellaneous 28

4 Intramolecular Oxime-Olefin Cycloaddition (IOOC) 30

4.1 (*H*)-Nitrones from Oximes 30
4.1.1 Pyrrolidines 31
4.1.2 Carbocycles 34
4.1.3 Fused Rings (Pyrrolizidines, Indolizidines, and Quinolizidines) . . . 35
4.1.4 Tetrahydrofurans 36
4.2 Role of Substituents in the Oxime-(*H*)-Nitrone Isomerization 36
4.3 Miscellaneous 37

5 Intramolecular Azide-Olefin Cycloaddition (IAOC) 39

5.1 Azides to Triazolines, Imines, and Aziridines 39
5.2 α-Azido Ethers to Δ^3-Oxazolines 40
5.3 IAOC in the Synthesis of Bioactive Materials 42

6 Intramolecular Nitrilimine Cycloaddition (INIC) 43

6.1 Generation and Reactivity of Nitrilimines 43
6.2 Intramolecular Cycloaddition 45

7 References 46

1 Introduction

Cycloaddition between a 1,3-dipole and an olefin [1], which takes place in a [4π + 2π] fashion and is analogous to the Diels-Alder reaction, has acquired a prominent place in organic synthesis [2]. Such cycloadditions have been traditionally employed for the construction of five-membered heterocycles. However, the versatility of 1,3-dipoles and dipolarophiles, the regio- and stereoselectivity during cycloaddition [3], and the scope for further transformation of the cycloadducts to a variety of multifunctional molecules [4] have elevated this class of reactions to an enviable methodology not only for the construction of functionalized normal ring carbocycles and heterocycles but also for the synthesis of complex natural products [5]. In this context, intramolecular 1,3-dipolar cycloadditions [6] are particularly well known as they involve concomitant formation of two rings fused to each other, from acyclic precursors in practically one single step. Since dipole and dipolarophile are in close proximity, such reactions are entropically favored and often proceed with a high degree of regio- and stereoselectivity [7].

Among the many recent applications to natural products, syntheses of pyrrolizidine and indolizidine alkaloids that take advantage of the 1,3-dipolar cycloaddition methodology have been reviewed [8]. The regio- and stereochemistry [9] as well as synthetic applications [10] of nitrile oxide cycloadditions have also been discussed.

This review covers primarily the results of intramolecular 1,3-dipolar cycloadditions reported by us in the past 15 years in perspective to closely related work by others.

2 Intramolecular Nitrile Oxide Cycloaddition (INOC)

Nitrile oxides are usually prepared via halogenation and dehydrohalogenation of aldoximes [11] or via dehydration of primary nitro alkanes (Scheme 1) [12]. However, it is important to note that nitrile oxides are relatively unstable and are prone to dimerization or polymerization, especially upon heating. 1,3-Dipolar cycloaddition of a nitrile oxide with a suitable olefin generates an isoxazoline ring which is a versatile synthetic intermediate in that it provides easy access to γ-amino alcohols, β-hydroxy ketones, β-hydroxy nitriles, unsaturated oximes, and a host of other multifunctional molecules (Scheme 1) [5a]. Particularly for the formation of β-hydroxy ketones, nitrile oxide-olefin cycloaddition serve as an alternative to the Aldol reaction.

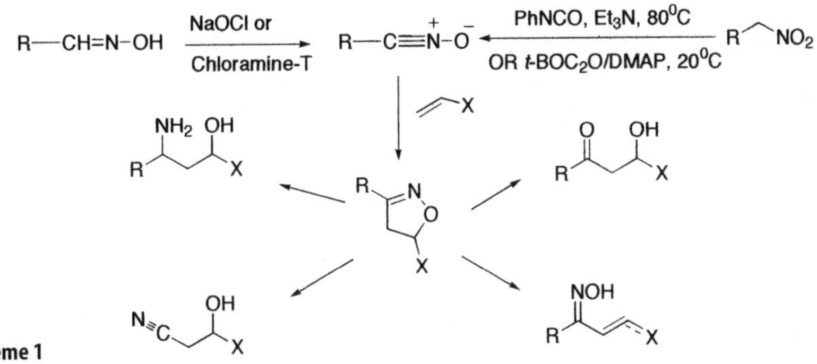

Scheme 1

2.1 Some General Factors Influencing the Cycloaddition

A substituent effect on the rate and stereoselectivity of INOC reaction has been observed (Eq. 1) [13]. Thus, *gem*-dicarboalkoxy and *gem*-dithioalkoxy groups were found to have profound accelerating effect on the cyclization (Entries g and h, Table 1). When C-3 in **2** was monosubstituted, good diastereoselectivity was observed depending on the relative size of the substituents (Ph > Me > CO_2Me).

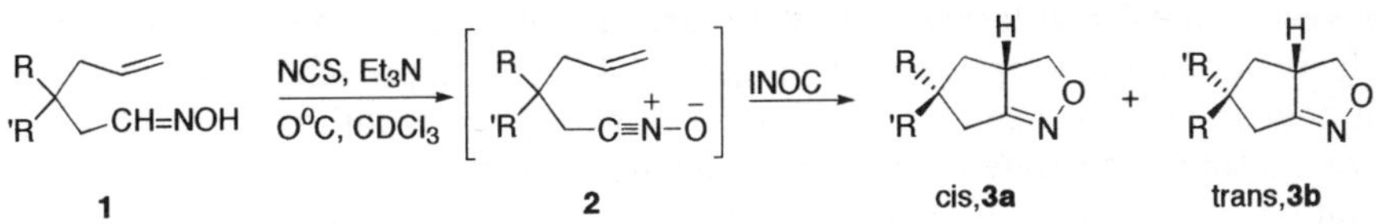

(1)

A dipolarophile bearing an ionic group and an associated counterion provides enhanced selectivity as has been recently demonstrated by Raposo and Wilcox [14]. Cycloaddition of benzonitrile **4** and the uncharged amine **5a** (a chiral phenylmaleimide derivative) in THF or chloroform provides a mixture of cycloadducts **6–9a** in 1:4:4:4 diastereomeric ratio (i.e., 8:5 in favor of the "methyl face" approach of the dipolarophile. The ortho-substituents of the

Table 1

1–3	Substituents	INOC $t_{1/2}$ (min)	INOC k_{rel}	**3a:3b**
a	R = R′ = H	990	1	
b	R = Me, R′ = H	624	1.6	87:13
c	R = R′ = Me	910	1.1	
d	R = R′ = $(CH_2)_4$	338	2.9	
e	R = Ph, R′ = H	248	4	90:10
f	R = CO_2Me, R′ = H	173	5.7	70:30
g	R = R′ = CO_2Me	46	21.5	
h	R = R′ = $S(CH_2)_3S$	<4	>247	

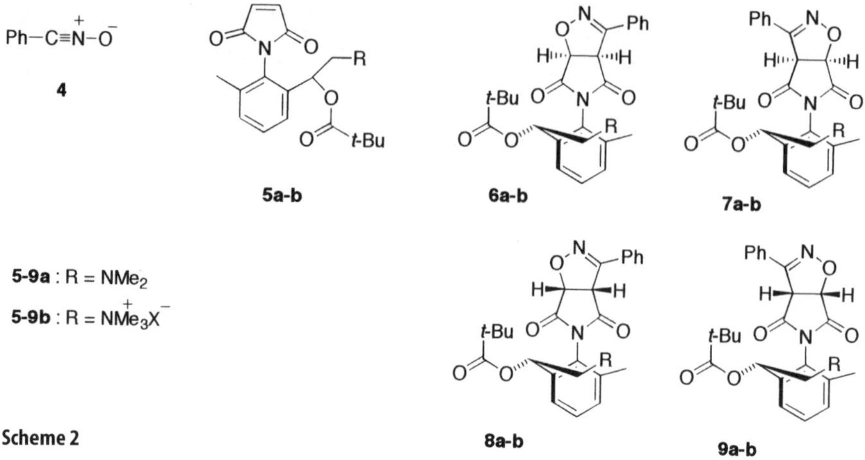

Scheme 2

phenyl ring raise the *N*-Ph rotational barrier to over 28 kcal/mol [14b] resulting in a difference in the two faces of the maleimide, Scheme 2) [14c]. However, when the cycloaddition was performed using the salt **5b**, the face selectivity was reversed for all solvents. The diastereomeric ratio was 0:20:1:2 (i.e., 20:3 in favor of the "salt face" approach product **7b**, chloroform as reaction solvent) revealing the direct influence of the charge in the reaction. Formation of **7b** as the major isomer despite the fact that the 1,3-dipole approaches from the more hindered side has been rationalized in terms of the hypothesis that **7b** arises from a transition state (TS) wherein the dipole is oriented to optimize electrostatic attraction to the ion pair [14d]. This is based on the original proposition by Huisgen [15] that the mechanism of 1,3-dipolar cycloadditions lies between a completely synchronous process and an alternative biradical or zwitterionic process, both the possibilities getting strongly affected by charged groups in the dipolarophile.

2.2 Nitrile Oxides from Oximes

Aldoximes can be oxidatively dehydrogenated to nitrile oxides using a variety of oxidants such as lead tetraacetate [16a], alkali hypohalites [11a], NBS in DMF followed by base treatment [16b], chloramine-T [11b], 1-chlorobenzotriazole [16c], mercuric acetate [16d], etc. However, we employed either NaOCl or chloramine-T for most of our INOC reactions. For instance, a piperidine ring fused to an isoxazoline as in **14** was constructed using the INOC methodology (Scheme 3) [17]. Monoalkylation of *N*-tosylallylamine **10** with the bromoacetal **11** provided the unsaturated acetal **12a**, which was hydrolyzed to the aldehyde **12b**. Oximation of **12b** followed by treatment of the oxime **13** with chloramine-T provided the isoxazoline **14** via spontaneous ring closure of a nitrile oxide intermediate.

The regioisomer **18** of isoxazoline **14** was also synthesized by an INOC reaction (Eq. 2). Homoallylamine **16** prepared by displacement of 4-bromo-1-butene

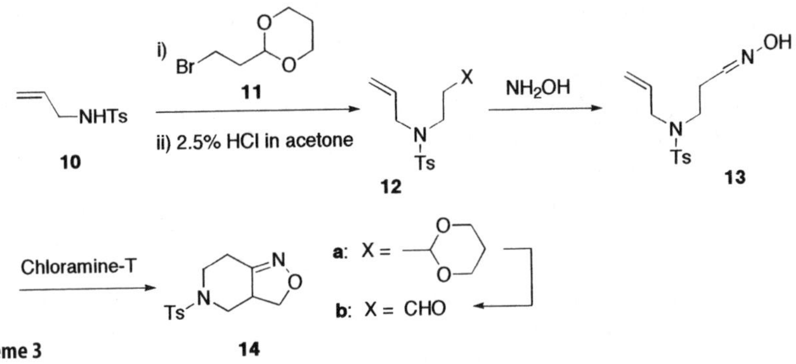

Scheme 3

(15) with *p*-toluenesulfonamide under phase transfer conditions followed by conversion into the unsaturated ester **17**, DIBAL reduction of **17**, oximation, and ring closure via a nitrile oxide provided **18** [17].

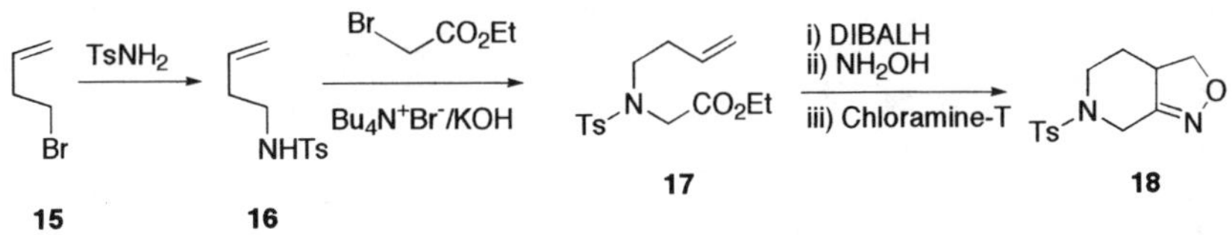

(2)

The versatility of the INOC reaction is evident from the synthesis of tetrahydrofurans fused to an isoxazoline **22a–f** (Eq. 3) [18]. α-Allyloxyaldoximes **21**, formed by the reduction of β-nitrostyrenes **19** with $SnCl_2 \cdot 2H_2O$ in the presence of an unsaturated alcohol **20**, are transformed to isoxazolines **22** in high yield on treatment with NaOCl via stereoselective ring closure of a nitrile oxide intermediate (Table 2).

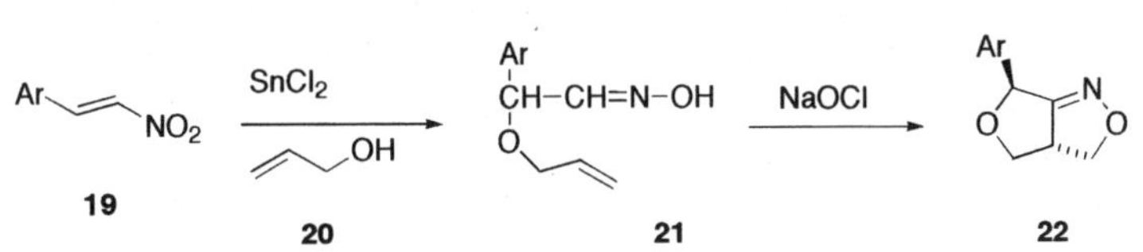

(3)

Table 2

	Ar	% Yield of **21**	*E*:*Z*	% Yield of **22**	*trans*:*cis*
a	Ph	59	6:1	86	4:1
b	4-MePh	53	2:1	90	4:1
c	4-MeOPh	68	2:1	87	4:1
d	3,4-(OCH_2)Ph	69	3:2	88	4:1
e	3,4-$(MeO)_2$Ph	54	10:1	84	4:1
f	3,4,5-$(MeO)_3$Ph	64	5:4	82	3:1

It is important to note that the *E* isomer of the oxime predominates over the *Z* isomer in all the cases **21 a–f** and the preferred stereochemistry for the isoxazolines **22 a–f** is *trans.*

Chiral tricyclic fused pyrrolidines **29 a–c** and piperidines **29 d–g** have been synthesized starting from L-serine, L-threonine, and L-cysteine taking advantage of the INOC strategy (Scheme 4) [19]. L-Serine (**23 a**) and L-threonine (**23 b**) were protected as stable oxazolidin-2-ones **24 a** and **24 b**, respectively. Analogously, L-cysteine **23 c** was converted to thiazolidin-2-one **24 c**. Subsequent *N*-allylation or homoallylation, DIBALH reduction, and oximation afforded the ene-oximes, **27 a–g**. Conversion of ene-oximes **27 a–g** to the desired key intermediates, nitrile oxides **28 a–g**, provided the isoxazolines **29 a–g**. While fused pyrrolidines **29 a–c** were formed in poor yield (due to dimerization of nitrile oxides) and with moderate stereoselectivity (as a mixture of *cis* (major) and *trans* (minor) isomers), corresponding piperidines **29 d–g** were formed in good yield and excellent stereoselectivity (as exclusively *trans* isomers, see Table 3).

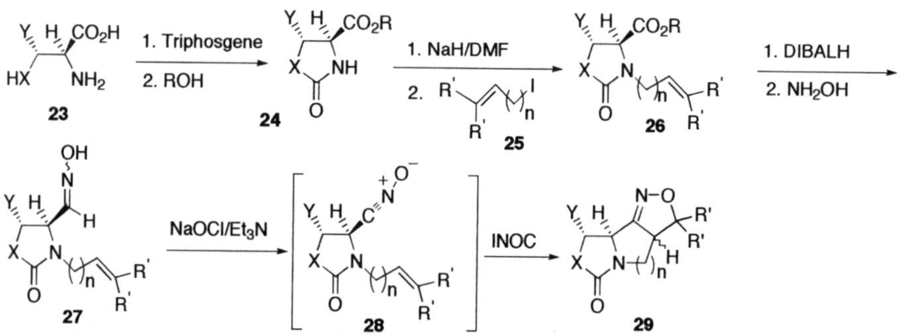

Scheme 4

The high degree of stereoselectivity in the piperidine ring formation has been attributed to the nitrile oxide adopting a chair-like TS as opposed to a flexible TS in the pyrrolidine ring formation. Subsequent reductive cleavage (Ra-Ni) of the tricyclic pyrrolidines provided inseparable mixture of bicyclic fused pyrrolidines bearing hydroxy and keto substituents. Analogous reduction of fused piperidines using Ra-Ni and $LiAlH_4$ was more stereoselective.

Table 3

23–29	X	Y	n	R′	% Yield of 29, *cis*	% Yield of 29, *trans*
a	O	H	1	H	30	12
b	O	Me	1	H	24	–
c	S	H	1	H	32	7
d	O	H	2	H	46	–
e	O	H	2	Me	52	–
f	S	H	2	H	54	–
g	S	H	2	Me	68	–

A nitrile oxide generated from a sugar derived aldoxime **30** underwent INOC reaction to the chiral pyranoisoxazoline **31** (Eq. 4) [20]. Reductive cleavage of isoxazoline **31** followed by acetylation provided the tetrasubstituted pyran **32.**

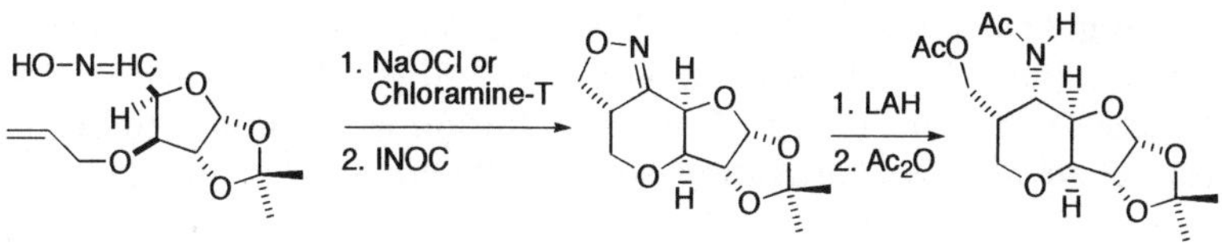

(4)

In the seven-step stereoselective total synthesis of ptilocaulin **44** [21], a potent antileukemic and antimicrobial agent isolated [22] from marine sponges, the oxime **36** was treated with NaOCl providing the tricyclic isoxazoline **38** in 89% yield without isolation of the nitrile oxide intermediate **37** (Scheme 5) [23]. Isoxazoline **38** was obtained as a mixture of four diastereomers and their ratio was

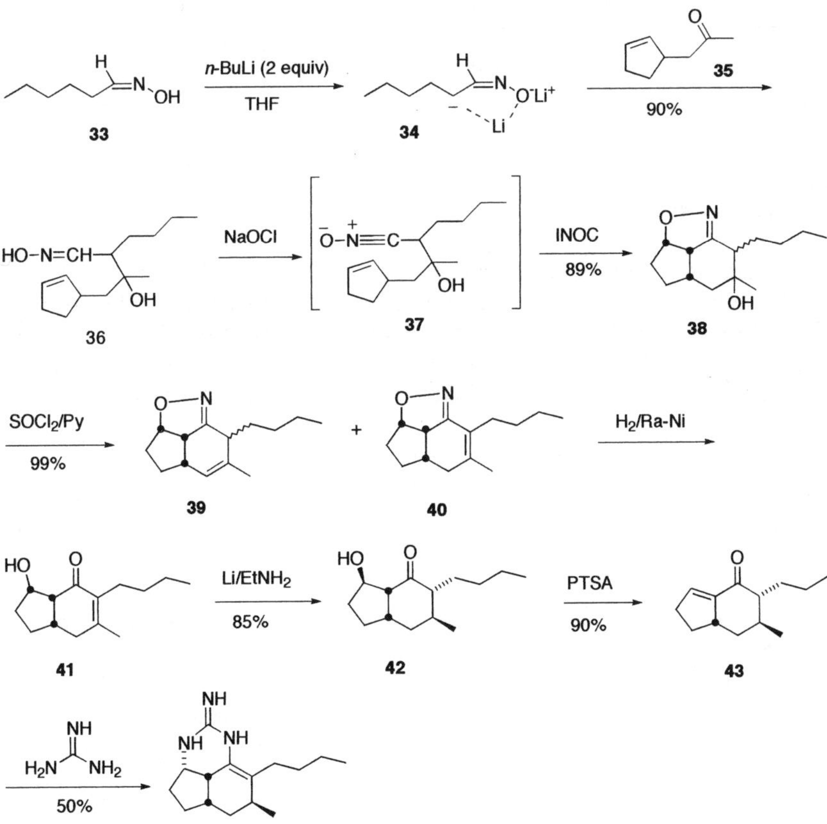

Scheme 5

identical to the isomer ratio of starting oxime **36**, suggesting that the INOC reaction had indeed occurred stereospecifically. The isomers were solely due to various configurations at C-7 and C-8, indicating that the three stereocenters formed during cycloaddition were homogeneous *cis,cis*.

A highly stereocontrolled synthesis of (+) testosterone **49** was accomplished wherein the A/B ring system was constructed via INOC reaction of **47** to isoxazoline **48** (Scheme 6) [24]. The cycloaddition was assumed to be taking place via a chair-like TS **47** providing isomerically pure isoxazoline **48** in 87% yield.

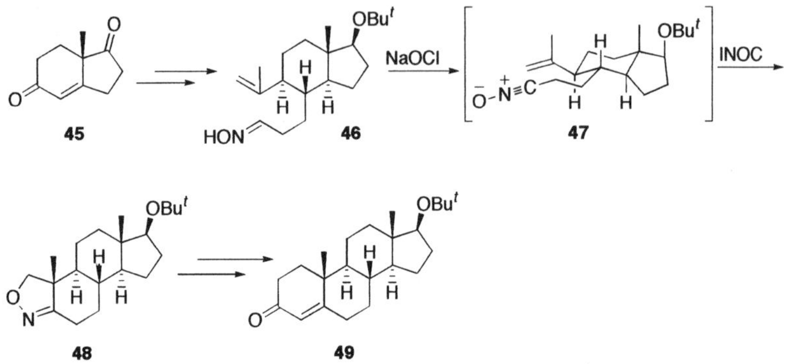

Scheme 6

2.2.1
α-Bromoaldoximes

α-Bromoaldoximes **55** are useful intermediates [25], particularly for the synthesis of vinylnitroso compounds (e.g., **53**) (Scheme 7) [26]. Nucleophilic displacement on **55** was considered as a general method to synthesize INOC precursors. However, unlike *α*-bromoketoximes, which can be prepared via bromination of ketones and oximation, *α*-bromoaldoximes are difficult to obtain partly because

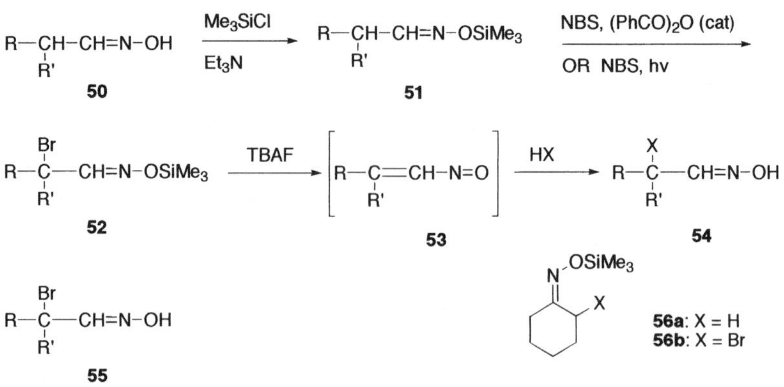

Scheme 7

Table 4. α-Bromination of *O*-trimethylsilyl aldoximes

50–55	a	b	c	d	e	f
R	H	Me	Et	*n*-Bu	Me	Ph
R′	H	H	H	H	Me	H
Yield of **52** (%)	30	95	95	90	100	70

α-bromination of aldehydes is a low yield reaction leading to unstable products [27]. Direct α-bromination of aldoximes (e.g., **50**) with a variety of brominating agents was also not successful. However, smooth bromination of the silylated oximes **51** was readily accomplished [28]. Thus reaction of **51b–e** with NBS proceeded in the presence of benzoyl peroxide in refluxing CCl_4 to produce the brominated products **52b–e** in high yield (see Table 4). Although thermal conditions proved unsuitable for the bromination of **51a** and **51f**, reasonable yields of **52a** and **52f** were obtained by photochemical irradiation. Ketoximes can be transformed into their bromo derivatives (e.g., **56b**) in a similar manner. Treatment of the *N*-trimethylsilyloxy α-bromoaldoxime **52** with F^- ion (TBAF) is best performed in the presence of other nucleophiles resulting in an overall nucleophilic functionalization α to the oximino center and providing **54**. This process appears to take place via the transient intermediacy of unsaturated nitroso compounds **53** [26b, 28].

The above methodology has been extremely useful for the synthesis of a variety of INOC precursors. For instance, treatment of *O*-trimethylsilyl α-bromoaldoximes **52b, e, f** with F^- ion in presence of unsaturated alcohols **57** produces oximino ethers **58** which can be readily oxidized using NaOCl (Scheme 8) [29]. The transient nitrile oxide intermediates formed undergo spontaneous cyclization to fused isoxazolines **59**. The preferred stereoisomer in the formation of the five-membered ring ethers is *trans* whereas in the six-membered ring ethers the *cis* isomer predominates (see Table 5). MM2 calculations helped rationalize the experimentally observed stereoselectivites (see Table 5).

In order to establish the generality of the reaction, the cycloaddition of a number of closely related systems has been investigated. Thus, treatment of **52b**

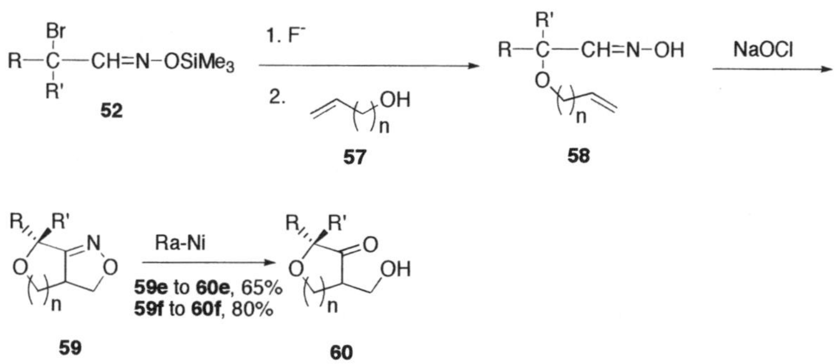

Scheme 8

Table 5. Experimental and MM2 Calculation Results of the INOC Reaction of Unsaturated Oximino Ethers **58**

52	58 to 59	n	R	R′	Isolated (%) Yield of **59**	*trans* kcal	*cis* kcal	ΔE kcal	*trans*:*cis* (Exp)
b	a	1	Me	H	71	28.06	28.44	0.38	2.4:1
e	b	1	Me	Me	95	–	–	–	single
f	c	1	Ph	H	88	37.54	38.50	0.96	>100:1
b	d	2	Me	H	78	20.08	18.53	1.55	<1:100
e	e	2	Me	Me	45	–	–	–	single
f	f	2	Ph	H	40	29.41	29.07	0.34	1:6

(R = Me, R′ = H) with cyclohexenol in the presence of F⁻ ion followed by NaOCl oxidation gave the tricyclic ether **61** in 65% yield (Scheme 9) [29]. The use of propargyl alcohol and propargyl thiol led, via the acetylenic oximes, to fused tetrahydrofuranoisoxazoles **62a** and **62b**, and tetrahydrothiopheno[3,4-*c*]isoxazole **62c**, respectively. Reaction of 1-butyn-4-ol with *O*-trimethylsilyl α-bromoaldoxime **52e** (R = R′ = Me) led to the tetrahydropyranoisoxazole **62d**.

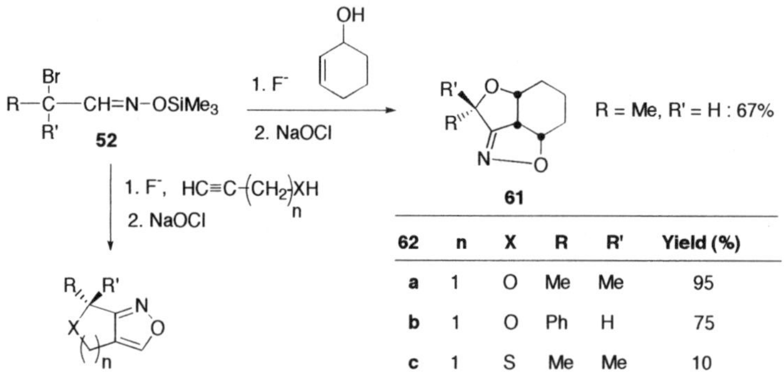

62	n	X	R	R'	Yield (%)
a	1	O	Me	Me	95
b	1	O	Ph	H	75
c	1	S	Me	Me	10
d	2	O	Me	Me	95

Scheme 9

The reaction of the α-bromo aldoxime **52e** (R = R′ = Me) with unsaturated alcohols has been extended to the heterocyclic systems furfuryl alcohols and 2-thiophene methanol [29b]. The furanyl and thiophenyl oximes **63a–c** were treated with NaOCl and the resulting heterocyclic nitrile oxides were found to undergo spontaneous intramolecular dipolar cycloaddition to produce the unsaturated tricyclic isoxazolines **64a–c** in high yield (Eq. 5). In these cases, the heterocyclic ring acts as the dipolarophile with one of the double bonds adding to the nitrile oxide [30].

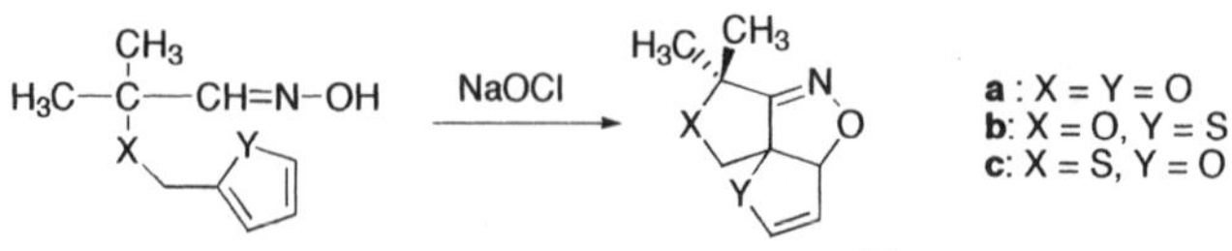

a: X = Y = O
b: X = O, Y = S
c: X = S, Y = O

(5)

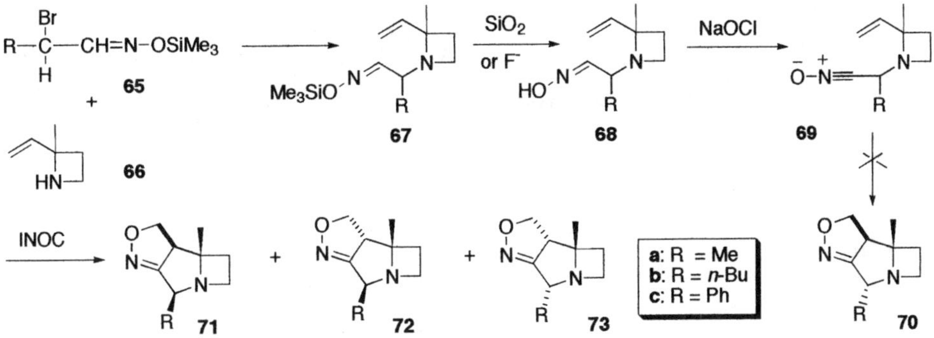

Scheme 10

A nitrile oxide-vinylazetidine system **69** has been studied aimed at assessing the stereochemical influence of a substituent far removed from the double bond (Scheme 10) [31]. Nucleophilic displacement of the halogen in the *O*-silylated α-bromoaldoximes **65** by the azetidine **66** led to the silylated oximes **67**. The latter were smoothly desilylated on chromatography over silica gel or in the presence of F^- ions to produce **68** as a mixture of diastereomers. The unsaturated oximes were converted via nitrile oxides **69** to the cycloaddition products by treatment with NaOCl. The major products were the stereoisomers possessing R and Me groups in a *cis* configuration (i.e., **71** and **72**), in which the *trans-cis* isomer **72** predominates over the all *cis* isomer **71** in a 2:1 ratio. Conspicuously absent in all three cases (R = Me, Bu, and Ph) was the *trans* isomer corresponding to **70**. Examination of molecular models indicated the apparent reason to be steric interaction between the α-oriented R group and the hydrogen at C-4. In the case of **65a** (R = Me) and **65b** (R = Bu) all the three isomers **71**–**73** were present, but **73** was absent when R was a larger substituent (i.e., Ph).

Consistent with the unfavorable interactions is the fact that the dimethyl derivative **74** failed to undergo the intramolecular cycloaddition since in this case one of the Me groups would necessarily interfere in the transition state for cyclization.

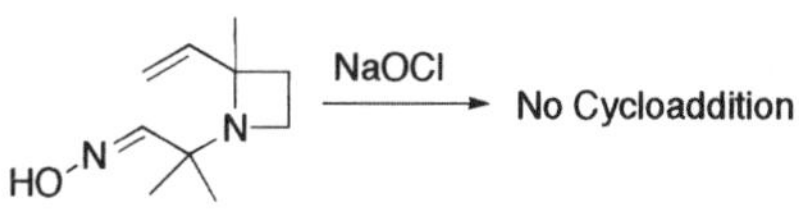

74 (6)

A one pot synthesis of isoxazolines **78a**–**f** involves base mediated 1,4-addition of malonate or alcohol **76** possessing an allylic substituent, conversion of the resulting nitronate to the α-chloroaldoxime (hydroxymoyl chloride **77**) and its subsequent dehydrohalogenation to the nitrile oxide intermediate which cyclizes to isoxazoline **78** (Eq. 7, Table 6) [32].

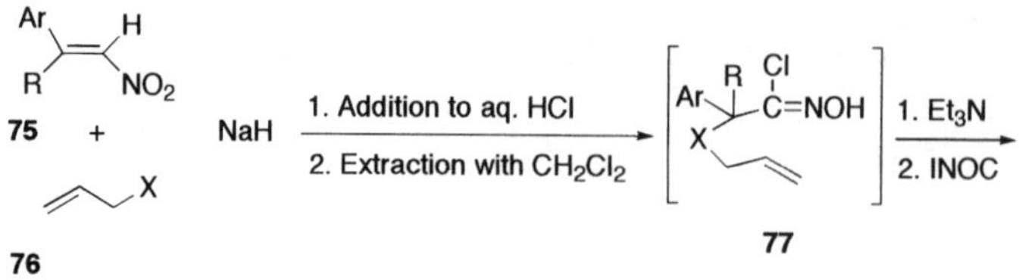

(7)

Table 6

78	Ar	R	X	% Yield of Isoxazoline **78**
a	Ph	H	CH(COEt)	60 (*cis*), 8 (*trans*)
b	4-Tolyl	H	CH(COEt)	75 (*cis*), 10 (*trans*)
c	2-Thienyl	H	CH(COEt)	60 (*cis*), 17 (*trans*)
d	2-Furyl	H	CH(COEt)	61 (*cis*), 24 (*trans*)
e	Ph	Ph	OH	64
f	4-F-Ph	H	OH	61

Similar workup conditions applied to the nitronate arising from 1,4-addition of unsaturated Grignard or Li reagents **80** (M = MgBr provided better yields) to β-Ar or β-hetero-Ar nitro alkenes **79** provided isoxazolines **82** in a practically one pot reaction sequence (Eq. 8, Table 7) [33].

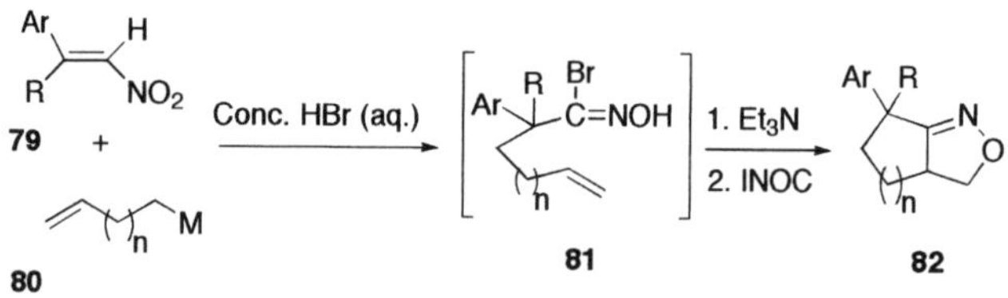

(8)

Table 7

82	Ar	R	n	% Yield of Isoxazoline **82**
a	Ph	Ph	1	95
b	Ph	Ph	2	92
c	Ph	H	2	88 [a,b]
d	2-Thienyl	H	2	85 [a,b]
e	*N*-Phenyl-3-indolyl	H	1	33 [c]

[a] From conversion of isolated **81** to **82**.
[b] Isomer ratio has not been reported.
[c] Only one isomer (*cis*) has been reported.

2.3 Nitrile Oxides from Nitro Alkanes

Another pathway investigated for the nitrile oxide cycloaddition was generation of the intermediate from a nitro alkane by treatment with PhNCO/Et_3N [12a] or *tert*-BOC_2O/DMAP (Scheme 1) [12b]. For instance, Michael addition of vinylazetidine **66** to nitroethene **83a** (R = H) and nitrostyrene **83b** (R = Ph) provided nitro alkanes **84a** and **84b**, respectively (see Eq. 9 and Table 8) [31]. Treatment of **84** with PhNCO/Et_3N led to a mixture of *cis* and *trans* tricyclic azetidines **85** and **86** (the remainder being the phenylurea **87**) without isolation of the nitrile oxide intermediate (see Table 8). Interestingly, both stereoisomers **85b** and **86b** had only the *cis* Ph-Me configuration despite the fact that **84b** was a 1:1 mixture of diastereomers. This may be due to the unfavorable interaction discussed in Sect. 2.2.1 (Scheme 10).

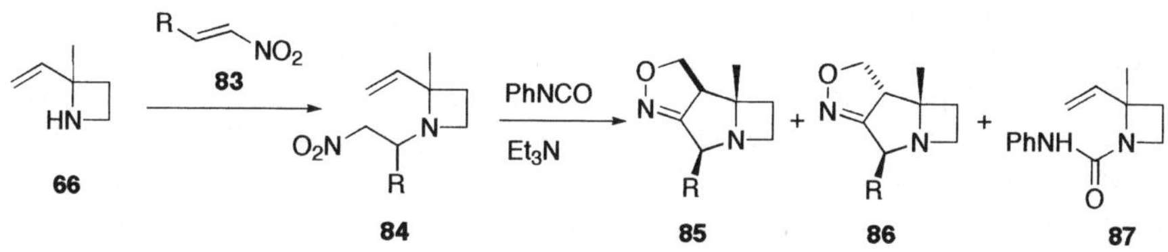

(9)

The stereochemical predictions for the intramolecular cycloadditions described in Scheme 10 and Eq. (9) in terms of MMX calculations are compared to experimental results in Table 9. The most striking trend from inspection of the data is the predominance of *trans* cycloadduct. The computed energy difference between the transition states **88** and **89** (Fig. 1) when R = H corresponds to 1.42 kcal/mol. The experimental preference for the *trans-cis* isomers **72a** and **86a** is correctly predicted by the MMX calculations. It seems reasonable to invoke $A^{1,3}$ strain which is present in the transition state **88** leading to the *cis* product **71a** to explain the *trans* over *cis* preference. Since the substituent R is

Table 8

84	R	84, Isomer Ratio	Yield (%)	85:86	85 + 86 Yield (%)
a	H	–	80	2:3	50
b	Ph	1:1	100	1:2	41

Table 9

Cycloadduct	R	Total *cis*	Energy *trans*	kcal/mol TE	*cis/trans* (Exp)
71a, 72a	Me	50.69	50.48	−0.21	1:2
85a, 86a	H	45.59	44.17	−1.42	2:3

Fig. 1

near the nitrile oxide center, a preference for the R group to be *cis* to the Me substituent is also noticed, which is presumably due to steric interactions on the α side between an azetidine H and the R group when the latter is α.

The INOC reaction of a series of 4-vinyl *N*-substituted β-lactams has also been investigated [31b, 34]. This turned out to be a good comparison with the INOC reaction of the azetidine systems described above. The synthesis of the nitrile oxides **93** involves treating 4-vinylazetidinone **90** with a dibromoalkane (Scheme 11). The resulting bromolactam **91** was converted to the corresponding nitro compound **92**, which was then converted to the nitrile oxide **93**. The nitrile oxides **93b, c** that were not isolated underwent spontaneous INOC reaction in good yield to six- and seven-membered rings **94b** and **94c**, respectively. Attempts to produce the analogous five- and eight-membered compounds were not successful; in the case of **92a** a retro-Michael addition took place to give vinyl lactam **90**, while in the case of **92d** polymerization of the nitrile oxide occurred. While ring closure of **93b** with formation of a six-membered ring occurred stereospecifically to produce **94b**, ring closure of **93c** to a seven-membered ring provided a 40:60 mixture of stereoisomers *cis*-**94c** and *trans*-**94c**. The stereochemical *cis* assignment to the Me and the methylene side chain is based on NOE experiments (ca. 8% enhancement).

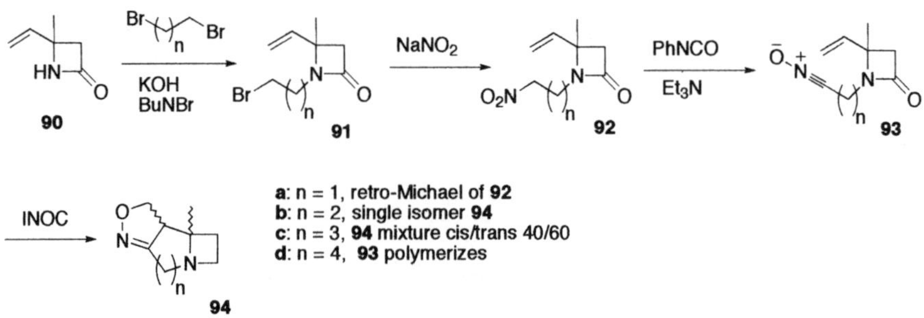

Scheme 11

The INOC reaction strategy has been applied in the synthesis of fused rings possessing functionality in the angular Me group (e.g., **100**, Scheme 12) [35]. The first step utilizes NaH-mediated monoalkylation of ketoester **95** using a dibromoalkane. After the Wittig reaction of the ω-bromoketone **96**, a nitro group was introduced which serves as the nitrile oxide precursor **97**. Sponta-

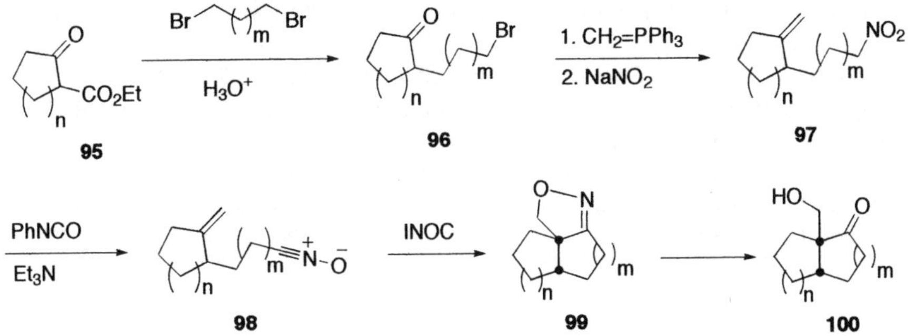

Scheme 12

neous ring closure of **98** upon generation from **97** to the tricyclic system **99** occurred when the ring in the methylenecycloalkane was six-membered (n = 2, see also Table 10). Thus, fused ring systems **99b** and **99d** (i.e., with n = 2 and m = 1 or 2) could be readily prepared. On the other hand, fusion to the methylenecyclopentane system **99** was unsuccessful and led to polymerization of **98** when n = 1 and m = 1, 2, or 3. Such a difference in reactivity could not be attributed to a large preference for an axial side chain in the cyclohexane system vs the more flexible cyclopentane system because MM2 calculations showed that the energy difference between axial and equatorial side chains in **98** is similar in magnitude regardless of the ring size.

Calculations reveal that both ground and TS energies are lower for **99b** and **99d**, which are formed readily in the INOC reaction, than for **99a**, **99c**, and **99e** for which ring closure was not observed (Table 10). Furthermore, a large energy difference between *cis*- and *trans*-fused rings in **99b** and **99d** in favor of the *cis*-fused isomers is obvious from the calculations. Raney Ni cleavage of the isox-

Table 10. Molecular mechanics calculations on the INOC reactions of nitrile oxides **98** (energies in kcal/mol)

			cis-99	*trans*-99	98	E_{TS}^{99}-E_{GS}^{98}
a	n = m = 1	GS	53.13		17.50	
		TS	29.07			11.57
b	n = 2, m = 1	GS	43.02	57.49	12.42	
		TS	20.84			8.42
c	n = 1, m = 2	GS	50.66		17.55	
		TS	30.70			13.13
d	n = m = 2	GS	36.47	50.53	13.11	
		TS	22.45			9.34
e	n = 1, m = 3	GS	54.57		18.21	
		TS	35.28			17.07
f	n = 2, m = 3	GS	48.17	49.45	13.77	
		TS	28.72			14.95

azoline **99d** led to ketol **100d** whose *cis*-decalin skeleton was confirmed by spectral characteristics.

Monoalkylation of *N*-tosylallylamine **10** with dibromoalkane **101** proceeded in 60–90% yield (Eq. 10; see also Scheme 3 and Eq. 2) [17]. The bromoalkylamines **102** were converted to nitro compounds **103**. In situ transformation of **103** into nitrile oxides led to spontaneous cycloaddition with formation of isoxazolines fused to 5-, 6-, and 7-membered ring heterocycles **104a–c**. Under very high dilution conditions, **103d** was converted to **104d**, an isoxazoline fused to an 8-membered azocine, in low (10%) yield.

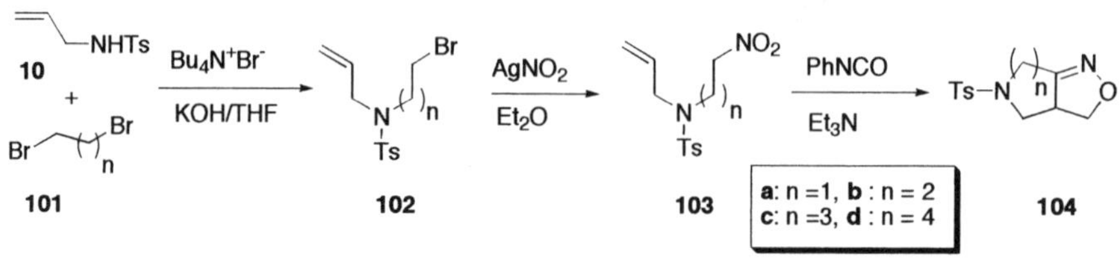

(10)

Using the same sequence, tricyclic quinolinoisoxazoline **108** was formed on intramolecular cycloaddition starting with aminocyclohexene derivative **105** via bromo alkene **106** and nitro alkene **107** (Scheme 13) [17].

The intramolecular cycloaddition of the norbornadiene-tethered nitrile oxides **110** (Eq. 11 and Table 11) was reported to be highly regio- and stereoselective, providing the exo cycloadduct **111** as the exclusive product out of the four possible regio/stereoisomers [36]. The cycloadduct **111** provides a stereoselective entry into tricyclic (e.g., **112**) and spirocyclic (e.g., **113**) frameworks.

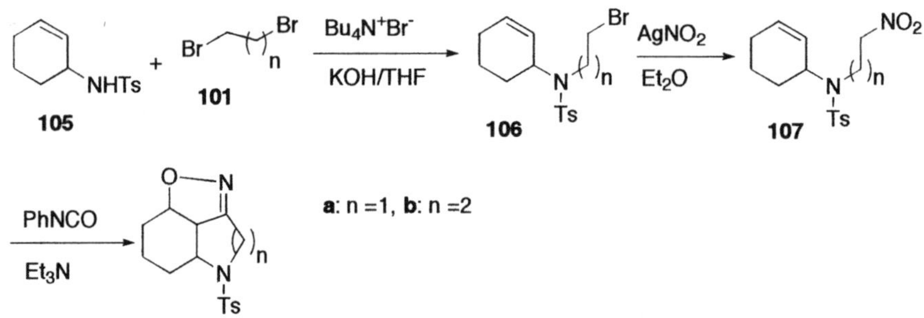

Scheme 13

A regio- and stereospecific INOC reaction of unsymmetrical silaketals **114**, synthesized in one pot from unsaturated alcohols, nitro ethanol, and dichlorosilanes, via the nitrile oxide **115** to isoxazolines **116** has been described (Scheme 14) [37a]. The intermolecular version of the cycloaddition, under similar conditions, proceeds with poor regio and stereoselectivity.

Table 11

109–113	n	X	R¹	R²	% Yield
a	1	CH_2	H	H	86
b	2	CH_2	H	H	75
c	3	CH_2	H	H	0
d	1	O	H	H	69
e	2	O	H	H	52
f	2	CH_2	H	OTBS	78
g	1	CH_2	Me	H	82
h	1	CH_2	Hex	H	83
i	1	CH_2	$SiMe_3$	H	0
j	1	CH_2	Br	H	69

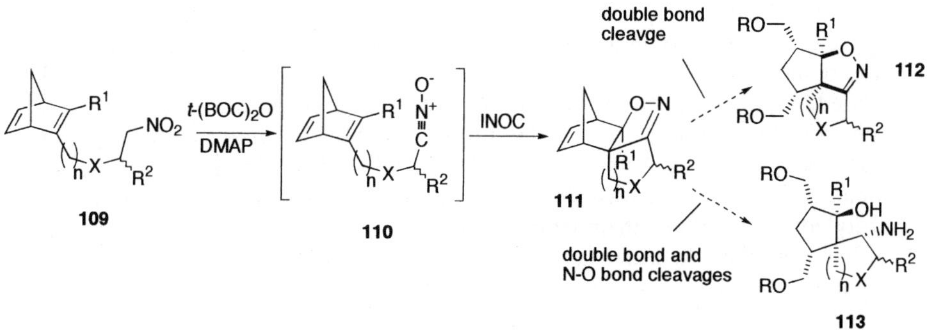

(11)

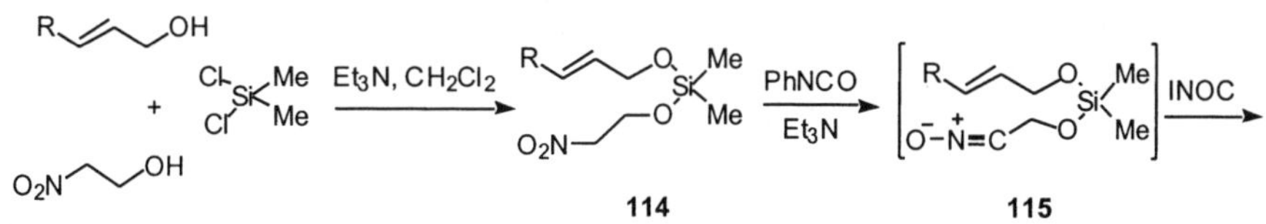

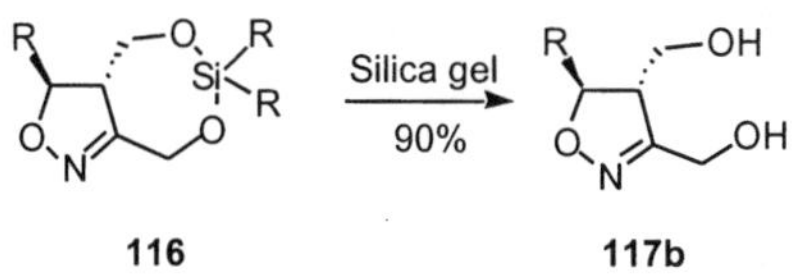

114-117	R	Yield (%) of 114	Yield (%) of 116
a	H	75	60
b	Ph	65	65
c	Me	67	70
d	CO_2Et	70	67

Scheme 14

A high level of diastereoselectivity was reported despite the length of the tether in the INOC reaction of silaketals (e.g., **118**) possessing an allylic substituent (Eq. 12) [37b].

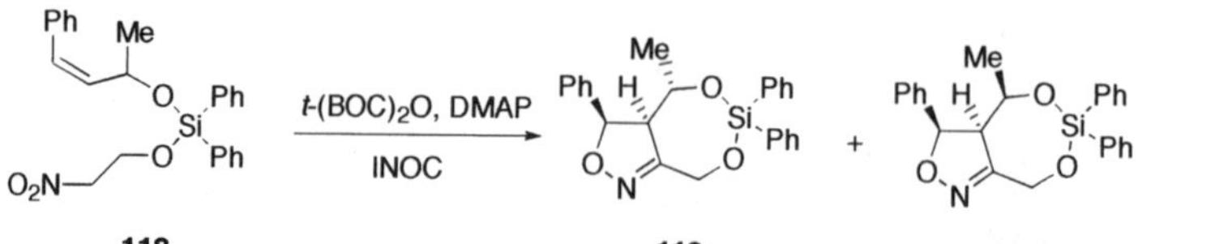

118 **119a** **119b** **119a**:**119b** = 96:4

(12)

Although the unsaturated nitrile oxides **124** can be prepared via the aldoxime route (see Scheme 8), the older procedure suffers from the disadvantage that a tenfold excess of allyl alcohol and two additional steps are required when compared to Scheme 15. Therefore, unsaturated nitro ether **123** that can be prepared by condensation of an aldehyde **120** and a nitro alkane followed by Michael addition of alcohol **122**, was a useful precursor to nitrile oxide **124** [38]. The nitrile oxide **124** spontaneously cyclized to ether **125**. This procedure is particularly suitable for the synthesis of tetrahydrofurans (**125a–h**) and tetrahydropyrans (**125i–k**) possessing Ar substituents in 72–95% yield (Table 12). The seven-membered ether **125l** was obtained only in 30% yield on high dilution. The acetylenic nitro ether **126** underwent INOC reaction to provide the isoxazole **127**.

In a related work, Enders and coworkers showed that the nitro alkanes **130a, b** obtained by the diastereoselective oxa Michael addition of (1*R*,2*S*)-(–)-*N*-formylnorephedrine **128** to nitro alkenes **129a, b** (Table 13) undergo diastereoselective INOC reaction (62–90% via nitrile oxides **131** providing access to enantiomerically pure *N*-protected amino diols (e.g., **133b**) in good overall yields (Scheme 16) [39]. While corresponding intermolecular cycloaddition of analogous optically active nitrile oxides proceeded with decreased stereoselectivity, attempted oxa-Michael addition-ISOC in one pot led to retro-Michael addition (ether cleavage).

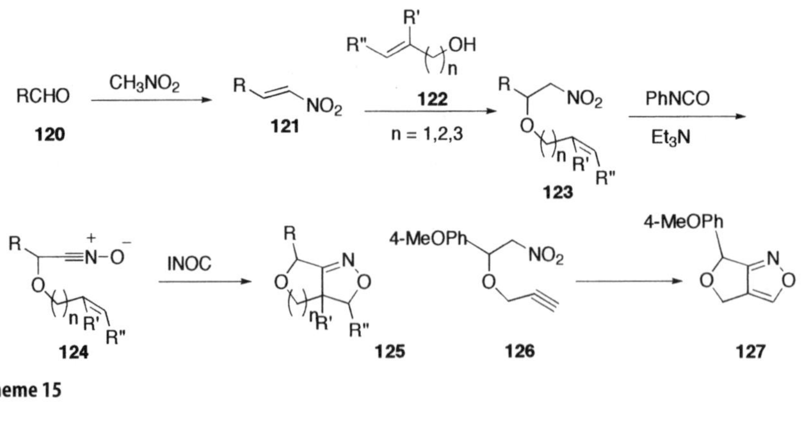

Scheme 15

Table 12

	n	R	R′	R″	Isolated Yield % of **125**	*trans*:*cis*
a	1	Me	H	H	90	2.5:1
b	1	Et	H	H	83	2.5:1
c	1	i-Pr	H	H	85	6:1
d	1	Ph	H	H	87	4:1
e	1	4-MeOPh	H	H	87	4:1
f	1	4-MeOPh	H	Me	79	3:1
g	1	4-MeOPh	H	Ph	78	3:1
h	1	4-MeOPh	Me	H	84	2:1
i	2	Me	H	H	72	1:6
j	2	4-MeOPh	H	H	95	1:6
k	2	2,4,6-Me_3Ph	H	H	80	<5:95
l	3	4-MeOPh	H	H	30	<1:99

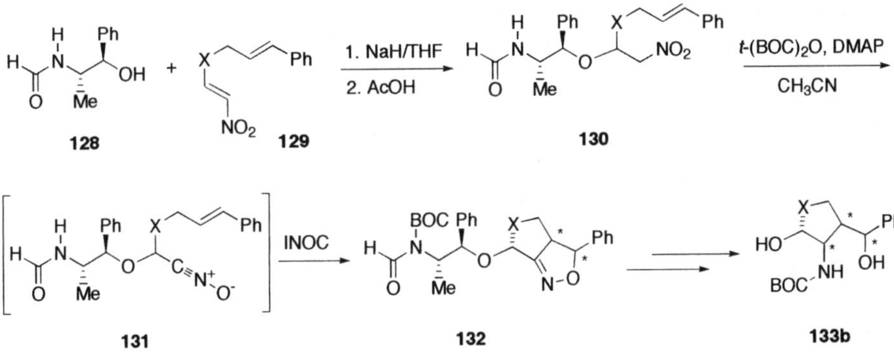

Scheme 16

Table 13

129–133	X	Yield (%) of **132**	de (%)
a	CH_2	62	62
b	CH_2O	74	95

As we found that furan and thiophene substituted oximes can be used as substrates for the INOC reactions (Eq. 5) [29b]; similarly, furan substituted nitro alkane **134** is also a good substrate for INOC reactions (Eq. 13) [40]. The furfuryl derivative **134**, prepared via Michael addition of furfuryl alcohol to 4-methoxy-β-nitrostyrene, was subsequently transformed without isolation of the intermediate nitrile oxide **135** to the triheterocyclic isoxazoline **136** as a 5:1 mixture of isomers in high yield.

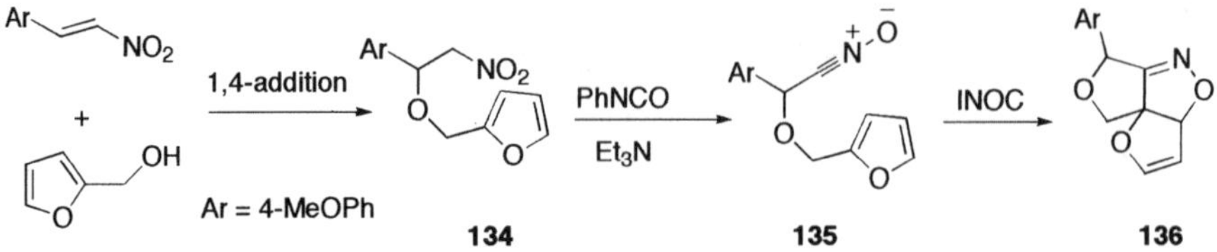

(13)

Such intramolecular nitrile oxide-heterocycle cycloadditions (INHC) also take place with nitrogen-containing heterocycles. For instance, indole was converted by monosubstitution of α,ω-dibromoalkanes under ambient conditions in moderate yields to the α,ω-bromoalkylheterocycles **137** [40] (Scheme 17). Substitution of the halide by a nitro group was followed by generation of nitrile oxides **139**. Although cycloaddition to **140** did not occur when n = 1, it proceeded smoothly at room temperature to cycloadduct **140** in high yield when n = 2. The corresponding reaction when n = 3 was less smooth and required reflux temperature and high dilution conditions.

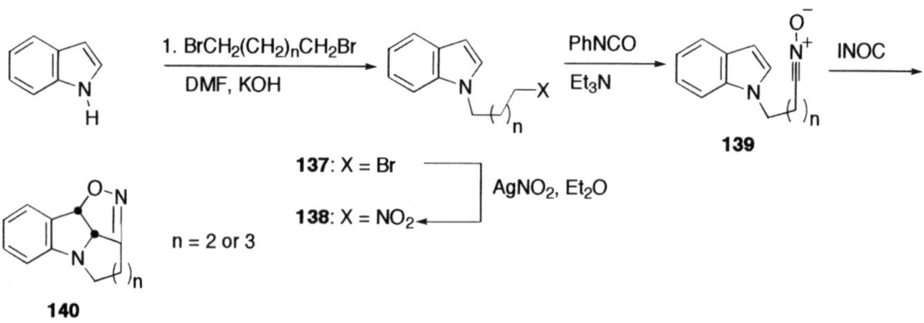

Scheme 17

Analogously, when the above sequence was followed with pyrrole, ring closure of **141** did occur (when n = 2 or 3) but the cycloadduct **142** underwent rearomatization to an oxime which added to phenylisocyanate providing carbamate **143** in moderate yield [40].

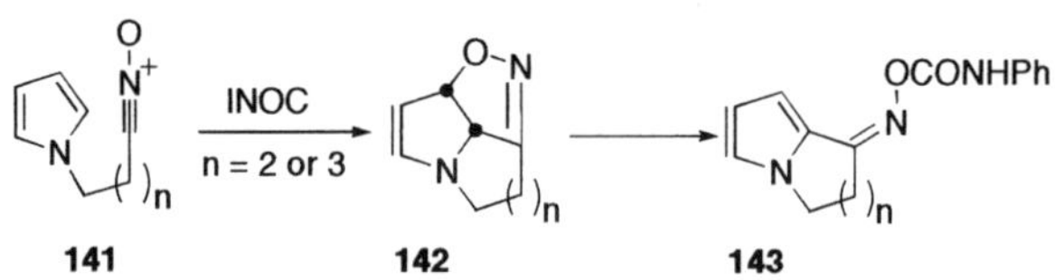

(14)

The INOC strategy involving nitro alkanes has been successfully applied for the synthesis of natural product analogues. Thus, the oxime derivative **147**, possessing the A/B ring of taxane diterpenes with an aromatized C ring, has been synthesized by an INOC reaction of nitrile oxide **145**, generated from nitro compound **144**, presumably via the cycloadduct **146**, as a single isomer in 94% yield (Scheme 18) [41].

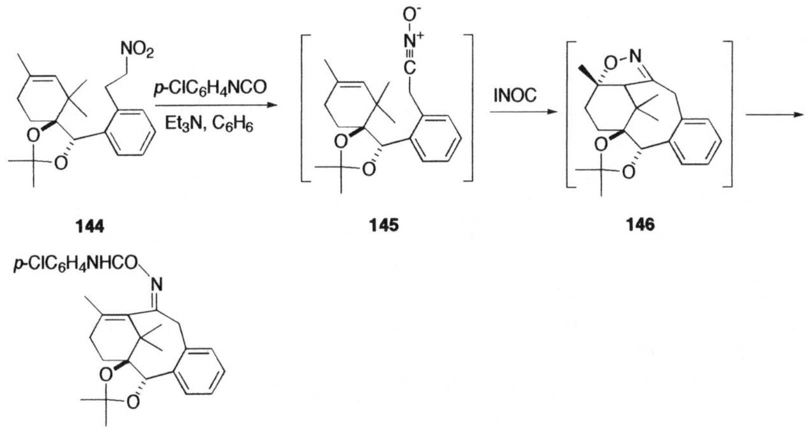

144 145 146

147

Scheme 18

The key step in the diastereoselective synthesis of model insect antifeedant **152** starting from α-cyclocitral **148** was the INOC reaction of oxime **149** or nitro alkane **150** to the isoxazoline **151** (Eq. 15) [42].

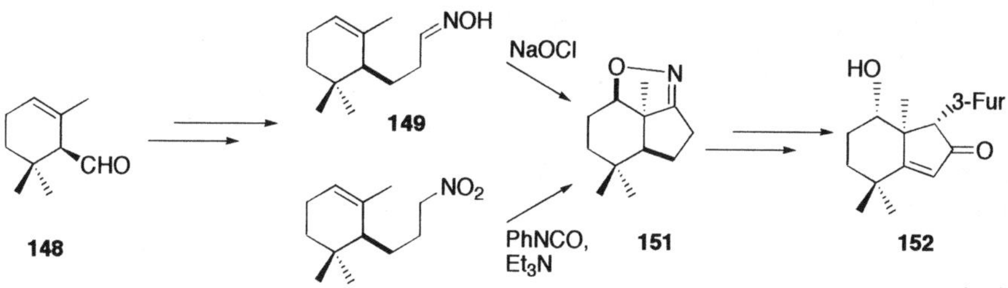

148 149 150 151 152

(15)

3
Intramolecular Silyl Nitronate-Olefin Cycloaddition (ISOC)

Although nitrile oxide cycloadditions have been extensively investigated, cycloadditions of silyl nitronates, synthetic equivalent of nitrile oxides in their reactions with olefins, have not received similar attention. Since we found that the initial cycloadducts, *N*-silyloxyisoxazolidines, are formed with high degree of stereoselectivity and can be easily transformed into isoxazolines upon treatment with acid or TBAF, intramolecular silylnitronate-olefin cycloadditions (ISOC) have emerged as a superior alternative to their corresponding INOC reactions [43]. Furthermore, adaptability of ISOC reactions to one-pot tandem sequences involving 1,4-addition and ISOC as the key steps has recently been demonstrated [44].

3.1 INOC vs ISOC

The unsaturated nitro compounds **153a–g** were first treated with PhNCO and Et_3N at room temperature (Scheme 19, Table 14) [43]. The nitrile oxides **154** thus generated cyclized spontaneously to a mixture of *trans* and *cis* isoxazolines **157** and **158**, respectively, in high yield but with moderate selectivity (see Table 14). For thioethers **157** or **158b–e** and **k**, this nitro route is the only entry, since attempts to oxidize the oxime failed due to sensitivity of the sulfur function. The greater selectivity in the ether over the thioether system may be due to the larger sulfur atom allowing for a more flexible transition state with smaller differences between the *trans* and *cis* isomer. While the stereoselectivity was reversed for cyclization to the carbocyclic systems **157–158f, g**, it is similar to that of the ether and the thioether systems in the case of **157–158h**

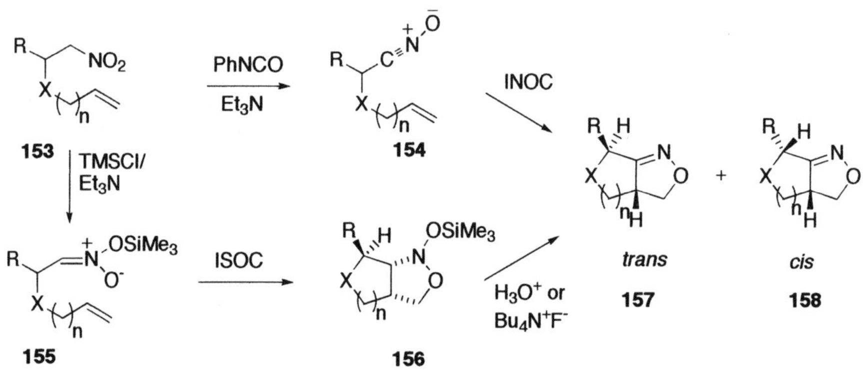

Scheme 19

Table 14

Entry	n	R	X	INOC Yield %	Reaction 157:158	ISOC Yield %	Reaction 157:158
a	1	Me	O	87	2.4:1	74	>99:1
b	1	Me	S	84	1:1	89	>99:1
c	1	i-Pr	S	90	1:1	84	>99:1
d	1	Ph	S	80	3:2	85	>99:1
e	1	4-MeOPh	S	63	3:2	87	>99:1
f	1	Me	$C(CO_2Me)_2$	81	1:7	84	>99:1
g	1	4-MeOPh	$C(CO_2Me)_2$	90	1:5	91	>99:1
h	1	Ph	CH_2	80	3.5:1	–[a]	–[a]
i	1	4-MeOPh	CH_2	88	3.5:1	–[a]	–[a]
j	2	Me	O	–[b]	1:5	–[b]	5:3
k	2	Me	S	–[b]	1:5	–[b]	5:3

[a] See Scheme 22.
[b] Not reported.

and **i**. Remarkably, in the ISOC reaction of **155a–g**, the *trans* stereoisomer was the sole product regardless of the *trans-cis* ratio obtained in the INOC route [43, 44].

Comparison of INOC and ISOC reactions for the construction of six-membered ethers **157–158j** and thioethers **157–158k** showed a reversal of stereoselectivity (see entries j and k).

3.2
One-Pot Reactions Involving Michael Addition and ISOC

One-pot tandem sequences involving 1,4-addition and ISOC as the key steps have been developed for the construction of N and O heterocycles as well as of carbocycles [44]. In this sequence, the nitronate arising from 1,4-addition to an α,β-unsaturated nitro alkene is trapped kinetically using trimethyl silyl chloride (TMSCl). The resulting silyl nitronate underwent a facile intramolecular 1,3-dipolar cycloaddition with the unsaturated tether (e.g., Schemes 20–22).

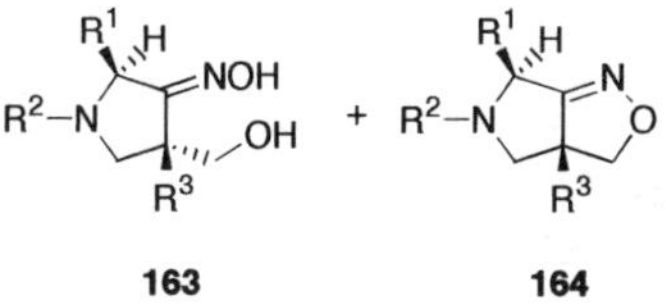

Scheme 20

Table 15

161–164	R^1	R^2	R^3	Workup	% Yield of **163**
a	*p*-MeOPh	allyl	H	F^-/OH^-	60
b	Ph	allyl	H	OH^-	63
c	*p*-MeOPh	c-hex	H	OH^-	52
d	*p*-MeOPh	Et	Me	F^-	66
e	*p*-MeOPh	Bn	H	OH^-	66
f	Me	allyl	H	OH^-	10–20[a]

[a] The main product is the bicyclic isoxazoline **164f**.

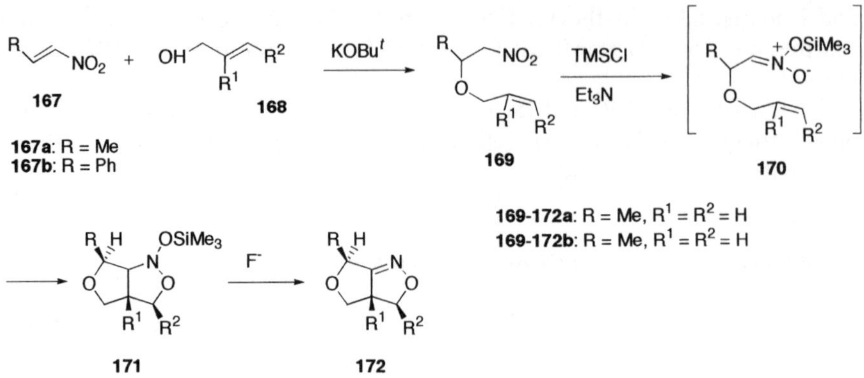

Scheme 21

Table 16

Entry	167	168	169–172	R^1	R^2	Tandem % Yield of **172**	Stepwise % Yield of **172**	% Yield of **169**
1	a	a	a	H	H	41	–	–
2	b	a	b	H	H	74	77	90
3	b	c	c	H	Me	79	68	84
4	b	d	d	Me	H	40	69	75
5	b	e	e	H	MeO	30	0[a]	79
6	b	f	f	-CH=	CH-O-	0[a]	0[a]	72

[a] Decomposition of the substrate

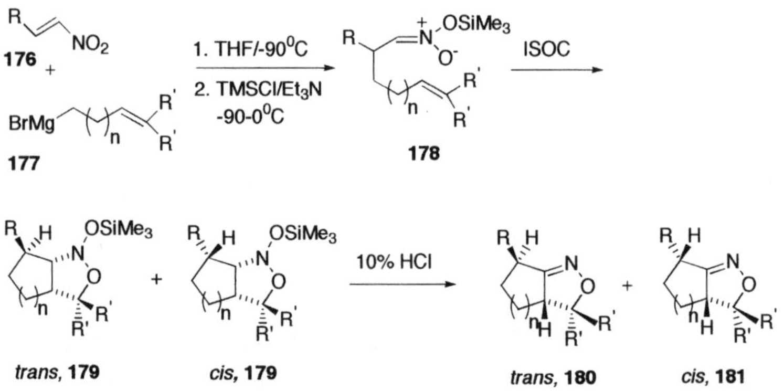

Scheme 22

In the synthesis of N heterocycles, this technique also overcomes competitive retro Michael addition that lowers the yield of 1,4-adduct in the Michael addition of amines to nitro olefins. Thus, a toluene solution of nitro olefin **159** was treated with allylamine **160**, Et_3N, and TMSCl under nitrogen at ambient tem-

Table 17

Entry	n	R	R′	% Yield **180** + **181**	*trans*:*cis* **180**:**181**
a	1	Ph	H	66	>99:1
b	1	4-MeOPh	H	43	>99:1
c	1	i-Pr	H	18	>99:1
d	1	Ph	Me	53	>95:5
e	2	Ph	H	62	1:2.6
f	2	4-MeOPh	H	58	1:2.6

perature for 14–24 h (Scheme 20) [44a]. The silyl nitronate **161**, generated in situ, underwent smooth ISOC reaction to the *N*-trimethylsilyloxyisoxazolidines **162**. Attempted desilylation of **162a** with aqueous HCl or anhydrous CF_3CO_2H followed by liberation of the salt provided a 3:1 mixture of hydroxy oxime **163a** and isoxazoline **164a** (see Table 15). While acid catalyzed elimination of Me_3SiOH from **162a** provides isoxazoline **164a**, base treatment to liberate the amine after acidification causes partial ring cleavage to oxime **163a**. Apparently, only a single isomer (*trans*) was observed for both **163a** and **164a**. The stereochemistry was established by NMR studies. Formation of the oxime-isoxazoline mixture can be avoided by treating the crude *N*-trimethylsilyloxyisoxazolidines **162** with TBAF/THF or more conveniently with methanolic KOH, affording exclusively the oxime (see Table 15). This basic *N*-O bond cleavage probably proceeds through an alkoxy nitroso intermediate **166** (Eq. 16) as is evident from the blue-green color which appears during the reaction.

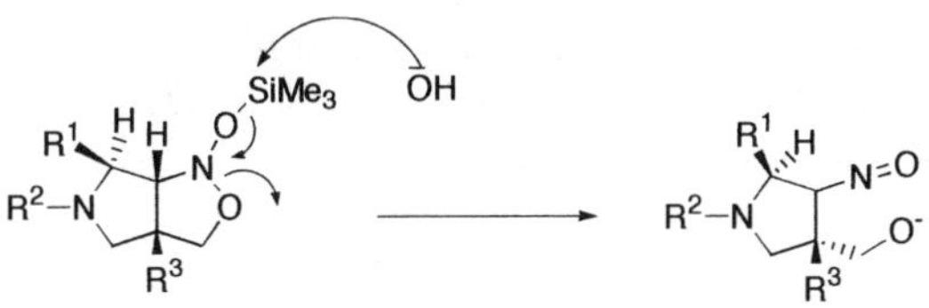

165 **166** (16)

The *N*-allylated compounds may serve as precursors to the *N*-unsubstituted pyrrolidines by a mild Pd(0) catalyzed deallylation procedure [45]. The deallylation may be more straightforward than hydrogenolysis of the *N*-benzylated product in which complications due to the presence of the oxime and the other Bn moiety (1,2 bond in the pyrrolidine ring) may arise during reduction. Attempts to prepare piperidine and azepine systems following the above procedure were unsuccessful.

ISOC reaction was employed to synthesize substituted tetrahydrofurans **172** fused to isoxazolines (Scheme 21) [44b]. The silyl nitronates **170** resulted via the nitro ethers **169** from base-mediated Michael addition of allyl alcohols **168** to nitro olefins **167**. Cycloaddition of **170** followed by elimination of silanol provided **172**. Reactions were conducted in stepwise and one-pot tandem fashion (see Table 16). A terminal olefinic Me substituent increased the rate of cycloaddition (Entry 3), while an internal olefinic Me substituent decreased it (Entry 4).

In the case of nitronates possessing ester or nitrile moieties as terminal olefin substituents, tandem Michael addition to produce substituted furans **174, 175** occurred faster than trapping of the nitronate anion by TMSCl (Eq. 17).

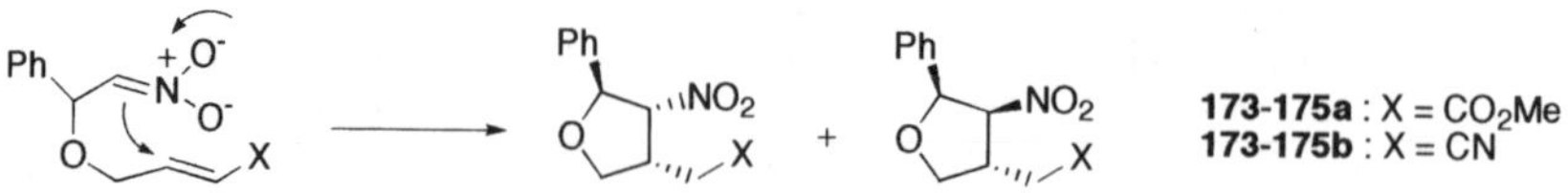

173 **174** **175** (17)

A one-pot tandem reaction sequence involving 1,4-addition of homoallyl or pentenyl Grignard reagent **177** to nitro olefin **176**, silylation of the resulting nitronate, followed by ISOC reaction of the silyl nitronate **178** and desilylation of the cycloadduct **179** has been developed for the synthesis of functionalized carbocycles (Scheme 22) [44c]. Whereas the ISOC reaction leading to cyclopentane rings fused to an isoxazoline (Entries a–d, Table 17) proceeded smoothly with a high degree of stereoselectivity when compared to analogous INOC reactions (see Scheme 19 and Table 14, Entries h and i), the corresponding cycloaddition to cyclohexane rings (Entries e and f, Table 17) was sluggish and less selective. It turned out that while the one-pot reaction sequence starting from β-Ar nitro olefins provided the isoxazolines in good overall yields, aliphatic nitro olefins did not perform well (see Table 17, Entry c) apparently owing to side reactions (reagent attacking the nitro group) [46] at the 1,4-addition stage.

The stereochemical course of the ISOC reaction has been rationalized in terms of analogous nitrone cycloadditions [47], assuming that in both cases the TSs leading to *N*-substituted isoxazolidines can have identical geometries. For instance, the formation of cyclopentanes fused to *N*-silyloxyisoxazolidines **179a–d** could take place via an exo TS **182a**, arising from an E-nitronate or via an endo TS **182b** which results from a Z-nitronate, both incidentally leading to *cis*-fused bicyclic isoxazolidines (Fig. 2). Those TSs (endo-E and *exo*-Z) leading

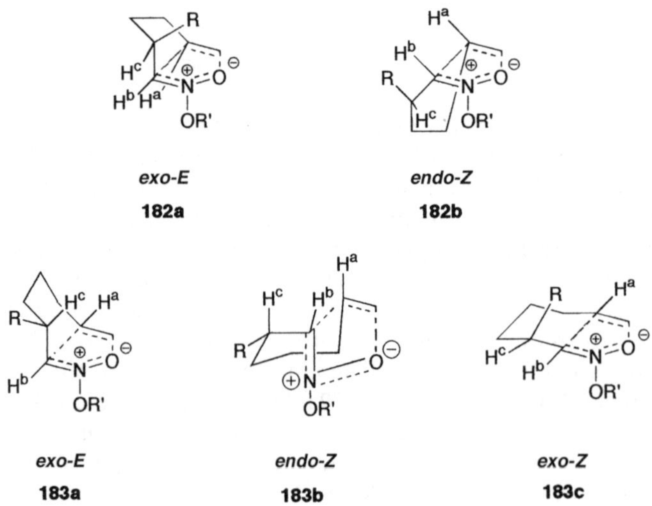

Fig. 2

to *trans*-fused systems, products of kinetic control, appeared energetically destabilized or geometrically unattainable [47, 48a].

Formation of *trans* isomers in overwhelming predominance in the ISOC reaction leading to five-membered rings (Entries a–d) has been ascribed to the orientation in which H^a, H^b, and R are on the exo face of TS **182b** (this avoids a possible $A^{1,2}$ strain between R and NO or between H^b and H^c [48b] that is presumably present in TS **182a**). Since elimination of silanol involving H^b in no way interferes with the orientation of H^a and R, a *trans* relationship between H^a and H^c is abundantly clear. This fully accords with the widely accepted view that approach of the dipole and dipolarophile takes place in two parallel planes [49] and that the endo TS is preferred in the absence of obvious steric effects [50]. Formation of approximately 5% *cis* isomer when the dipolarophile terminus is disubstituted is accountable in terms of the cycloaddition taking place via TS **182a**.

As far as the cycloaddition to the six-membered rings is concerned, the stereochemistry is dictated by chairlike transition states, consistent with the NMR data. In principle, an *exo*-E TS **183a** or an *endo*-Z TS **183b** could lead to the *cis* isomers **181e** and **181f** (H^a and H^c *cis* diaxial). However, there is a subtle difference that while the transformation of **183b** to **181e,f** takes place via *cis*-fused *N*-silyloxyisoxazolidines, **179e, f** (products of thermodynamic control) the intermediate *N*-silyloxyisoxazolidines **179e, f** would be *trans*-fused (products of kinetic control) if the TS is **183a.** TS **183b** appears preferred not only in terms of the superior FMO overlap but also in terms of the reaction conditions employed (60 °C, 15 h). Formation of *trans* isomers **180e, f** (which is assumed to take place via an *exo*-Z TS **183c** despite the fact that R is pseudo-axial and presumably experiences $A^{1,2}$ strain with NO group) in considerable amounts is in all probability a consequence of heating the reaction mixture which was required to complete the cycloaddition. Prolonged heating is known to have a detrimental effect on the selectivity, since equilibration of cycloadducts [50c, 51], presumably via cycloreversions and readditions, takes place readily at higher temperatures [52].

3.3 Sequential 1,3-Dipolar Cycloadditions

A strategy involving sequential 1,3-dipolar cycloadditions has been reported for the synthesis of novel bis-isoxazolo substituted piperidines **192a** and **192b** (Eqs. 18 and 19) [53]. It consists of the Michael addition of an unsaturated alkoxide **185** to *β*-nitrostyrene **184** followed by an INOC or ISOC reaction to provide isoxazolines **187–189** (Eq. 18 and Table 18). A polymer supported acyl chloride

Table 18

a,b	n	R	INOC Yield %	a:b	ISOC Yield %	a:b
187	1	H	88	9:1	[a]	[a]
188	1	Ph	75	6:1	82	~100:0[b]
189	2	H	77	3:97	40	1:1

[a] Not reported.
[b] Trace of **188b** was isolated.

scavenger was employed to remove excess alkoxide thereby increasing the efficiency of and yield in the Michael addition step. Addition of allylmagnesium bromide to the isoxazoline intermediate **187–189** followed by DCC coupling of the resulting isoxazolidine **190** with nitroacetic acid provided the nitro alkene **191** (Eq. 19). Finally, INOC reaction of both **191a** and **191b** proceeds in 65% yield in a stereoselective fashion providing **192a** and **192b**, respectively.

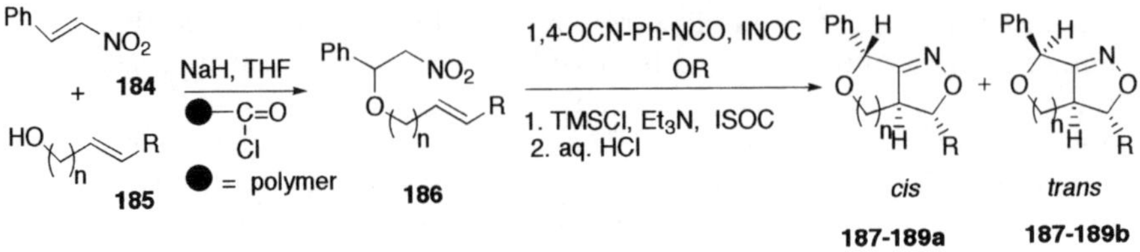

(18)

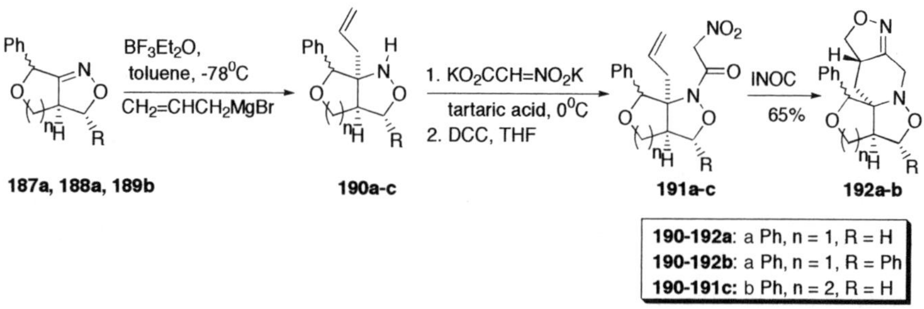

(19)

3.4 Miscellaneous

Potential precursors to stereoselective INOC and ISOC reactions (e.g., **195** and **196**, respectively) have been prepared via stereoselective conjugate additions of several allylic alcohols (e.g., **194**, X = O) and an allylic thiol (e.g., **194**, X = S) to a chiral (*E*)-nitro alkene (e.g., **193**) that was derived from (*R*)-2,3-isopropylidene

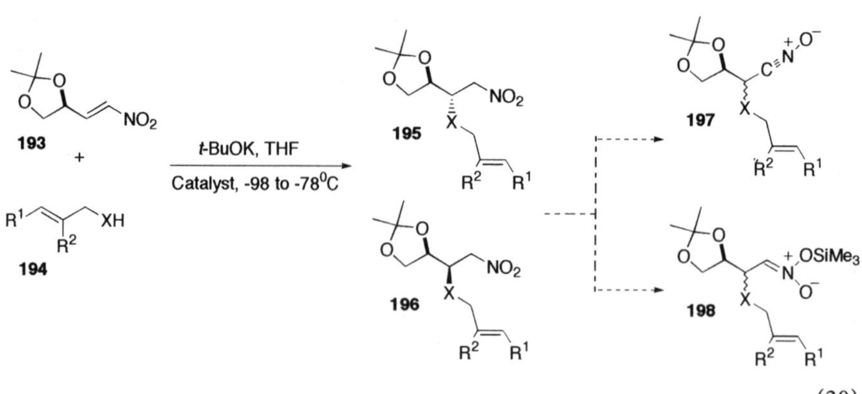

(20)

Table 19

Entry	R^1	R^2	X	Catalyst	**195** (*anti*):**196** (*syn*)	% Yield of **195** + **196**
1	H	H	O	–	70:30	54
2	H	CH_3	O	–	72:28	70
3	H	H	O	DMAP[a]	73:27	86
4	H	CH_3	O	DMAP[a]	75:25	82
5	H	H	O	CuCN[a]	82:18	76
6	CH_3	H	O	CuCN[a]	80:20	80
7	H	H	O	CuBr[a]	80:20	79
8	H	CH_3	O	CuI[a]	85:15	81
9	H	H	S	CuI[a]	78:22	74
10	CH_3	H	O	CuI[a]	85:15	82
11	CH_3	H	O	CuI[b]	87:13	81
12	H	H	O	CuCN[a]/CuI[a]	92:8	84
13	CH_3	H	O	CuCN[a]/CuI[a]	93:7	83
14	H	H	O	CuI[a]/ZnI_2[a]	95:5	87

[a] 1.2 equiv.
[b] 3 equiv.

glyceraldehyde (Eq. 20 and Table 19) [54a]. The ISOC ring closure of **198** has just been achieved [54b].

High levels of diastereocontrol in an ISOC reaction were induced by a stereogenic carbon center that bears a Si substituent (Scheme 23) [55]. For instance, conversion of nitro alkenes (e.g., **199**) to β-siloxyketones (e.g., **203**) has been accomplished via a key ISOC reaction-reduction sequence with complete control of 1,5-relative stereochemistry. The generality of the ISOC reaction of a silyl nitronate with a vinylsilane was demonstrated with seven other examples. Corresponding INOC reaction proceeded with lower stereoselectivity.

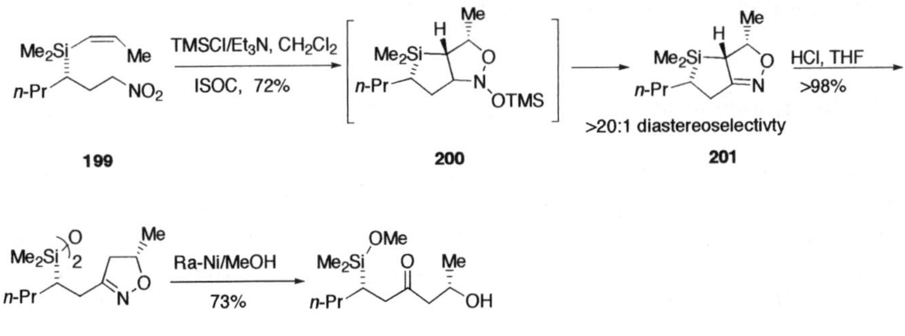

199 **200** **201** **202** **203**

Scheme 23

Stereocontrolled syntheses of the natural iridolactone, (+)-iridomyrmecine **207** (Eq. 21) and the alkaloid (–)-actinidine **213** (Scheme 24) have been accomplished based on an ISOC reaction as the key step [56, 57]. ISOC reaction of **204** followed by fluorodesilylation of the cycloadduct **205** provided the hydroxy-oxime **206** which was transformed to iridolactone **207** [56].

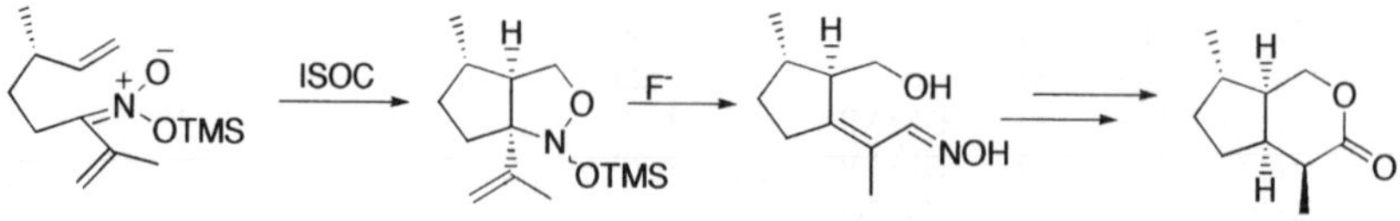

204 **205** **206** **207** (21)

Thioethers **210** are smoothly formed upon cyclization of silyl nitronates **209**, generated in situ from the nitro compounds **208**, on treatment with N,*O*-bis(trimethylsilyl)acetamide (BSA, Scheme 24) [57]. Fluorodesilylation of **210** gave the *N*-oxide **212**, presumably via highly reactive aldehyde **211**, which was reduced to the target compound actinidine **213** in an overall 27% yield.

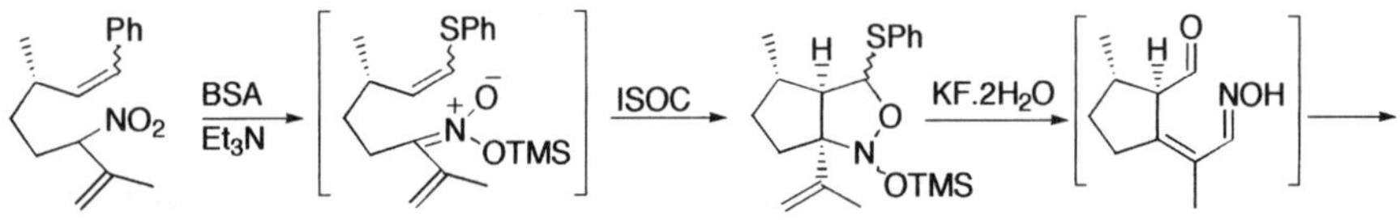

208 **209** **210** **211**

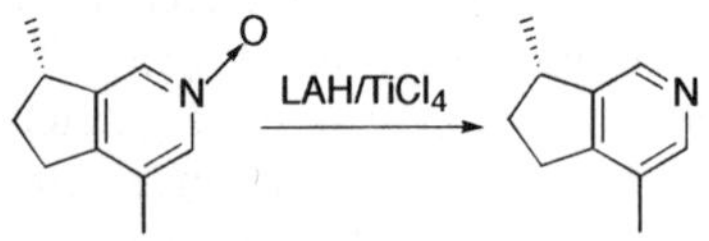

212 **213**

Scheme 24

4 Intramolecular Oxime-Olefin Cycloaddition (IOOC)

4.1 (*H*)-Nitrones from Oximes

Intramolecular nitrone cycloadditions often require higher temperatures as nitrones react more sluggishly with alkenes than do nitrile oxides and the products contain a substituent on nitrogen which may not be desirable. Conspicuously absent among various nitrones employed earlier have been NH nitrones, which are tautomers of the more stable oximes. However, Grigg et al. [58a] and Padwa and Norman [58b] have demonstrated that under certain conditions oximes can undergo addition to electron deficient olefins as Michael acceptors, followed by cycloadditions to multiple bonds. We found that intramolecular oxime-olefin cycloaddition (IOOC) can occur thermally via an *H*-nitrone and lead to stereospecific introduction of two or more stereocenters. This is an excellent procedure for the stereoselective introduction of amino alcohol functionality via *N*-O bond cleavage.

4.1.1
Pyrrolidines

Allylamines containing an aldoxime chain **215a** undergo smooth intramolecular cycloaddition to the pyrrolidinoisoxazolidines **216** in 65–100% yield simply on heating at 80–110 °C or even upon standing for long periods of time at room temperature (Scheme 25) [59]. The ring closure proceeded stereospecifically to generate three adjacent stereogenic centers that provide an entry into functionalized pyrrolidines. For instance, LAH reduction of **216a** and **216b** led to pyrrolidines **217a** and **217b** in 75% and 82% yield, respectively, each possessing stereospecifically positioned amino alcohols that do not bear a substituent on the amine function as would have resulted from a nitrone cycloaddition (Table 20). The advantage of IOOC reactions over INOC reactions for the stereospecific introduction of amino alcohol functionality was demonstrated by the fact that LAH reduction of isoxazoline **218a** provided an isomeric mixture of **217a**. Apparently the IOOC reaction proceeded via a thermal equilibration of the oxime **215a** to its nitrone tautomer **215b** via proton transfer from oxygen to nitrogen in the oxime function (see **215b**).

It was possible to effect IOOC reaction leading to six-membered rings, e.g., **220** in low yield (ca. 20%) by heating the reaction mixture at 110 °C (Eq. 22) [59]. In fact, Oppolzer and Keller [60] had previously reported the IOOC reaction of **219** to **220** in 20% yield by heating at 110 °C. Furthermore, the scope of these oxime-olefin cycloadditions has been extended to ketoximes, e.g., **221**. The latter was prepared by amination of α-bromoacetophenone with allylamine **214a**. Heating of **221** at 110 °C for 8 h led to cycloaddition with formation of the fused pyrrolidine **222** in 88% yield. As in Scheme 25, only one

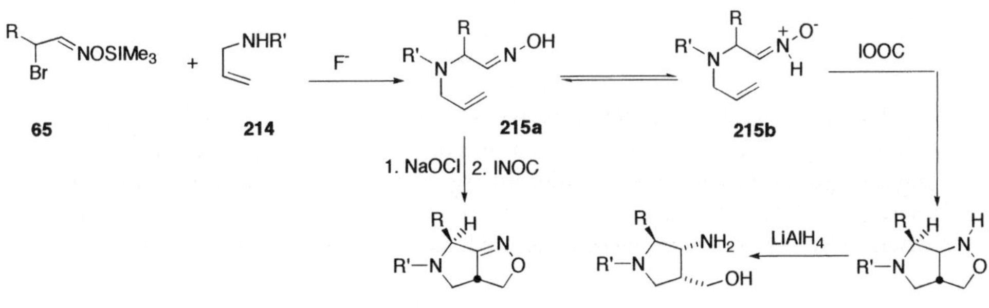

Scheme 25

Table 20

215–218	a	b	c	d	e	f
R′	Ph	Ph	c-Hex	c-Hex	Allyl	Allyl
R	Et	Me	Et	Me	Et	Me

stereoisomer was formed and LAH reduction led stereospecifically to amino alcohol **223** [59].

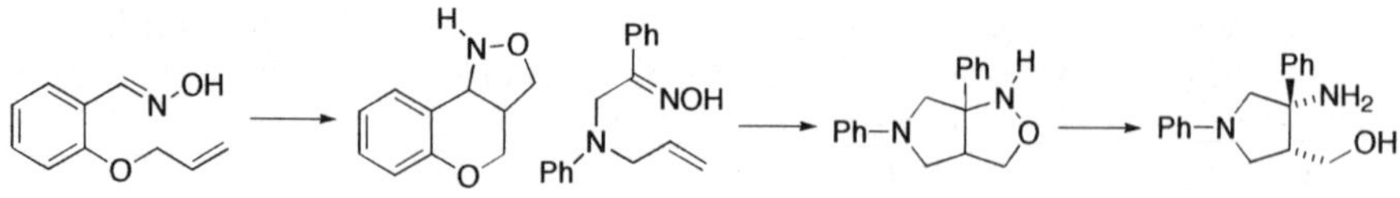

219 **220** **221** **222** **223**

(22)

Pyrrolidines fused to isoxazolidines **225** have been synthesized from oximes **224** in order to evaluate the stereoselectivity and the conformation of the resultant cycloadducts (Eq. 23) [61]. It has been shown that unshared orbital interactions on neighboring hetero atoms such as those present in isoxazolidines are responsible for a considerable energy barrier toward ring inversion [62]. The ring system in **225** differs from the all carbon bicyclo[3.3.0]octane in that the otherwise consecutive carbon framework is interrupted by the presence of a N atom in ring A (pyrrolidine) as well as by the relatively high-energy *N*-O bond in ring B (isoxazolidine). It was thought that this could predispose these molecules to a preferred conformation which may be ascertained by NMR studies, a situation not feasible in the all carbon system. Thus, further insight into the conformational preferences of such fused five-membered rings would be possible:

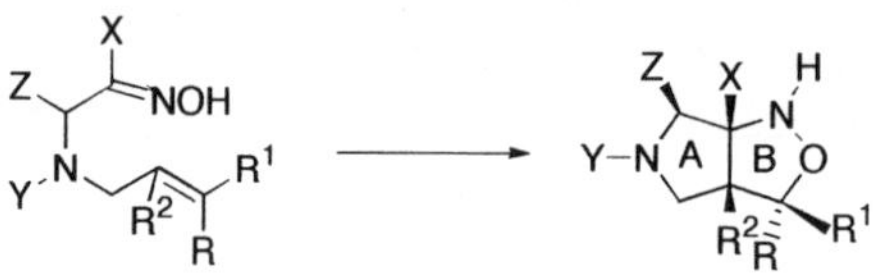

224 **225**

(23)

The unsaturated oximes **224** (see Table 21) were readily prepared by *N*-alkylation of allyl amines with α-bromoketones or *O*-silyl-α-bromoaldoximes. Heating the oximes **224** in toluene under an argon atm at 110–180 °C smoothly led to isoxazolidines **225** in good yields with *cis* ring junction stereochemistry. Even when three stereocenters were generated, as in **225g–l**, a single stereoisomer

Table 21

224, 225	a	b	c	d	e	f	g	h	i	j	k	l
X	H	Me	Me	H	Me	Ph	Ph	H	H	H	H	H
Y	Allyl	Me	cHex	Et	Ph	Ph	Ph	Ph	Ph	H	Allyl	Et
Z	H	H	H	H	H	H	H	Et	Et	Et	Et	Et
R	H	H	H	H	H	H	H	H	Me	H	H	H
R^1	H	H	H	H	H	H	Me	Me	Me	H	H	H
R^2	H	H	H	Me	H	H	H	H	H	H	H	Me

was isolated with the side chain always *cis* to the adjacent bridgehead substituent. In the case of **224h**, the cycloaddition took place with the stereospecific generation of four consecutive stereocenters. The reaction proceeded equally well for aldoximes and ketoximes. Furthermore, the presence of terminal (γ) Me substituents on the allyl amine enhanced the reaction rate (see **225g–i**) while a Me substituent on the β-carbon retarded the cycloaddition.

Compounds **225a–f** showed interesting dynamic phenomena on the NMR time scale with broad lines at room temperature and appearance of two sets of sharp peaks at –50 °C corresponding to conformers **226** and **227** (Fig. 3). By contrast, **225g–l** exist essentially as one conformer. These results show that the presence of a Me substituent adjacent to the O atom in ring B and *syn* to the ring junction hydrogen (see **225g**) prejudices the molecule in favor of conformer **226**, thus placing the Me substituent pseudoequatorially (cf. **226**, R^1 = Me). Similarly, a single β-substituent in the A ring (pyrrolidine) at position 8, *syn* to the ring junction hydrogen (**225j–l**) favors conformer **227** (cf. Z = Et) in which the A ring substituent Z can assume a pseudoequatorial position. The former effect dominates the latter when both rings are substituted (cf. **225h**), giving the product with both side chains in a β-orientation, but the preferred conformation is **226** and the Et group is forced into a pseudoaxial position. An additional Me group on C-4 (cf. **225i**) doesn't change these observations. Equilibrium measurements indicated a free energy of conversion of 13.2–13.4 kcal/mol, apparently a manifestation of the *N*-inversion in the isoxazolidine ring. NMR studies also showed that the *N*-H proton in the isoxazolidines **225** prefers an axial orientation. Furthermore, MM2 force field calculations with AMBER charge provided a good correlation between calculated and experimental vicinal coupling constants.

The calculations reveal a 0.31 kcal/mol difference between conformers **226a** and **227a** which corresponds to an equilibrium ratio of 32:68. This is in excellent agreement with experimental findings (**226a**:**227a**=33:66). The calculations also show that the lowest energy conformers of cycloadducts **225b** and **225d** correspond, respectively, to **226b** (60%) and **226d** (58%). Compounds **225h** and **225l** were found to exist predominantly (90%) as conformer **226** whereas structures **225j** and **225k** clearly favor conformer **227** (>90%) in which the Et substituent is found in the pseudoequatorial position.

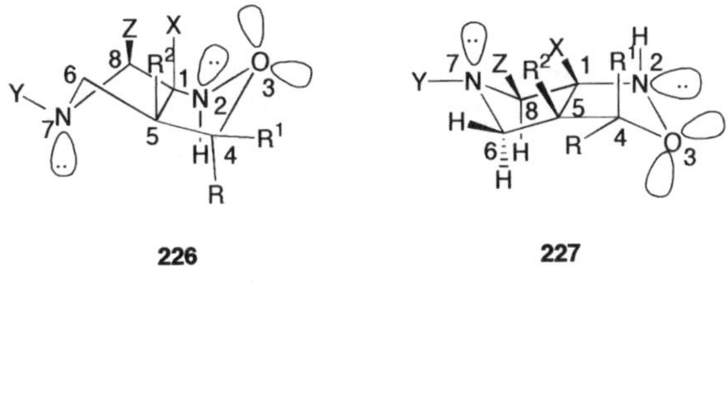

Fig. 3 **226** **227**

4.1.2
Carbocycles

It was also of interest to apply such IOOC reactions to formation of carbocyclic rings. Oxime olefins **230a–e**, formed in good yield via reaction of **229** with *O*-silyl-α-bromoaldoximes **228** in the presence of F⁻ ions, cyclized in a sealed tube at 190 °C to provide **231a–e** (Eq. 24, Table 22) [63]. Reduction of **231a** provided amino alcohol **232a** in 68% yield. Amino alcohol **232e** was converted stereospecifically to the fused β-lactam **233**.

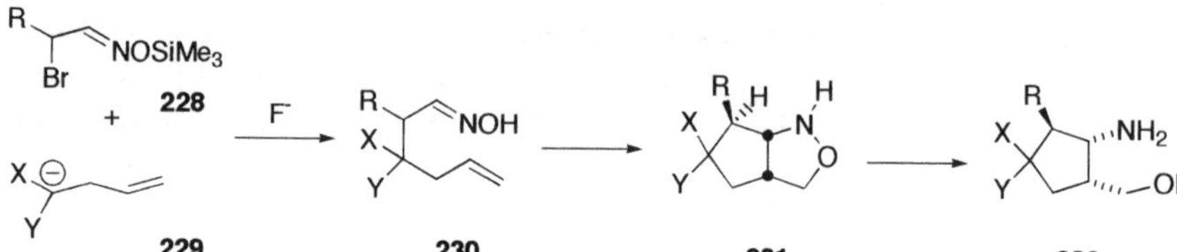

229 **230** **231** **232**

(24)

Table 22

	X	Y	R	% Yield of **231**
a	CO_2Me	CO_2Me	Et	81
b	CO_2Me	COMe	Et	63
c	H	H	Et	55
d	H	H	Ph	61
e	H	H	SO_2Ph	–

The IOOC reactions were extended to cyclization of ketoxime **234** which provided the isoxazolidine **235** stereospecifically in 75% yield (Eq. 25).

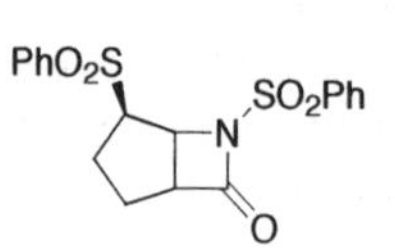

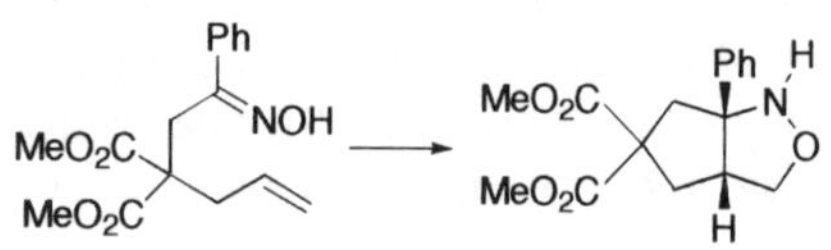

(25)

233 **234** **235**

Formation of cyclohexane fused isoxazolidine **237** from oxime **236** was less selective, providing a mixture of isomers in 51% yield (Eq. 26). On the other hand, citronellal oxime **238** led stereospecifically and in 80% yield to the fused cyclohexane **239** when heated at 190 °C for 5 h.

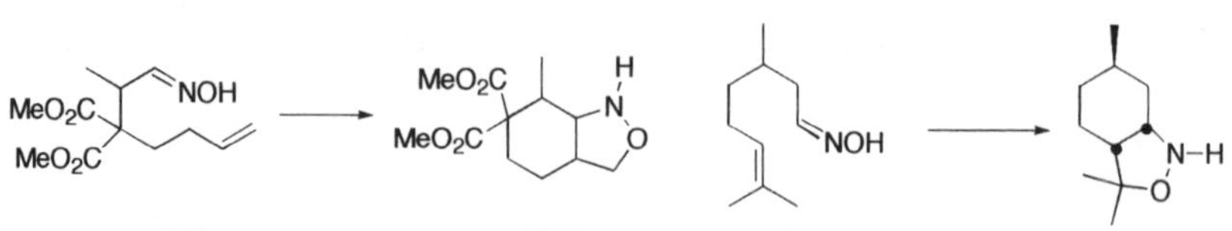

236 **237** **238** **239** (26)

4.1.3
Fused Rings (Pyrrolizidines, Indolizidines, and Quinolizidines)

The utility of IOOC reactions in the synthesis of fused rings containing a bridgehead N atom such as pyrrolizidines, indolizidines, and quinolizidines which occur widely in a number of alkaloids has been demonstrated [64]. Substrates **242a–d**, that possess properly positioned aldoxime and alkene functions, were prepared from proline or pipecolinic acid **240** (Eq. 27). Esterification of **240** and introduction of unsaturation on N by *N*-alkylation produced **241** which was followed by conversion of the carbethoxy function to an aldoxime **242.** IOOC reaction of **242** led to stereoselective formation of various tricyclic systems **243.** This versatile method thus allows attachment of various unsaturated side chains that can serve for generation of functionalized five- or six-membered (possibly even larger) rings.

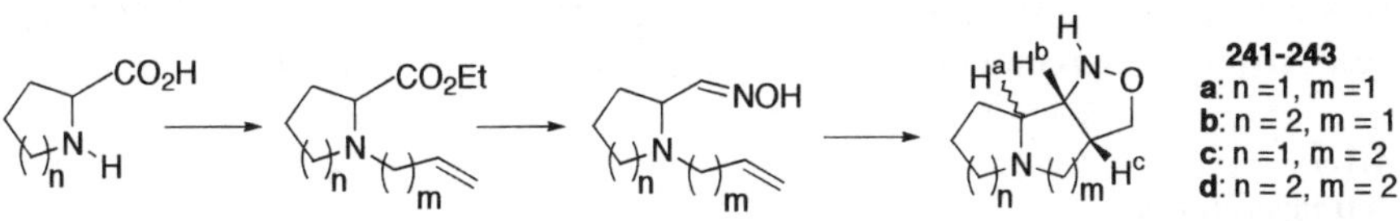

240 **241** **242** **243** (27)

In an alternative approach, the isomeric unsaturated pyrrolidine or piperidine aldoximes **245a** and **245b** were prepared and subjected to IOOC reaction affording **246a** and **246b**, respectively (Eq. 28). Esterification of **240** followed by *N-tert*-BOC protection and DIBALH reduction provided aldehyde **244** (X = O) which was subjected to Wittig olefination. Introduction of a two carbon aldoxime chain on N in **244** (X = CH_2) was carried out by alkylation with Et α-bromoacetate after deprotection of the N atom in **244.** Reduction and oximation led to **245.**

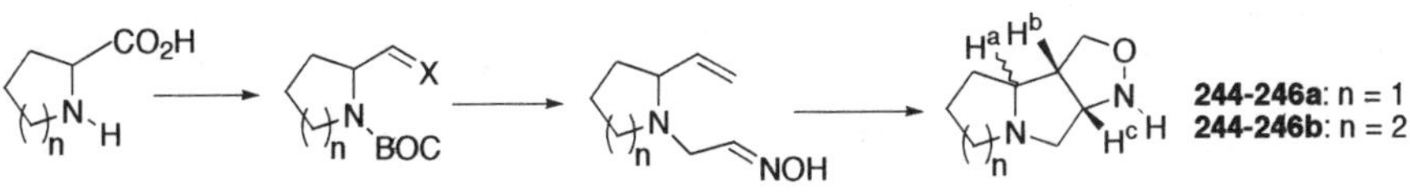

240 **244** **245** **246** (28)

Although the thermal IOOC reactions can sometimes be accomplished at 80 °C [64], the oxime olefins **242a–d** and **245a, b** required heating in toluene at 180 °C in a sealed tube. Thus tricyclic pyrrolidines **243a** (*anti*) and **246a** (*anti*), the indolizidines **243b** (*anti*:*syn* = 75:25), **243c** (*syn*), and **246b** (*anti*), and the quinolizidine **243d** (*syn*) were isolated in 60–75% yield. It is noteworthy that the IOOC products **243a, 243c, 243d, 246a,** and **246b** were obtained stereochemically pure. Indolizidine **243b,** on the other hand, was obtained as a mixture.

It is clear from the foregoing that ring closure to five-membered rings fused to the isoxazolidine, regardless whether part of a pyrrolizidine or of an indolizidine system, led mainly to the *cis-anti* isomers (see **243a, b** and **246a, b**) while formation of six-membered fused ring (either as part of an indolizidine or of a quinolizidine system) produced the *cis-syn* stereoisomer (see **243c** and **243d**). That the ring junction between the isoxazolidine and the newly formed five- or six-membered ring is always *cis* is indicated by coupling constants and examination of molecular models.

Molecular mechanics calculations revealed a 2.4 kcal difference between the two diastereomeric transition states for isoxazolidine **243a**, but only a 1.16 kcal difference for isoxazolidine **243b**. This accounts for the 75:25 mixture of isomers obtained from **243b**, while a single diastereomer was produced from **243a**. In both cases, the lower energy isomer corresponds to the *anti* diastereomer. The calculations also showed that the lowest energy *anti* conformer of **246a** is about 1.62 kcal lower in energy than the *syn* isomer which fits with the generality that formation of a five-membered ring fused to the isoxazolidine should have the *anti* configuration. In the six-membered ring formation, the energy differences between *syn* and *anti* isomers of **243c** and **243d** were 3.29 kcal and 2.63 kcal, respectively, in agreement with the isolation of a single product.

4.1.4 *Tetrahydrofurans*

The IOOC route was followed for the synthesis of tetrahydrofurans possessing a γ-amino alcohol moiety **247** (Eq. 29) [18]. Aldoximes **21a–f** (see also Eq. 3 and Table 2), when heated in benzene in a sealed tube at 110–120 °C for 6 h, underwent smooth intramolecular cycloaddition to the tetrahydrofuranoisoxazolidines **246a–f** in 70–83% yield (Eq. 29). This ring closure proceeded stereospecifically to generate three adjacent stereogenic centers. LAH reduction of **246b** resulted in isolation of stereospecifically functionalized tetrahydrofuran derivative **247b** in 75% yield.

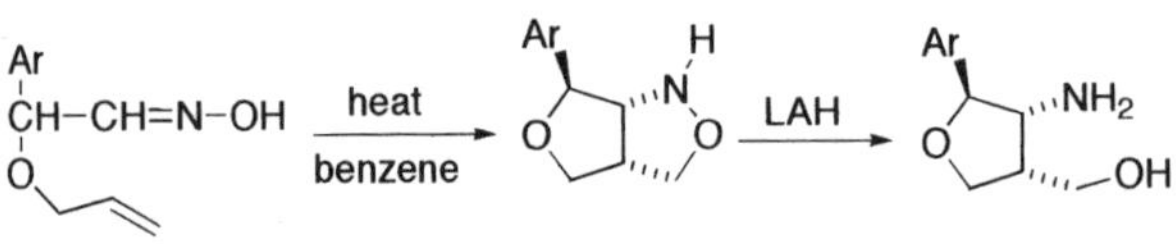

21 **246** **247**

247a: 70%, **b**: 72%, **c**: 77%
d: 70%, **e**: 83%, **f**: 74%

(29)

4.2 Role of Substituents in the Oxime-(*H*)-Nitrone Isomerization

From the above, it is interesting to note that while the temperature required to effect IOOC of an oxime possessing amine N is 80 °C and an oxime possessing ether O is 110 °C, it is much higher (190 °C) for corresponding C compounds. However, it is not clear if (and if yes, to what extent) the presence of the unshared electron pair on the amine N (as in **216** (Scheme 25), **224** (Eq. 23), **242** (Eq. 27), and **245** (Eq. 28)) or on the ether O (as in **219** (Eq. 22) and **21** (Eq. 29)) exercises an assisting effect in the proton transfer from O to N.

Interestingly, a facile oxime – *H*-nitrone isomerization in the pyridopyrimidine systems **248** has been attributed to the favorable proton transfer from the

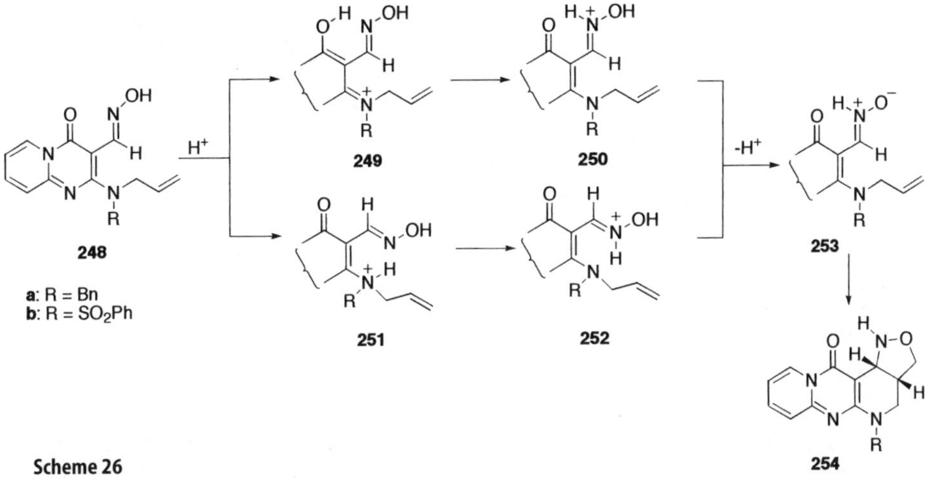

Scheme 26

protonated alkenamino and/or carbonyl moieties **249** or **251** to the lone pair of the imine N producing, after proton loss from **250** or **252**, the *H*-nitrone intermediate **253** (Scheme 26) [65]. Thus, cyclization of **248a** to isoxazolidine **254a** was 45.3 times faster than cyclization of oxime **248b** to isoxazolidine **254b** in dioxane at 68.8 °C indicating the catalytic role played by the alkenamino N. This is further supported by the fact that only the oxime with *E*-configuration underwent the oxime-*H*-nitrone isomerization. Similar intramolecular assistance of an alkenamino N and carbonyl group has been observed in the hydrazone-azomethine imine isomerization as well [66].

4.3 Miscellaneous

The structure-reactivity relationship between a 19-Me- and 19-nor-5,10-seco-steroid has been investigated using IOOC and intramolecular nitrone cycloaddition taking into account various stereochemical aspects (Schemes 27 and 28) [67]. The E-19-nor-5,10-seco-ketone **255a**, on treatment with hydroxylamine hydrochloride (R′ = H), undergoes IOOC via **256a** to a single isoxazolidine **257**

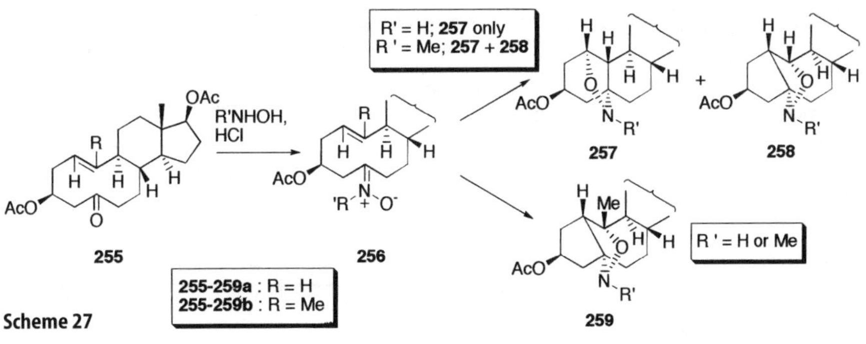

Scheme 27

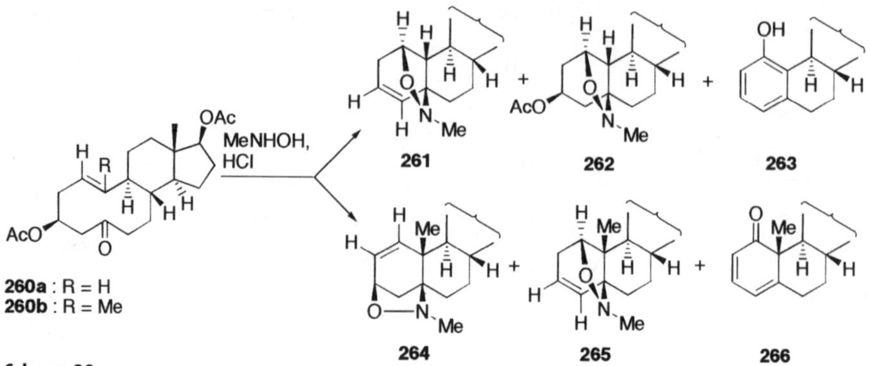

Scheme 28

(R′ = H, Scheme 27). On the other hand, reaction of **255a** with *N*-methylhydroxylamine hydrochloride produces a mixture of two regioisomers **257** and **258** (R′ = Me). When the E-1(10)-unsaturated 5-oxo-5,10-secosteroid **255b** was treated with hydroxylamine hydrochloride (R′ = H) or *N*-methylhydroxylamine hydrochloride (R′ = Me), isoxazolidine **259** was formed regio- and stereoselectively in high yield via intramolecular 1,3-dipolar cycloaddition of the nitrone intermediate **256** (R′ = H or Me).

The nitrone arising from reaction between (*Z*)-19-nor-5,10-secosteroidal ketone **260a** and *N*-methylhydroxylamine hydrochloride undergoes transannular 1,3-dipolar cycloaddition to give isoxazolidines **261** and **262** and an aromatic derivative **263** originating from **261** (Scheme 28). Corresponding reaction of **260b** produces two types of structurally different isoxazolidines **264** and **265** as well as the dienone **266**.

We have seen that substituted chiral pyrrolidines that display glycosidase inhibitory properties and are synthetic aza sugar analogs can been synthesized based on the INOC route starting from naturally occurring amino acids and their enantiomers (Scheme 4) [19]. Similarly, the ene-oximes **267a–c** prepared as in Scheme 4 undergo IOOC reaction (at 165 °C providing the tricyclic compounds **268a–c** as single diastereomers) (Scheme 29) [68]. The ring closure pro-

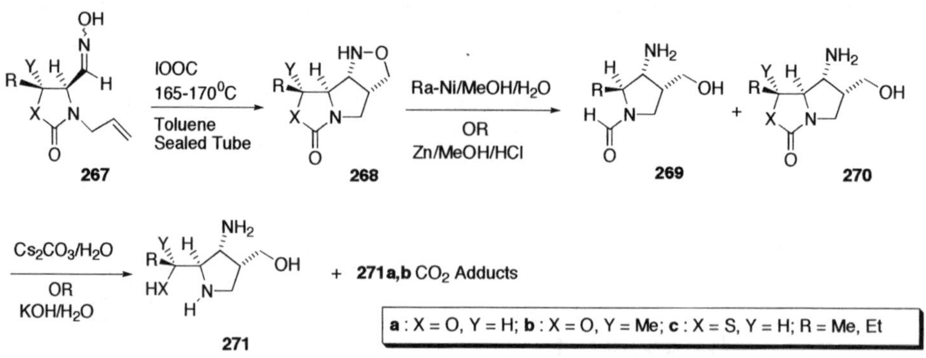

Scheme 29

duced solely an *anti-syn* fused ring system consistent with the stereoselectivity of IOOC ring closures [68, 69]. While the isoxazolidine ring of **268a, b** underwent *N*-O bond reduction with Ra-Ni to produce **270a, b**, the thiazolidine ring underwent desulfurization as well to give pyrrolidine **269**. However, the alternative reduction of **268c** with Zn gave amino alcohol **270c**. *O*-N deprotection of **270a–c** required overnight reflux in the presence of catalytic amounts of Cs_2CO_3 [70a] and KOH [70b].

In the case of **271a** and **271b**, covalently bound adducts of **271a, b** and CO_2 (e.g., carbamic acid derivatives) were detected. This was absent in the case of thiol derivative **271c**. The concentration of this material could be reduced by heating (a solution to 90 °C for 1 h) and increased by exposing to a CO_2 atmosphere.

Since the substrates of glycosidases are natural sugars that possess the D-configuration, it was desirable to prepare the D-enantiomer of aza sugar analogs L-**271** which, indeed, turned out to be the most active. The D-**271a** was prepared following Scheme 29 with a slight modification in that the deprotection (Ra-Ni) and hydrolysis (Cs_2CO_3) steps were switched, with virtually no change in the overall yield [68].

5 Intramolecular Azide-Olefin Cycloaddition (IAOC)

5.1 Azides to Triazolines, Imines, and Aziridines

1,3-Dipolar cycloaddition of azides with olefins provides a convenient access to triazolines, cyclic imines, and aziridines and hence is a valuable technique in heterocyclic synthesis. For instance, tricyclic *β*-lactams **273–276** have been synthesized using the intramolecular azide-olefin cycloaddition (IAOC) methodology (Scheme 30) [71].

The bromoalkene **91** was converted to an azido alkene **272** in quantitative yield utilizing a polymeric azide reagent [72] which was superior to NaN_3. Azido alkene **272a** underwent ring closure on heating in benzene (14 h) exclusively to the *cis* tricyclic *β*-lactam **273a**. In the formation of the seven-membered ring, both *cis* and *trans* isomers **273b** and **274b**, respectively, were isolated with

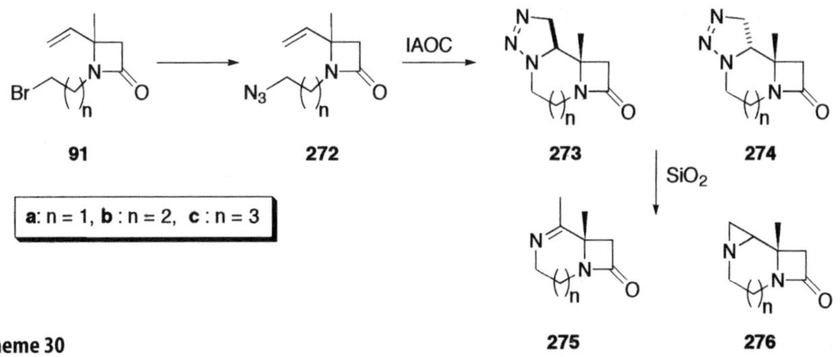

Scheme 30

great preference for the *cis* isomer **273b** (ratio of **273b** to **274b** = 9:1). Reluctance to undergo IAOC reaction (refluxing in toluene was necessary) and a reduced selectivity were observed in the formation of an eight-membered ring (ratio of **273c** to **274c** = 6:4).

Whereas thermolysis or photolysis [73] of triazolines **273, 274** resulted in a mixture of aziridine, imine, and polymeric material, a smooth chemoselective transformation of these triazolines was achieved by treatment with silica gel. Interestingly, the fused six-membered ring **273a** gave imine **275a** exclusively, while the fused seven-membered rings **273b** and **274b** led to a 7:3 mixture of imine **275b** and aziridine **276b**. The fused eight-membered ring triazoline **274c** (separated from **273c**) was converted on silica gel to aziridine **276c** exclusively. The *cis* triazoline **273c** furnished an isomeric aziridine, hence the triazoline decomposition is stereospecific. The decomposition of triazolines **273, 274** depended on the ring size, becoming difficult for larger rings.

The bromoalkylallylamines **102** (see also Eq. 10) serve as precursors to unsaturated azides **277** which can undergo IAOC reactions to novel heterocycles **278a–c** (Eq. 30) [74]. The transformation to **278** was best performed by direct heating of **102** with NaN_3 in DMSO-H_2O (Eq. 30). The isolated yields of triazolinopiperazine **278a**, triazolinodiazepine **278b**, and triazolinodiazocine **278c** were 25%, 45%, and 5%, respectively.

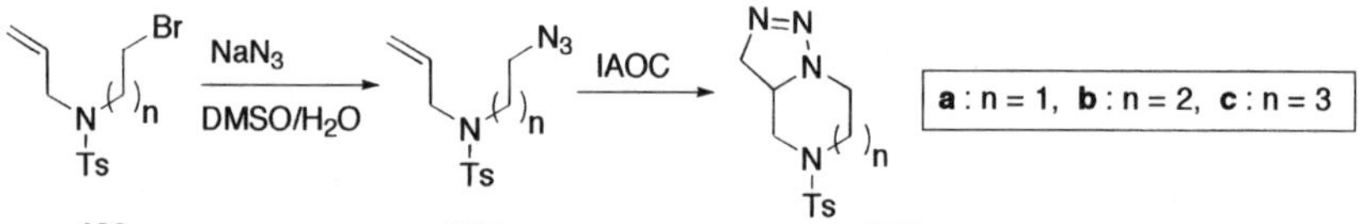

102 **277** **278** (30)

5.2
α-Azido Ethers to Δ^3-Oxazolines

α-Azido ethers are potential substrates for azide-olefin cycloaddition and can be conveniently prepared from aldehydes and ketones (Scheme 31). Thus, $TiCl_4$ promoted addition of hydrazoic acid (HN_3) to aldehydes in presence of an alcohol produces α-azido ethers **279** [75]. Similarly, simple ketones can be converted to Me α-azido alkyl ethers **280** by means of HN_3 and Me orthoformate in the presence of *p*-toluenesulfonic acid.

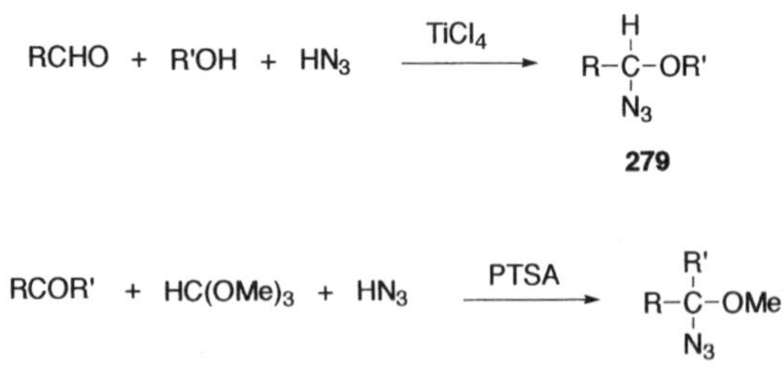

Scheme 31

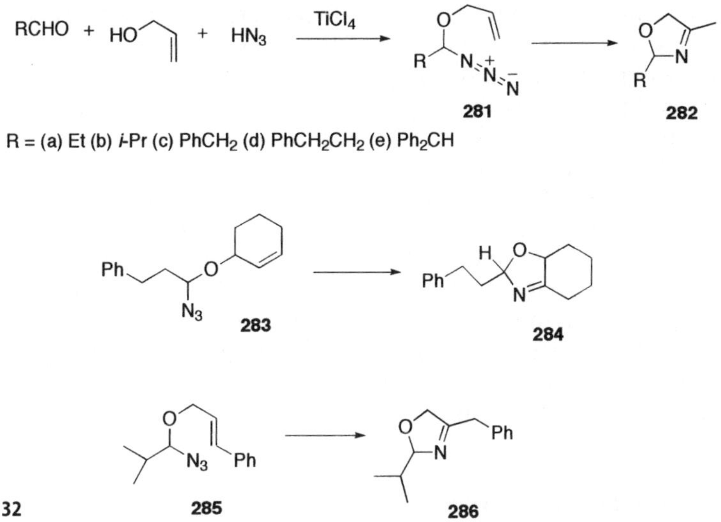

Scheme 32

Using the above procedures, allyl α-azido alkyl ethers of type **281** were prepared by employing an unsaturated alcohol such as allyl alcohol [76] (Scheme 32). The reaction of an aldehyde with allyl alcohol and HN_3 in a ratio of 1:3:9 carried out in the presence of $TiCl_4$ as catalyst provided azido ethers **281**, **283**, and **285** in 70–90% yield. The ratio of reagents is critical to ensure a high yield of azido ether and to prevent formation of acetal and diazide side products [75]. Thermolysis of azido alkenes **281**, **283**, and **285** in benzene (the solvent of choice) for 6–20 h led to 2,5-dihydrooxazoles **282**, **284**, and **286**, respectively, in 66–90% yield.

In order to determine whether oxazolines are formed via an independent nitrene pathway or via triazolines **287** (Fig. 4), the thermolysis of azido alkene **281 b** was followed by ^{1}H NMR in C_6D_6 at 70 °C. First, formation of both oxazoline **282 b** and triazoline **287** was observed at partial conversion. After 3 h of heating only oxazoline **282 b** was present. The same behavior was observed for **281 a**, **281 c**, and **281 e**. On the other hand, if a mixture of **281 b** and **287** after 50% conversion was chromatographed on silica gel, only oxazoline **282 b** and the fused aziridine **288** (*cis* + *trans*) were isolated in addition to starting material. The aziridine **288** was stable in refluxing benzene for 4 h and was not converted to **282 b**. The azido alkene **281 b** could be converted to triazoline **287** as a 2.9:1 mixture of *cis* and *trans* isomers, without appreciable presence of oxazoline

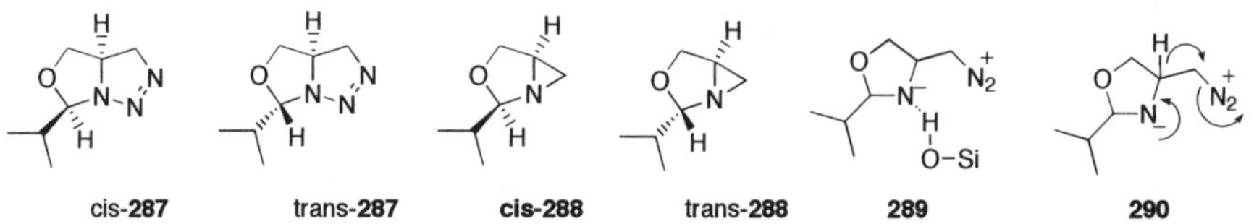

Fig. 4

282b or aziridine **288**, by heating at 50 °C in $CDCl_3$ for 1 h and following the reaction by ^1H NMR (up to 80% conversion). Isolation of pure **287** was not possible because it was unstable to chromatography.

The overall pathway for the conversion of the unsaturated azido ether **281** to 2,5-dihydrooxazoles **282** involves first formation of the dipolar cycloaddition product **287**, which thermolyzes to oxazoline **282** or is converted by silica gel to oxazolinoaziridine **288**. While thermolysis or acid-catalyzed decomposition of triazolines to a mixture of imine and aziridine is well-documented [71, 73], this chemoselective decomposition, depending on whether thermolysis or exposure to silica gel is used, is unprecedented. It is postulated that acidic surface sites on silica catalyze the triazoline decomposition via an intermediate resembling **289**, which prefers to close to an aziridine **288**. On the other hand, thermolysis of **287** may proceed via **290** (or the corresponding diradical) in which hydrogen migration is favored over ring closure.

5.3 IAOC in the Synthesis of Bioactive Materials

IAOC strategy has been extremely useful for the synthesis of several biologically active compounds. For instance, (±)-crinane **293**, a biologically active *Amaryllidaceae* alkaloid, has been synthesized involving the IAOC of **291** as the key C-N bond forming reaction (Eq. 31) [77].

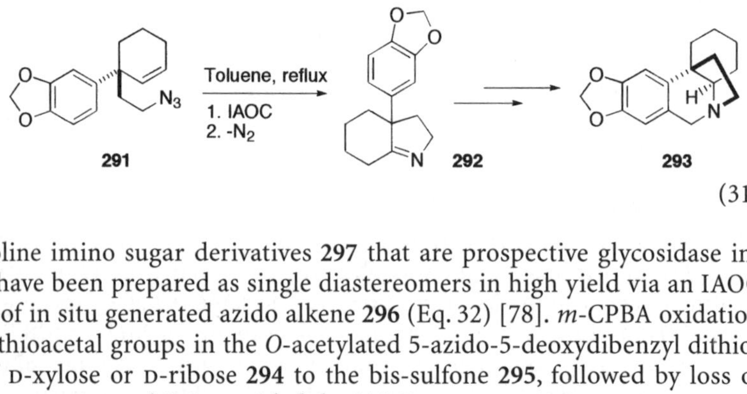

(31)

Triazoline imino sugar derivatives **297** that are prospective glycosidase inhibitors have been prepared as single diastereomers in high yield via an IAOC reaction of in situ generated azido alkene **296** (Eq. 32) [78]. *m*-CPBA oxidation of the dithioacetal groups in the *O*-acetylated 5-azido-5-deoxydibenzyl dithioacetal of D-xylose or D-ribose **294** to the bis-sulfone **295**, followed by loss of HOAc between C-1 and C-2 provided the IAOC precursor **296**.

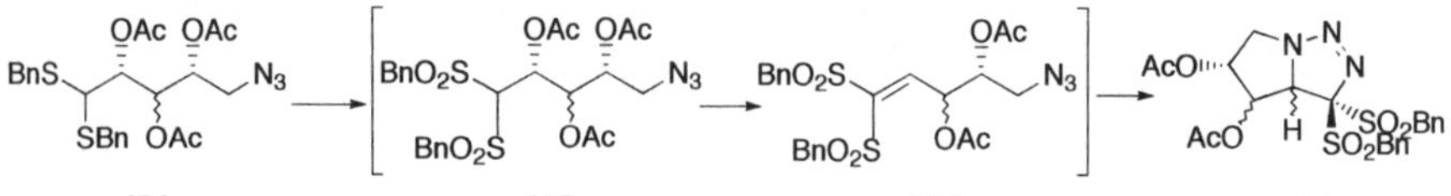

(32)

Intramolecular cycloaddition between an azide and an unsaturated ester (see **300**) was the key step in the synthesis of triazole carboxylic acids **302a, b**, prospective anionic sugar mimics (Eq. 33) [79].

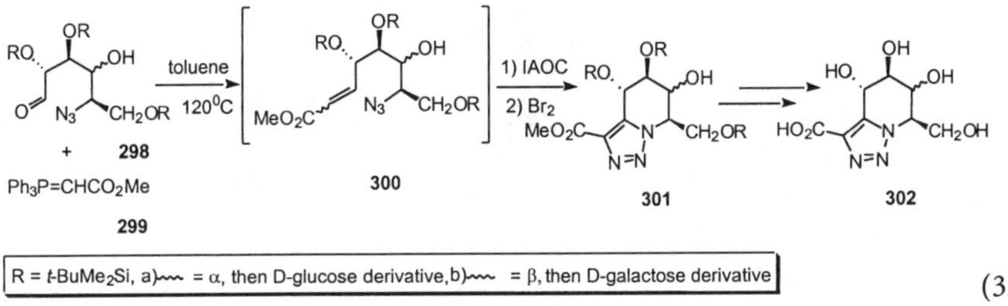

(33)

Azaspirocyclic ketoaziridines **304** (X = Cl or OTBS), potential intermediates for the total synthesis of antitumor alkaloid cephalotaxine **305**, have been prepared in 26% (X = Cl) and 76% (X = OTBS) yields, respectively, via an IAOC reaction of azide **303** (Eq. 34) [80].

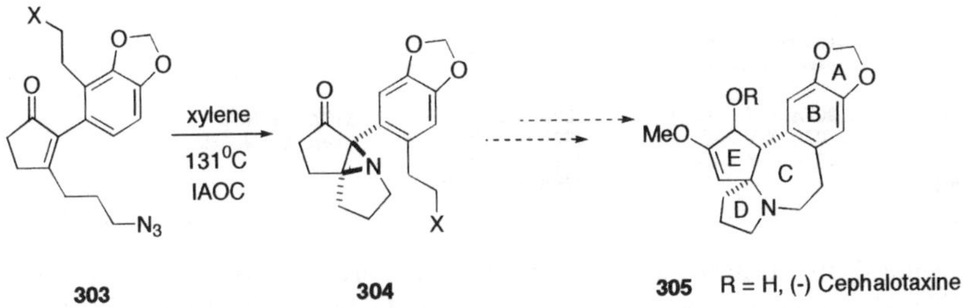

(34)

6 Intramolecular Nitrilimine Cycloaddition (INIC)

6.1 Generation and Reactivity of Nitrilimines

Nitrilimines are interesting 1,3-dipoles in that their reactions with olefins and acetylenes result in the formation of pyrazolines and pyrazoles, respectively. A simple and versatile procedure for the smooth generation of nitrilimines **306** by treatment of aldehyde hydrazones **311** with chloramine-T has been reported (Scheme 33) [81]. The method is applicable to aliphatic and aromatic aldehyde hydrazones and is superior to lead tetraacetate mediated dehydrogenation [82] of aldehyde hydrazones in terms of yield and reaction conditions. Other procedures for generation of nitrilimine intermediates include thermolysis or photolysis of either 2,5-diphenyltetrazole **307a** [83], oxathiadiazolines **307b** [84], 1,3,4-oxadiazolin-2-ones **307c** [85], sydnones **312** [86], or the sodium salt of α-nitroaldehyde hydrazones **310a** or pyridinium betaines **310b** [87]. The dehydrohalogenation of *N*-phenylbenzhydrazonyl halides **309** by triethylamine has been elaborated as a valuable source of nitrilimines **306** [88]. Nitrilimines **306** can also be formed by reaction of α-azobenzylhydroperoxide **309d**, an auto-oxidation product of aldehyde hydrazone, with triethylamine [89].

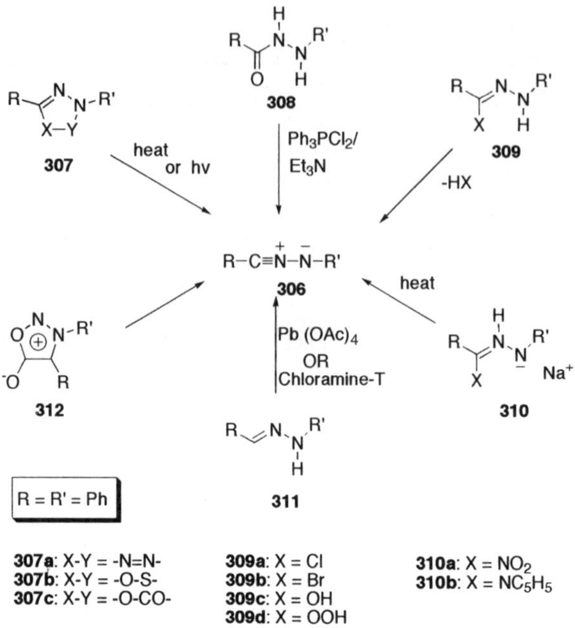

Scheme 33

In general, nitrilimines are generated in the presence of a suitable dipolarophile. Thus, heating an equimolar mixture of hydrazone **313**, alkene **314**, and chloramine-T trihydrate in ethanol under reflux for 3 h provided pyrazolines **315** in 68–90% yield [81]. The cycloaddition in all the cases was regiospecific as indicated by NMR (Table 23).

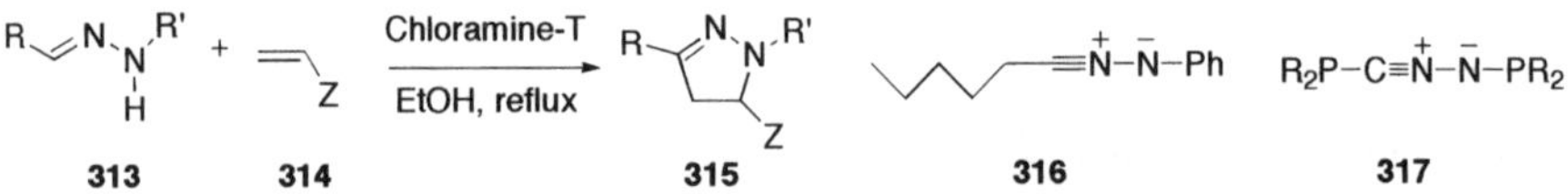

(35)

The only hydrazone previously converted to a nitrilimine is benzaldehyde phenylhydrazone **313a** which on treatment with $Pb(OAc)_4$ in the presence of acrylonitrile (**314a**) provided pyrazoline **315a** in 27% yield [90]. In all other

Table 23

313	a	a	a	b	b	c	d	e
R	Ph	Ph	Ph	i-Pr	i-Pr	*p*-MeOPh	*p*-MePh	*p*-MePh
R′	Ph	Ph	Ph	Ph	Ph	Ph	*p*-NO_2Ph	Ph
314	a	b	c	a	b	a	b	b
Z	CN	CO_2Et	Ph	CN	COEt	CN	COEt	COEt
315	a	b	c	d	e	f	g	h
% Yield	90	70	58	80	68	70	76	80

cases, $Pb(OAc)_4$ led mainly to formation of diacylhydrazides. By comparison, the procedure reported in Eq. (35) using chloramine-T gave **315a** in 90% yield [81]. Furthermore, the transient existence of nitrilimines **306** and **316** have been confirmed by ^1H NMR. The only nitrilimine that has been isolated so far is **317** whose stability has been attributed to steric factors [91].

The cycloaddition between a nitrilimine **319** and an aroyl substituted heterocyclic ketene aminal **318** has been found to be stepwise, involving an initial nucleophilic addition of **318** to **319** followed by intramolecular cyclocondensation of the intermediate **320** providing fully substituted pyrazole **321** (Eq. 36) [92]. When Ar′ was the 2,4-dinitrophenyl group, the intermediate **320** was isolable and required forcing conditions (xylene, reflux, 10 h) to undergo cyclization:

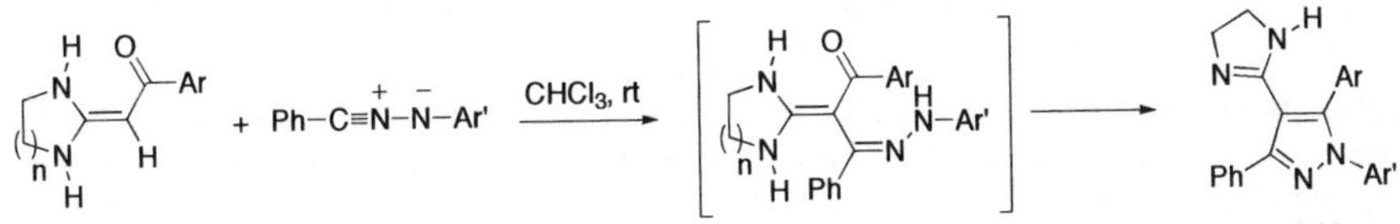

318 **319** **320** **321**

(36)

6.2 Intramolecular Cycloaddition

The applicability of the procedure shown in Eq. (35) for intramolecular cycloaddition was demonstrated by cyclization of *O*-allyloxybenzaldehyde hydrazone **322** to **323**, albeit in low (20%) yield (Eq. 37) [81].

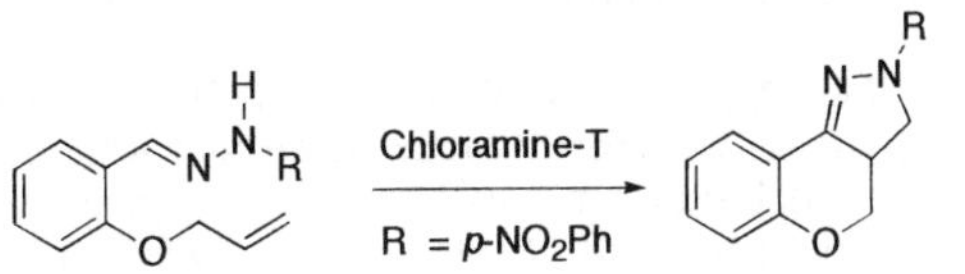

322 **323**

(37)

Alkenyl-substituted diarylnitrilimines **325**, generated by photolysis or thermolysis of corresponding tetrazoles **324**, undergo a regioselective INIC reaction to yield fused 2-pyrazolines (Scheme 34) [93]. Similarly, with alkynyl derivatives

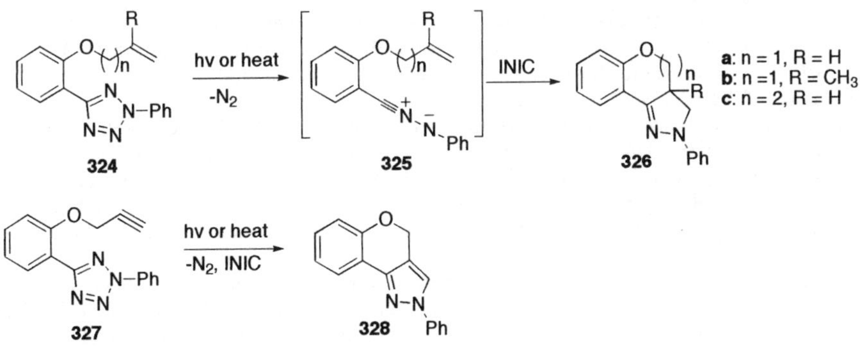

324 **325** **326** **327** **328**

Scheme 34

327, the corresponding pyrazoles **328** have been isolated. The nitrilimine **325a** has been characterized at –190 °C by UV spectroscopy.

Irradiation of 3,4-diarylsydnones **329** possessing an allyl or alkenyloxy substituent provided fused dihydropyrazoles **331** presumably via decarboxylation of sydnones **329** to nitrilimine **330** and the latter's INIC reaction (Eq. 38) [94].

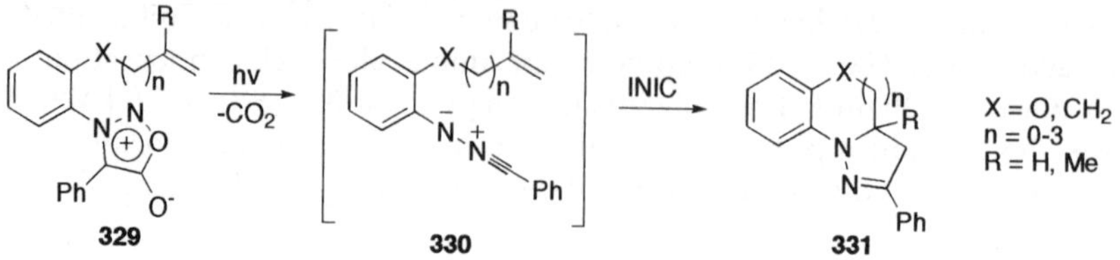

(38)

Acknowledgements. We are grateful to the US-Israel Binational Science Foundation, to the Marcus Center for Pharmaceutical and Medicinal Chemistry and to the Research Authority at Bar-Ilan University for support of this work.

7
References

1. a) Huisgen R (1963) Angew Chem Int Ed Engl 2:565; b) Huisgen R (1963) Angew Chem Int Ed Engl 2:633
2. Padwa A (ed) (1984) 1,3-Dipolar cycloaddition chemistry. Wiley, New York
3. Hassner A, Murthy KSK, Maurya R, Dehaen W, Friedman O (1994) J Heterocycl Chem 31:687
4. Curran DP (1988) In: Curran DP (ed) Advances in cycloaddition. JAI, London, vol 1, p 129
5. a) Kozikowski AP (1984) Acc Chem Res 17:410; b) Oppolzer W (ed) (1985) Symposia-in-print Number 21, Tetrahedron 41:3447; c) Jäger V, Müller R, Leibold T, Hein M, Schwarz M, Fengler M, Jaroskova L, Pärzel M, Le Roy P-Y (1994) Bull Soc Chim Belg 103:491
6. a) LeBel NA, Whang JJ (1959) J Am Chem Soc 81:6334; b) Padwa AW (1976) Angew Chem Int Ed Engl 15:123; c) Oppolzer W (1977) Angew Chem Int Ed Engl 16:10
7. Brieger G, Bennett JN (1980) Chem Rev 80:63
8. Broggini G, Zecchi G (1999) Synthesis 905
9. Kamimura A (1992) Yuki Gosei Kagaku Kyokaishi 50:808
10. Kanemasa S, Tsuge O (1990) Heterocycles 30:719
11. See for example: a) Grundman C, Dean JM (1965) J Org Chem. 30:2809; b) Hassner A, Rai KML (1989) Synthesis 57
12. a) Mukaiyama T, Hoshino T (1960) J Am Chem Soc 82:5339; b) Basel Y, Hassner A (1997) Synthesis 309; c) see also: Shimizu T, Hayashi T, Shibafuchi H, Teramura K (1986) Bull Chem Soc Jpn 59:2827 for Et chloroformate/Et_3N/DMAP method
13. Jung ME, Vu BT (1996) Tetrahedron Lett 37:451
14. a) Raposo C, Wilcox CS (1999) Tetrahedron Lett 40:1285; b) in DMSO at 150 °C; private communication; c) **6a** is the minor isomer because it requires the most hindered TS where the phenyl group must circumvent the dimethylamino group; d) The disappearance of **6b** has been attributed to a higher steric hindrance due to the new methyl group on the amine nitrogen and/or to electrostatic repulsion attending non-favorable approach of the dipole in the presence of the charge
15. Huisgen R (1988) In: Curran DP (ed) Advances in cycloaddition. Jay Press, London, vol 1, p 1
16. a) Just G, Dahl K (1968) Tetrahedron 24:5251; b) Grundmann C, Richter R (1968) J Org Chem 33:476; c) Kim JN, Ryu EK (1990) Synth Commun 20:1373; d) Rai KML, Linganna N, Hassner A (1992) Org Prep Proc Int 24:91

17. Hassner A, Rai KML (1990) Heterocycles 30:817
18. Hassner A, Rai KML, Dehaen W (1994) Synth Commun 24:1669
19. Falb E, Nudelman A, Gottlieb HE, Hassner A (2000) Eur J Org Chem 645
20. Mukhopadhyay R, Datta S, Chattopadhyay P, Bhattacharjya A, Patra A (1996) Ind J Chem 35B:1190
21. For two earlier and lengthier syntheses, see: a) Snider BB, Faith WC (1984) J Am Chem Soc 106:1443; b) Rousch WR, Walts AE (1984) J Am Chem Soc 106:721; c) Walts AE, Rousch WR (1985) Tetrahedron 41:3623
22. Harbour GC, Tymiak AA, Rinehart KL Jr, Shaw PD, Hughes RG Jr, Mizak SA, Coats JH, Zurenko GE, Li LH, Kuentzel SL (1981) J Am Chem Soc 103:5604
23. a) Hassner A, Murthy KSK (1986) Tetrahedron Lett 27:1407; b) Murthy KSK, Hassner A (1991) Isr J Chem 31:239; c) Hassner A, Murthy KSK, Maurya R, Dehaen W, Friedman O (1994) J Heterocyclic Chem 31:687
24. Ihara M, Tokunaga Y, Taniguchi N, Fukumoto K (1991) Tetrahedron 47:6635
25. a) Masaki M, Fukui K, Ohta M (1967) J Org Chem 32:3564; b) Smith JH, Heidema JH, Kaiser ET (1972) J Am Chem Soc 94:9276; c) Hassner A, Alexanian V (1979) J Org Chem 44:3861; d) Oppolzer W, Battig K, Hudlicky T (1981) Tetrahedron 37:4359
26. a) Corey EJ, Melvin LS, Haslaugh MF (1975) Tetrahedron Lett 3117; b) Gilchrist TL (1983) Chem Soc Rev 12:53; c) Denmark SE, Dappen MS (1984) J Org Chem 49:798
27. a) Stevens CL, Gillis BT (1957) J Am Chem Soc 79:3448; for formation of α-bromoaldehydes via silylenol ethers see: Reuss RH, Hassner A (1974) J Org Chem 39:1785
28. Hassner A, Murthy KSK (1987) Tetrahedron Lett 28:683
29. a) Padwa A, Chiacchio U, Dean DC, Schoffstall AM, Hassner A, Murthy KSK (1988) Tetrahedron Lett 29:4169; b) Hassner A, Murthy KSK, Padwa A, Chiacchio U, Dean DC, Schoffstall AM (1989) J Org Chem 54:5277
30. For other examples of dipoles adding across a π-bond of a furan see: Heinze I, Eberbach W (1988) Tetrahedron Lett 2051
31. a) Hassner A, Murthy KSK (1987) Tetrahedron Lett 28:4097; b) Hassner A, Murthy KSK, Padwa A, Bullock WH, Stull PD (1988) J Org Chem 53:5063
32. Yao C-F, Yang C-S, Fang H-Y (1997) Tetrahedron Lett 38:6419
33. a) Yao C-F, Kao K-H, Liu J-T, Chu C-M, Wang Y, Chen W-C, Lin Y-M, Lin W-W, Yan M-C, Liu J-Y, Chuang M-C, Shiue J-L (1998) Tetrahedron 54:791; b) see also: Liu J-T, Lin W-W Jang J-J, Liu J-Y, Yan M-C, Huang C, Kao K-H, Wang Y, Yao C-F (1999) Tetrahedron 55:7115; Yan M-C, Liu J-Y, Lin W-W, Kao K-H, Liu J-T, Jang J-J, Yao C-F (1999) Tetrahedron 55:12,493 for the use of alkyl chloroformate for the generation of nitrile oxides and in the intramolecular alkoxycarbonyl nitronate-olefin cycloaddition
34. Murthy KSK, Hassner A (1987) Tetrahedron Lett 28:97
35. Hassner A, Amarasekara AS, Padwa A, Bullock WH (1988) Tetrahedron Lett 29:715
36. Yip C, Handerson H, Jordan R, Tam W (1999) Org Lett 1:791
37. a) Dogbéavou R, Breau L (1997) Synlett 1208; b) Marrugo H, Dogbéavou R, Breau L (1999) Tetrahedron Lett 40:8979
38. Hassner A, Dehaen W (1991) Chem Ber 124:1181
39. Enders D, Haertwig A, Runsink J (1998) Eur J Org Chem 1793
40. Dehaen W, Hassner A (1991) J Org Chem 56:896
41. Hirai Y, Nagaoka H (1997) Tetrahedron Lett 38:1969
42. Fernández-Mateos A, Coca GP, González RR, Hernández CT (1996) J Org Chem 61:9097
43. Dehaen W, Hassner A (1990) Tetrahedron Lett 31:743
44. a) Gottlieb L, Hassner A (1995) J Org Chem 60:3759; b) Hassner A, Friedman O, Dehaen W (1997) Liebigs Ann 587; c) Namboothiri INN, Hassner A, Gottlieb HE (1997) J Org Chem 62:485
45. Garro-Helion F, Merzouk A, Guibe F (1993) J Org Chem 58:6109
46. Namboothiri INN, Hassner A (1996) J Organomet Chem 518:69
47. Armstrong P, Grigg R, Heaney F, Surendrakumar S, Warnock WJ (1991) Tetrahedron 47:4495

48. a) Lebel NA, Post ME, Whang JJ (1964) JACS 86:3759; b) We thank Prof. J. Leighton of Columbia University for this suggestion
49. a) Huisgen R, Grashey R, Hauck H, Seidl H (1968) Chem Ber 101:2043; b) Huisgen R (1976) J Org Chem 41:403; c) Leroy G, Nguyen MT, Sana M (1978) Tetrahedron 34:2459
50. a) Tufariello JJ, Ali SkA (1978) Tetrahedron Lett 19:4647; b) Tufariello JJ, Puglis JM (1986) Tetrahedron Lett 27:1265; c) Burdisso M, Gandolfi R, Grunanger P, Rastelli A (1990) J Org Chem 55:3427; d) Jung ME, Gervay J (1990) Chemtracts 3:284
51. a) Lebel NA, Banucci EG (1971) J Org Chem 36:2440; b) Burdisso M, Gandolfi R, Grunanger P (1989) Tetrahedron 45:5579
52. a) Bianchi G, Micheli CD, Gandolfi R (1979) Angew Chem Int Ed Engl 18:721; b) Bianchi G, Gandolfi R (1984) In: Padwa A (ed) 1,3-Dipolar cycloaddition chemistry. Wiley, New York, vol 2, chap 14
53. Huang KS-L, Lee EH, Olmstead MM, Kurth MJ (2000) J Org Chem 65:499
54. a) Cheng Q, Oritani T, Hassner A (2000) Synth Commun 30:293; b) Cheng Q, Hassner A (2000) (unpublished results)
55. Young DGJ, Gomez-Bengoa E, Hoveyda AH (1999) J Org Chem 64:692
56. Stepanov AV, Veselovsky VV (1997) Russ Chem Bull 46:1606
57. Stepanov AV, Lozanova AV, Veselovsky VV (1998) Russ Chem Bull 47:2286
58. a) Armstrong P, Grigg R, Warnock WJ (1987) J Chem Soc Chem Commun 1325; Grigg R (1987) J Chem Soc Rev 16:89; b) Padwa A, Norman BH (1988) Tetrahedron Lett 29:2417
59. Hassner A, Maurya R, Mesko E (1988) Tetrahedron Lett 29:5313
60. Oppolzer W, Keller K (1970) Tetrahedron Lett 1117
61. Hassner A, Maurya R, Friedman O, Gottlieb HE, Padwa A, Austin D (1993) J Org Chem 58:4539
62. Raban M, Kost D (1984) Tetrahedron 40:3345
63. Hassner A, Maurya R (1989) Tetrahedron Lett 30:5803
64. a) See [7] for a recent review; b) Hassner A, Maurya R (1989) Tetrahedron Lett 30:2289; c) Hassner A, Maurya R, Padwa A, Bullock WH (1991) J Org Chem 56:2775
65. Gotoh M, Sun B, Hirayama K, Noguchi M (1996) Tetrahedron 52:887
66. Sun B, Adachi K, Noguchi M (1996) Tetrahedron 52:901
67. Mihailovic M, Rajkovic MM, Lorenc LB, Pavlovic VD, Milovanovic AZ, Tinant, B, Declercq J-P (1996) Tetrahedron 52:11995
68. a) Hassner A, Falb E, Nudelman A, Albeck A, Gottlieb HE (1994) Tetrahedron Lett 35:2397; b) Falb E, Bechor Y, Nudelman A, Hassner A, Albeck A, Gottlieb HE (1999) J Org Chem 64:498
69. Chiacchio U, Corsaro A, Pistara V, Rescifina A, Romeo G, Romeo R (1996) Tetrahedron 52:7875
70. a) Ishizuka T, Kunieda T (1987) Tetrahedron Lett 28:4185; b) Abd Elall EHM, Al Ashmawy MI, Mellor JM (1987) J Chem Soc Perkin Trans 1 2729
71. Hassner A, Murthy K (1987) Tetrahedron Lett 28:97
72. Hassner A, Stern M (1986) Angew Chem Int Ed Engl 25:478
73. For a review of triazoline decomposition by thermolysis and photolysis see: Scheiner P (1970) In: Thyagarajan (ed) Selective organic transformations. Wiley, vol 1, pp 327–362
74. Rai KML, Hassner A (1990) Heterocycles 30:817
75. Hassner A, Fibiger R, Amarasekara AS (1988) J Org Chem 53:22
76. Hassner A, Amarasekara AS, Andisik D (1988) J Org Chem 53:27
77. Schkeryantz JM, Pearson WH (1996) Tetrahedron 52:3107
78. Norris P, Horton D, Giridhar DE (1996) Tetrahedron Lett 37:3925
79. Krülle TM, de la Fuente C, Pickering L, Aplin RT, Tsitsanou KE, Zographos SE, Oikonomakos NG, Nash RJ, Griffiths RC, Fleet GW (1997) Tetrahedron: Asymmetry 8:3807
80. Molander GA, Hiersemann M (1997) Tetrahedron Lett 38:4347
81. Rai KML, Hassner A (1989) Synth Commun 19:2799
82. Gladstone WAF, Aylward JB, Norman ROC (1969) JCS Chem Commun 2587
83. a) Huisgen R, Seidel M, Sauer J, Mcfarland JW, Wallbillich G (1959) J Org Chem 24:892; b) Huisgen R, Sauer J, Seidel M (1961) Chem Ber 94:2503

84. Reimlinger H, Vandewalle JJM, King GSD, Lingier WRF, Merenyi R (1970) Chem Ber 103:1918
85. a) Sauer J, Mayer KK (1968) Tetrahedron Lett 325; b) Marky M, Meier H, Wunderli A, Hemigartner H, Schmid H, Hansen HJ (1978) Helv Chim Acta 61:1477
86. a) Ollis, WD, Ramseden CA (1976) Adv Heterocycl Chem 19:1; b) Gotthardt H, Reiter F (1979) Chem Ber 112:1206, 1635; c) Pfoertner KH, Foricher J (1980) Helv Chim Acta 63:653
87. Shawali AS, Hasseneen HM, Sherif SM (1980) J Heterocycl Chem 17:1745
88. a) Huisgen R, Seidel M, Wallbillich G, Knupfer H (1962) Tetrahedron 17:3; b) Hasseneen HM, Mousa HAM, Abed NM, Shawali AS (1988) Heterocycles 27:695; c) Hasseneen HM, Ead HA, Elwan NM, Shawali AS (1988) Heterocycles 27:2857; d) El-Abadellah MM, Hussein AQ, Kamal MR, Al-Adhami KH (1988) Heterocycles 27:917; e) Hegarty AF, Cashman MP, Scott FL (1972) JCS Perkin Trans 2 44; f) Baruah AK, Prajapathi D, Sandhu JS (1988) Tetrahedron 44:6137
89. Tezuka T, Otsuka T (1988) Chem Lett 1751
90. Gladstone WAF, Aylward JB, Norman ROC (1969) J Chem Soc 2587
91. Sicard G, Baceiredo A, Bertrand G (1988) J Am Chem Soc 110:2663
92. Liu B, Wang M-X, Huang Z-T (1999) Tetrahedron Lett 40:7399
93. Meier vH, Heimgartner H (1985) Helv Chim Acta 68:1283
94. Meier vH, Heimgartner H (1986) Helv Chim Acta 69:927

4-Acetoxy- and 4-Cyano-1,3-Dioxanes in Synthesis

Christopher J. Sinz · Scott D. Rychnovsky

Department of Chemistry, University of California, Irvine, California, 92627-2025, USA
E-mail: srychnov@uci.edu

Alkylations of 4-cyano-1,3-dioxanes (cyanohydrin acetonides) represent a highly practical approach to *syn*-1,3-diol synthesis. Herein we present a comprehensive summary of cyanohydrin acetonide chemistry, with particular emphasis on practical aspects of couplings, as well as their utility in natural product synthesis. Both 4-acetoxy-1,3-dioxanes and 4-lithio-1,3-dioxanes have emerged as interesting *anti*-1,3-diol synthons. The preparation and utility of these two synthons are described.

Keywords: Cyanohydrin acetonide alkylations, Reductive decyanations, Oxocarbenium ions, Reductive lithiation

1	**Introduction**	52
2	**4-Cyano-1,3-Dioxanes in Synthesis**	52
2.1	Background	53
2.1.1	Alkylation/Reductive Lithiation of Glucopyranosyl Sulfones	53
2.1.2	Stork Cyanohydrin Alkylations	54
2.2	Preparation of 4-Cyano-1,3-dioxanes	54
2.3	Cyanohydrin Acetonide Alkylations and Reductive Decyanations	55
2.3.1	Alkylation Selectivity	55
2.3.2	Reductive Decyanation Selectivity	56
2.3.3	Practical Aspects of Cyanohydrin Acetonide Alkylations	58
2.3.4	Challenging Electrophiles	60
2.4	Cyanohydrin Acetonides in Natural Product Synthesis	61
2.5	Radical Nitrile Transfer Reactions	72
3	**4-Acetoxy-1,3-dioxanes in Synthesis**	72
3.1	Background	72
3.2	Preparation of 4-Acetoxy-1,3-dioxanes	73
3.3	Nucleophilic Additions to 4-Acetoxy-1,3-dioxanes	74
3.3.1	Allyl/Crotyl Metal Reagents	74
3.3.2	Alkyl Zinc Reagents	77
3.3.3	Alkynyl Metal Reagents	80

Topics in Current Chemistry, Vol. 216

4	**4-Lithio-1,3-dioxanes in Synthesis**	82
4.1	Background	82
4.2	Kinetic and Thermodynamic Stability of 4-Lithio-1,3-dioxanes	83
4.2.1	Experimental Studies	83
4.2.2	Theoretical Studies	86
4.3	Synthetic Applications	87
5	**Alternative Approaches to 1,3-Diol Synthesis**	90
6	**Conclusions**	90
7	**References**	91

1 Introduction

Polyol chains are common features in highly oxygenated natural products. The synthesis of such chains is best effected by a convergent strategy. We have found that reactive synthons derived from 1,3-dioxanes have unique advantages for these targets. First, reactive intermediates at the 4-position of a 1,3-dioxane are subject to anomeric effects and often show very high levels of stereoselectivity. Second, the products of coupling reactions using these intermediates are protected by the 1,3-dioxane structure, and these protecting groups can often be taken through the synthesis without further manipulation, except for a final deprotection. The highly selective coupling reactions avoid tedious diastereomer separations, and the coupling of protected segments avoids complex reprotections that can add many steps to a synthetic sequence. These synthons have been used in the rapid and efficient synthesis of a number of complex natural products.

We have investigated a variety of synthons incorporating reactive intermediates at the 4-position of a 1,3-dioxane. The cyanohydrin acetonides couple as anionic intermediates, but the configuration of the resulting diol is introduced through a radical reduction. Cyanohydrin acetonide synthons lead to *syn*-1,3-diols. Organometallic couplings with 4-acetoxy-1,3-dioxanes proceed through oxocarbenium ion intermediates, and are related to a number of methods used in the synthesis of *C*-glycosides. These synthons lead to *anti*-1,3-diols. Finally, 4-lithio-1,3-dioxanes are anionic intermediates that can be manipulated to produce either *syn*- or *anti*-1,3-diols. The strengths and weaknesses of each of these synthons will be discussed below.

2 4-Cyano-1,3-Dioxanes in Synthesis

The intriguing structural complexity and often potent biological activity exhibited by many polyacetate-derived natural products have generated a continuing interest in the development of efficient, stereoselective approaches to the assembly of repeating 1,3-diol subunits. One such method is the alkylation and reduc-

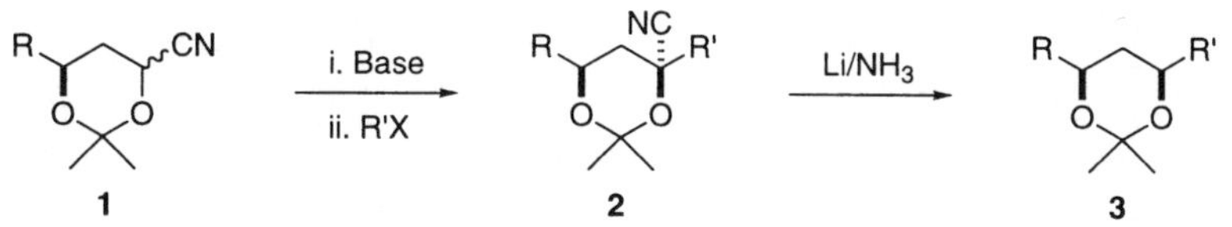

Scheme 1

tive decyanation of 4-cyano-1,3-dioxanes **1** (cyanohydrin acetonides) (Scheme 1). The aim of this section is to highlight the utility of these *syn*-1,3-diol synthons in natural product synthesis, as well as to describe some practical considerations regarding cyanohydrin acetonide alkylations and reductive decyanations.

2.1 Background

2.1.1 *Alkylation/Reductive Lithiation of Glucopyranosyl Sulfones*

Beau and Sinäy described a method which laid the groundwork for cyanohydrin acetonide alkylations [1]. Their strategy involved alkylation and reductive desulfonylation of glucopyranosyl sulfones **4**. In this one-pot procedure, low temperature alkylation and subsequent reductive desulfonylation with lithium naphthalenide generated β-*C*-glycosides with good selectivity (> 10:1 β:α) and in moderate to good yield (Eq. 1).

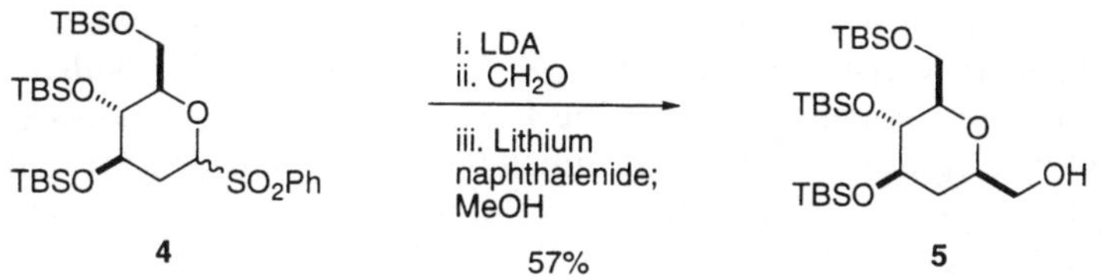 (1)

The anomeric configuration is set in the reductive lithiation step, which proceeds via a radical intermediate. Hyperconjugative stabilization favors axial disposition of the intermediate radical, which after another single electron reduction leads to a configurationally stable α-alkoxylithium intermediate. Protonation thus provides the β-anomer. The authors were unable to determine the stereoselectivity of the alkylation step, due to difficulty with isolation. However, deuterium labeling studies pointed to the intervention of an equatorially disposed α-alkoxylithium 7 (thermodynamically favored due to the reverse anomeric effect) which undergoes alkylation with retention of configuration (Eq. 2).

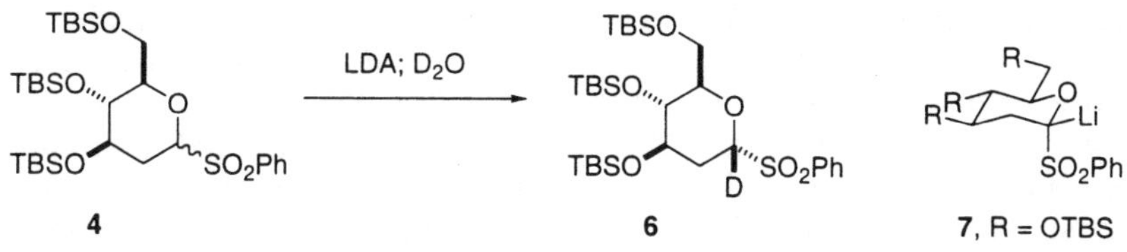 (2)

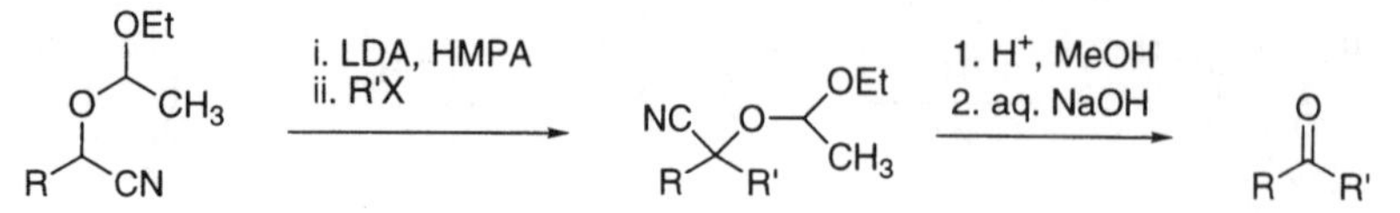

8 9 10

Scheme 2

2.1.2
Stork Cyanohydrin Alkylations

Stork first demonstrated the utility of protected cyanohydrins as acyl anion equivalents in 1971 [2]. The acetal-protected cyanohydrin **8** was transformed into the corresponding anion with LDA in THF/HMPA, which was then alkylated with a range of alkyl halides, including secondary bromides (Scheme 2). A mild acidic hydrolysis yielded a cyanohydrin, which provided the ketone after treatment with base. The Stork cyanohydrin alkylation and its variants have become important methods in natural product synthesis [3, 4].

2.2
Preparation of 4-Cyano-1,3-Dioxanes

4-Cyano-1,3-dioxanes are typically prepared from the corresponding β-hydroxy esters [5]. Temporary protection of the alcohol as a trimethylsilyl ether, reduction of the ester with DIBALH, and treatment of the resulting aldehyde with a slight excess of TMSCN and a catalytic quantity of KCN/18-crown-6 complex [6] give a β-trimethylsilyloxy cyanohydrin. Acetonide formation with acetone and 2,2-dimethoxypropane generates the 4-cyano-1,3-dioxane as a nearly 1:1 mixture of diastereomers. It should be noted that the cyanohydrin formation can be run neat or in dichloromethane. Conversion of ester **11** (Eq. 3) to cyanohydrin acetonide **12** proceeded in 87% overall yield [7, 8]; for a wide range of β-hydroxy esters, the yield for this sequence is greater than 70%, and only a single chromatographic purification is required.

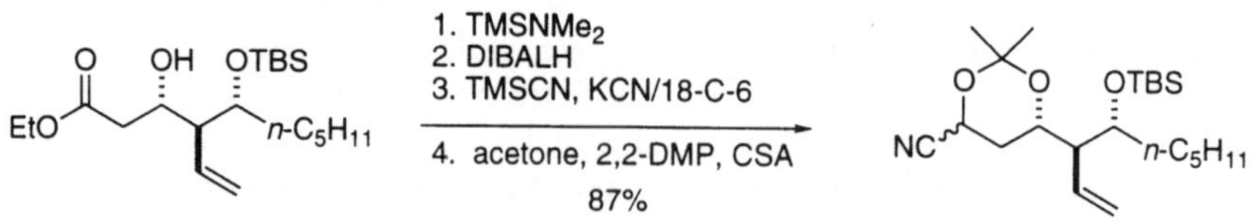

11 12 (3)

A 1,3-diol sometimes represents a more convenient precursor to cyanohydrin acetonides. For these instances, an alternate procedure has been developed. Selective oxidation of a 1,3-diol with TEMPO/NaOCl generates a sensitive β-hydroxy aldehyde (see also Sect. 3.2). The neat β-hydroxy aldehyde is prone to dimerization, but can be handled in solution without significant dimerization. Conversion to the cyanohydrin acetonide is accomplished in a manner similar to that described above. Following this procedure, diol **13** was converted to 4-cyano-1,3-dioxane **14** in 71% overall yield [9] (Eq. 4). Yields as high as 90% have been realized for this sequence [10].

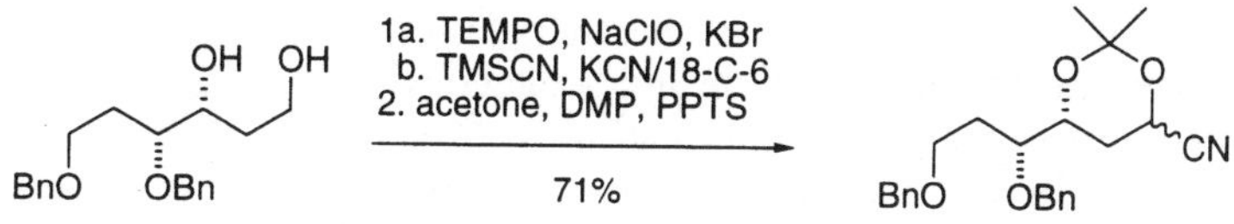

13 **14** (4)

2.3 Cyanohydrin Acetonide Alkylations and Reductive Decyanations

2.3.1 *Alkylation Selectivity*

Though the *syn*-1,3-diol relationship is ultimately established in the reductive decyanation (*vide infra*), the alkylation is itself highly selective. Selectivities are typically greater than 100:1 in favor of the axial nitrile. This selectivity can be rationalized by a chair-like intermediate **15** (Scheme 3) for which equatorial alkylation is highly favored on steric grounds. Approach of the electrophile from an axial trajectory leads to a *syn*-pentane-like interaction.

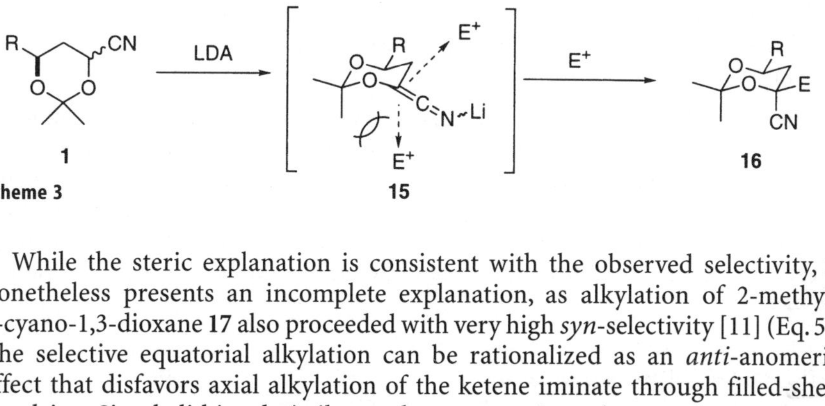

1 **15** **16**

Scheme 3

While the steric explanation is consistent with the observed selectivity, it nonetheless presents an incomplete explanation, as alkylation of 2-methyl-4-cyano-1,3-dioxane **17** also proceeded with very high *syn*-selectivity [11] (Eq. 5). The selective equatorial alkylation can be rationalized as an *anti*-anomeric effect that disfavors axial alkylation of the ketene iminate through filled-shell repulsion. Simple lithiated nitriles are known to exist as ketene iminates, but it would be easy to rationalize the preference for equatorial alkylation by considering the relative stability of hypothetical equatorial and axial alkyllithium reagents, *vide infra*. Preferential equatorial alkylation was also observed by Beau

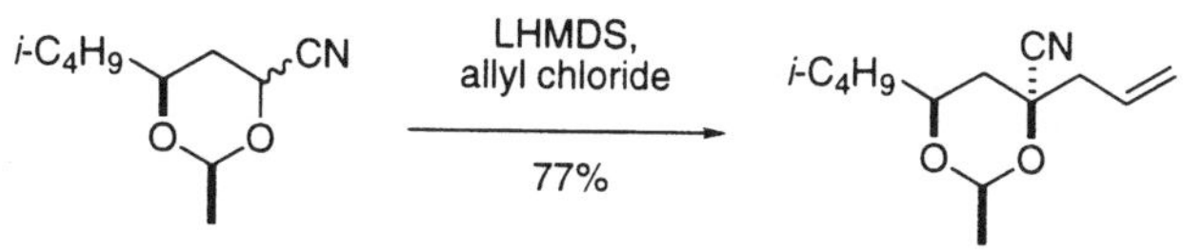

17 **18** (5)

and Sinäy in their glucopyranosyl sulfone alkylations (Eq. 1). The sulfone alkylations are likely to involve a discrete equatorial alkyllithium intermediate in which the sterically demanding sulfone adopts an axial configuration [1].

Leahy demonstrated that unsaturation at the 5-position of a 4-cyano-1,3-dioxane can lead to a reversal in selectivity [12] (Eq. 6). Alkylation of cyanohydrin acetonide **19** with benzyl bromide generated a 9:1 mixture of **20** and **21**, with the *anti*-isomer **20** predominating, in 57% overall yield. An alkylithium intermediate in which overlap with the methylidene π^* orbital favors the axial configuration could account for this anomalous selectivity.

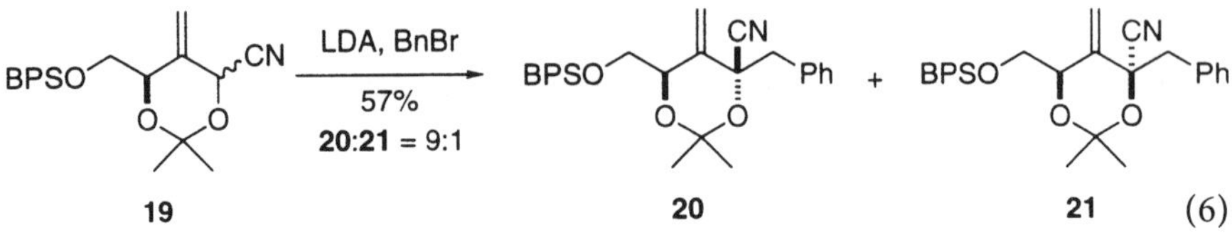

(6)

2.3.2
Reductive Decyanation Selectivity

The *syn*-1,3-diol acetonide is ultimately established by reductive decyanation. These reactions proceed with exceptionally high selectivity. The selectivity observed in reductive decyanations could in principle have two origins:

1) The reaction simply proceeds with retention of configuration, and the diastereomer ratio reflects that of the alkylation.
2) The selectivity arises from axial protonation.

Rychnovsky demonstrated that the latter explanation is correct; in reductive decyanations, the intermediate radical equilibrates to the most stable (axial) radical, and this equilibration determines the stereochemical outcome. Reductive decyanation of a 52:48 mixture of cyanohydrin acetonides **22** provided the *syn*-product **25** with 99:1 selectivity (Scheme 4). *Ab initio* calculations revealed a ca. 3.5 kcal/mol enthalpy difference between the axial and equatorial radical

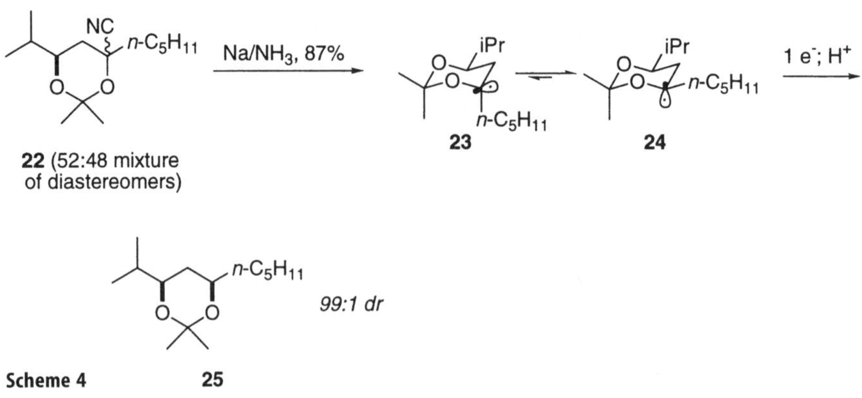

Scheme 4

intermediates [13]. These results are consistent with earlier observations by Beau and Sinäy [1] and Cohen [14] in reductions of sulfones and sulfides, respectively. Husson has demonstrated that conformationally constrained α-aminonitriles undergo reductive decyanation with high selectivity [15, 16].

Alternative conditions for reductive decyanations can be used. The allylic ether in compound **26**, an intermediate in a total synthesis of (–)-roxaticin, was prone to reduction when treated with lithium in liquid ammonia. Addition of the substrate to an excess of lithium di-*tert*-butylbiphenylide in THF at –78 °C, and protonation of the alkyllithium intermediate provided the reduced product **27** in 63% yield, as a single diastereomer (Eq. 7). α-Alkoxylithium intermediates generated in this manner are configurationally stable at low temperature, and can serve as versatile synthons for carbon-carbon bond forming processes (see Sect. 4).

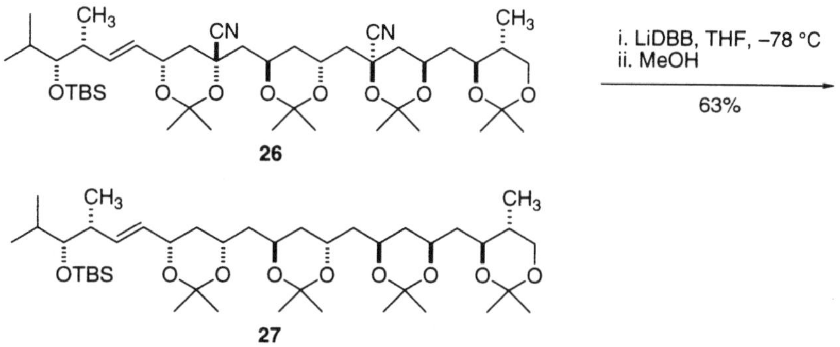

(7)

During the course of the development of our group's alkylation/reductive decyanation strategy, a very reliable method for distinguishing between *syn*- and *anti*-1,3-diols was discovered [17, 18]. The acetonide methyl groups reliably display diagnostic ^{13}C-NMR chemical shifts, allowing for stereochemistry to be determined simply by inspection (Fig. 1). Evans later extended the ^{13}C-NMR chemical correlation to polypropionate chains [19, 20].

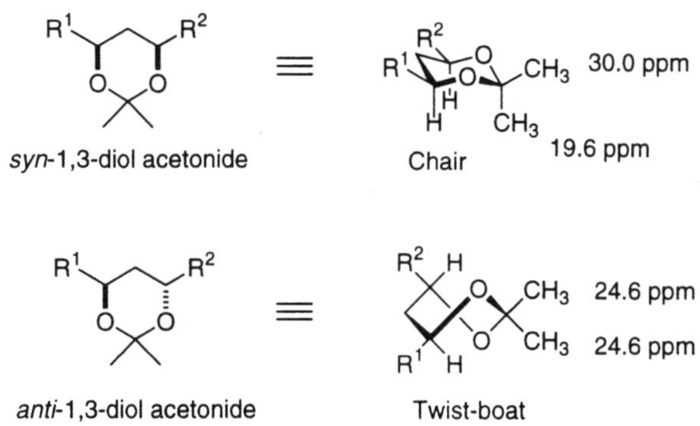

Fig. 1. Determination of 1,3-diol stereochemistry by the ^{13}C acetonide method

2.3.3
Practical Aspects of Cyanohydrin Acetonide Alkylations

Lithiated cyanohydrin acetonides are potent nucleophiles. Reactive electrophiles like butyl bromide work well (Eq. 8). Less reactive electrophiles like β-alkoxy- and β-silyloxy bromides (Eqs. 9 and 10) also smoothly participate in alkylations. Increased steric bulk near the reacting center of the cyanohydrin acetonide is well tolerated (Eq. 11) [21].

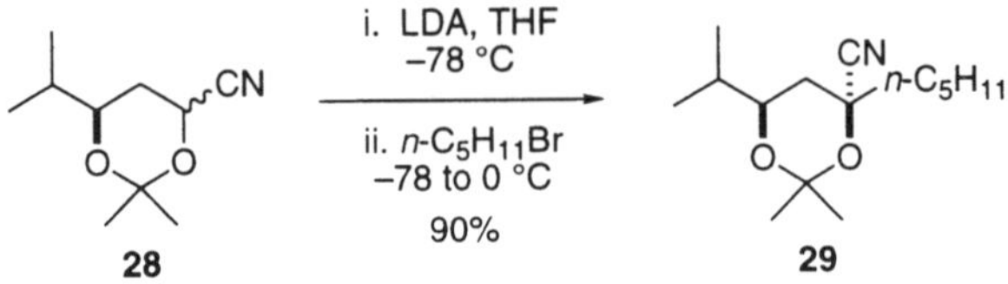

(8)

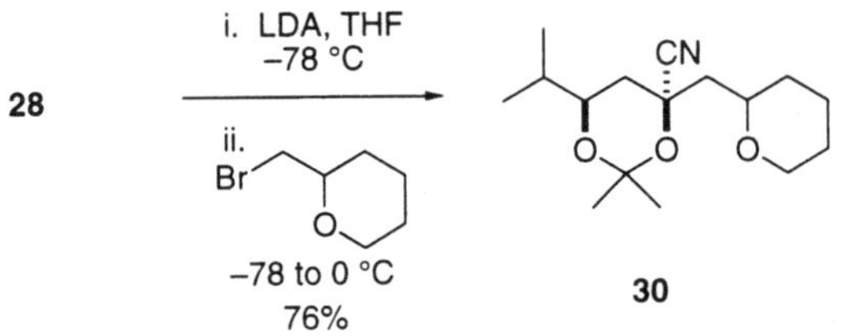

(9)

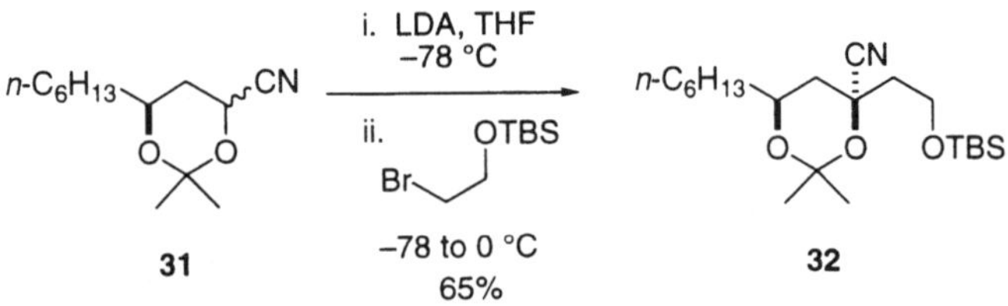

(10)

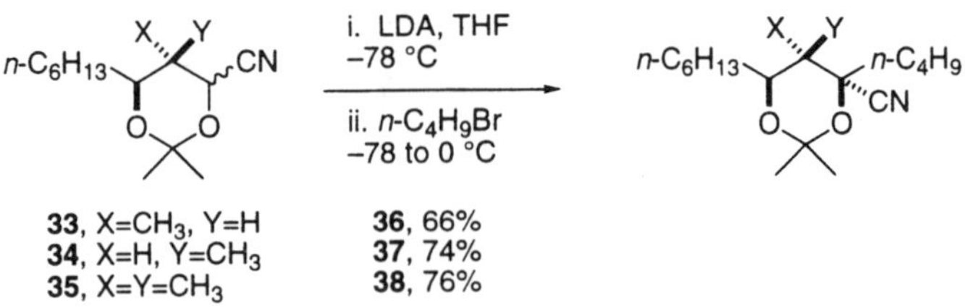

(11)

A range of amide bases can be employed. Typically LDA is used, but in certain complex cases, $LiNEt_2$ was found to be more effective. One exceptional case involves the ostensibly simple alkylation of a cyanohydrin acetonide with allyl chloride (Eq. 12). Here, use of LDA gave essentially none of the desired product **39**, whereas KHMDS or LHMDS gave excellent yields [5].

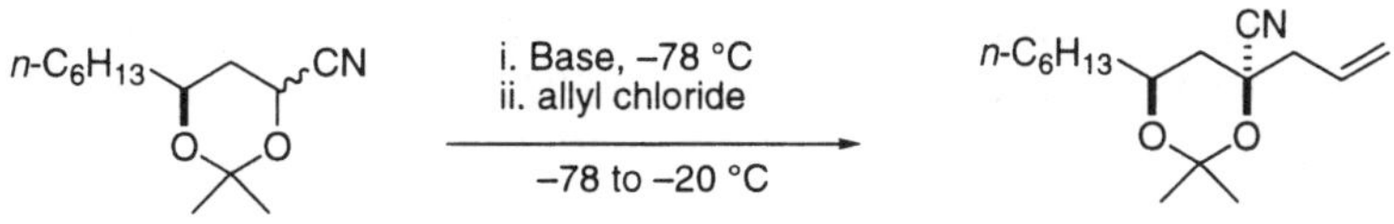

31 **39**

Base	% Yield
LDA	5
KHMDS	73
LHMDS	82

(12)

The difference in reactivity between the anions generated from LDA and LHMDS is difficult to rationalize, but nonetheless reproducible. The same effect has been observed with substituted allyl halides and propargyl halides. Instability of the product under the reaction conditions may account for this phenomenon. Thus, for alkylation of allylic and propargylic halides, LHMDS and KHMDS are the bases of choice.

Cyanohydrin acetonide alkylations do not require strict adherence to a general protocol. Originally, the deprotonation was performed at –78 °C for 1–2 h, followed by the optional addition of DMPU (DMPU has an accelerating effect on alkylations, but in some cases was found to be deleterious [8]) and the electrophile, and subsequent warming to –20 °C. Highly reactive electrophiles can undergo alkylation at –78 °C, but less reactive electrophiles, such as β-alkoxyalkyl halides, require warming to –40 °C or above for complete reaction. A temperature profile study for a typical alkylation showed that nitrile anions are stable at –20 °C or below, but begin to decompose around 0 °C [5].

Reports by Takahashi [22] and Stork [23] on the intramolecular cyclizations of protected cyanohydrin anions pointed to the possibility of generating the cyanohydrin acetonide anions in the presence of the electrophile. This procedure would be particularly advantageous on a small scale, as any adventitious water could be simply titrated away. In a recently developed procedure [10], excess base (LDA or $LiNEt_2$) is added until TLC indicates complete consumption of the cyanohydrin acetonide. This procedure has been exploited in an ongoing total synthesis [24] of dolabelide A [25] (Eq. 13). In this case, the "premix" conditions have served equally well in gram and milligram scale couplings to provide **42** in excellent yield. In contrast to some other segment coupling reactions,

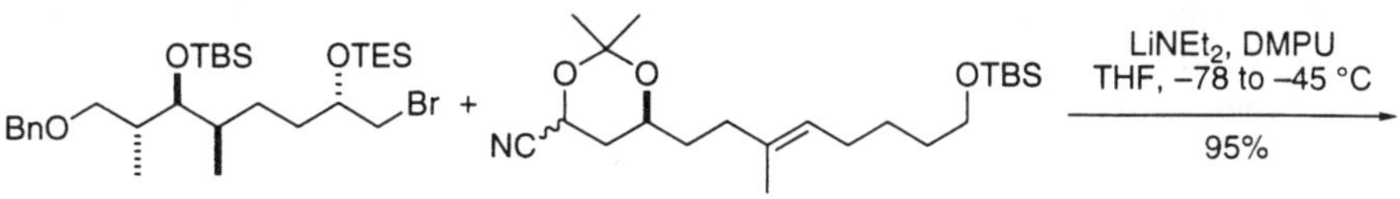

40 **41**

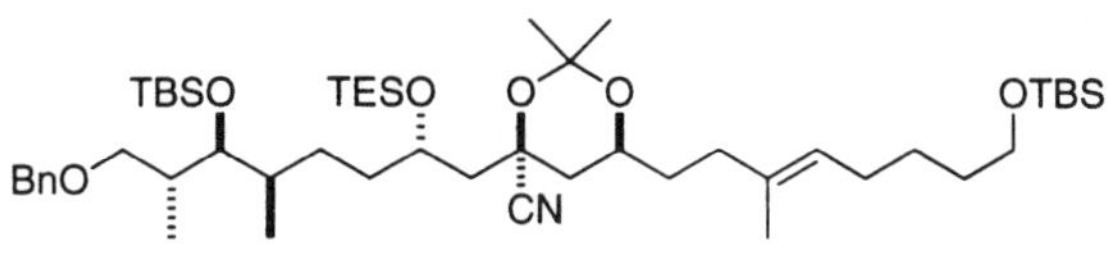

42

(13)

a large excess of the nucleophile is not required. In the example shown above, as little as 1.8 equiv. of the cyanohydrin acetonide can be used, without lowering the yield.

2.3.4
Challenging Electrophiles

Most often, the application of cyanohydrin acetonide couplings to a natural product synthesis calls for coupling with a primary alkyl halide. This has proven successful in every instance. However, on occasion, alkylations of more hindered epoxides or hindered alkyl halides are desirable. These reactions are less dependable.

A novel strategy for remote stereocontrol was employed in an approach to the spiroacetal fragment of milbemycin b1 [26]. Treatment of spiroacetal **43** with lithium(pyrrolidide), followed by addition of epoxide **44** provided **45** in 72% yield (Scheme 5). The stereogenic centers at C24 and C25 in **43** lock the spiroacetal rings in a conformation that defines the C19 axial and equatorial positions. Reductive decyanation established the configuration at C19 with concomitant removal of the terminal benzyl ether. Cleavage of the methyl acetal and spiroacetalization gave the target **46** in 57% yield from **45**.

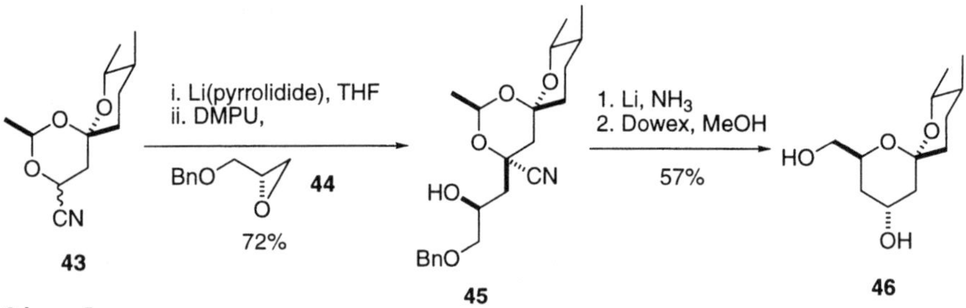

Scheme 5

While alkylation of terminal epoxides is reliable, attempted alkylations of 1,2-disubstituted epoxides have proved capricious. An unsuccessful approach to the swinholides, which called for the alkylation of cyanohydrin **47** with epoxide **48**, is one such example. In the event, alkylation cleanly produced imidate **49**, rather than the expected product **50** [27] (Eq. 14).

An alternative coupling strategy for a related fragment of the swinholides highlights the challenges of alkylating secondary, β-branched alkyl halides. Model studies hinted at the feasibility of this approach; alkylation of cyanohydrin **51** with a γ-branched iodide **52** proceeded smoothly [28] (Eq. 15). However, extension to a more complex system was unsuccessful (Eq. 16). The target compound **55** could be isolated in only 16% yield, due in large part to competitive formation of **56** by dehydrohalogenation [28].

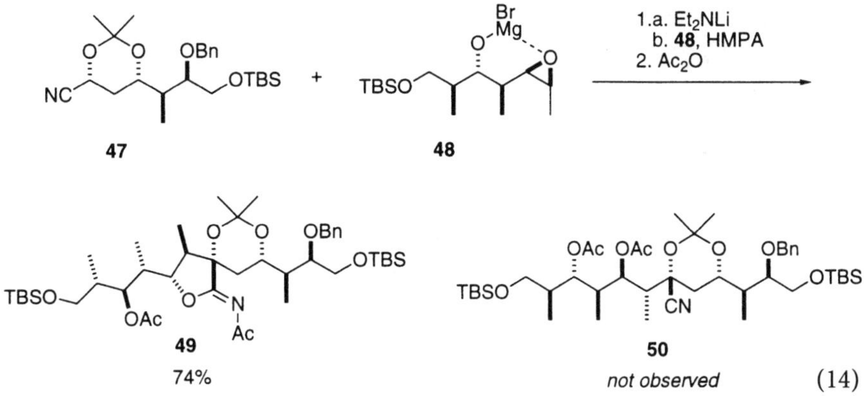

(14)

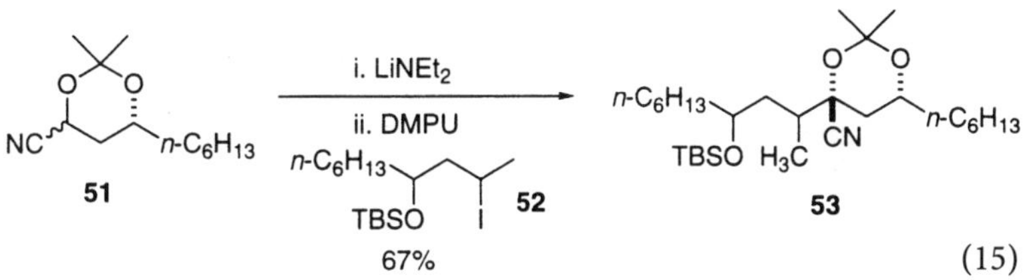

(15)

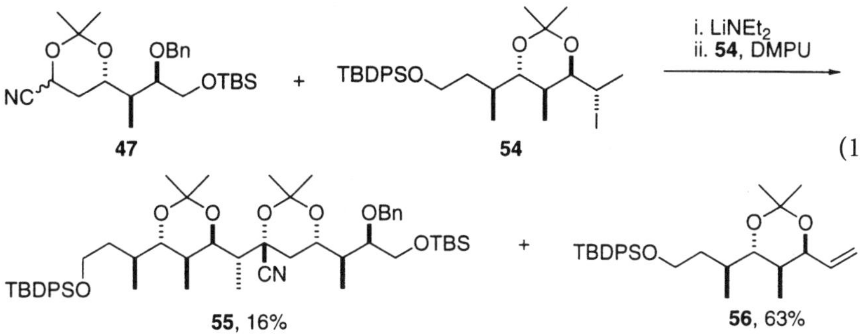

(16)

2.4
Cyanohydrin Acetonides in Natural Product Synthesis

The cyanohydrin acetonide method has been a valuable tool in natural product synthesis. The first reported demonstration of this strategy was the total synthesis of (–)-roxaticin [29]. In this approach, treatment of cyanohydrin **57** with an excess of the C_2-symmetrical dibromide **58** provided **59**, without overalkylation (Scheme 6). A second alkylation involving cyanohydrin **60** gave **61** in excellent yield. (–)-Roxaticin was accessed in ca. 10 steps from tetraacetonide **62**.

In another application of the cyanohydrin acetonide method, cyanohydrin acetonide **64** (Fig. 2) was developed as a common precursor to both the nucleophilic and electrophilic components of a convergent coupling [30]. Orthogonal

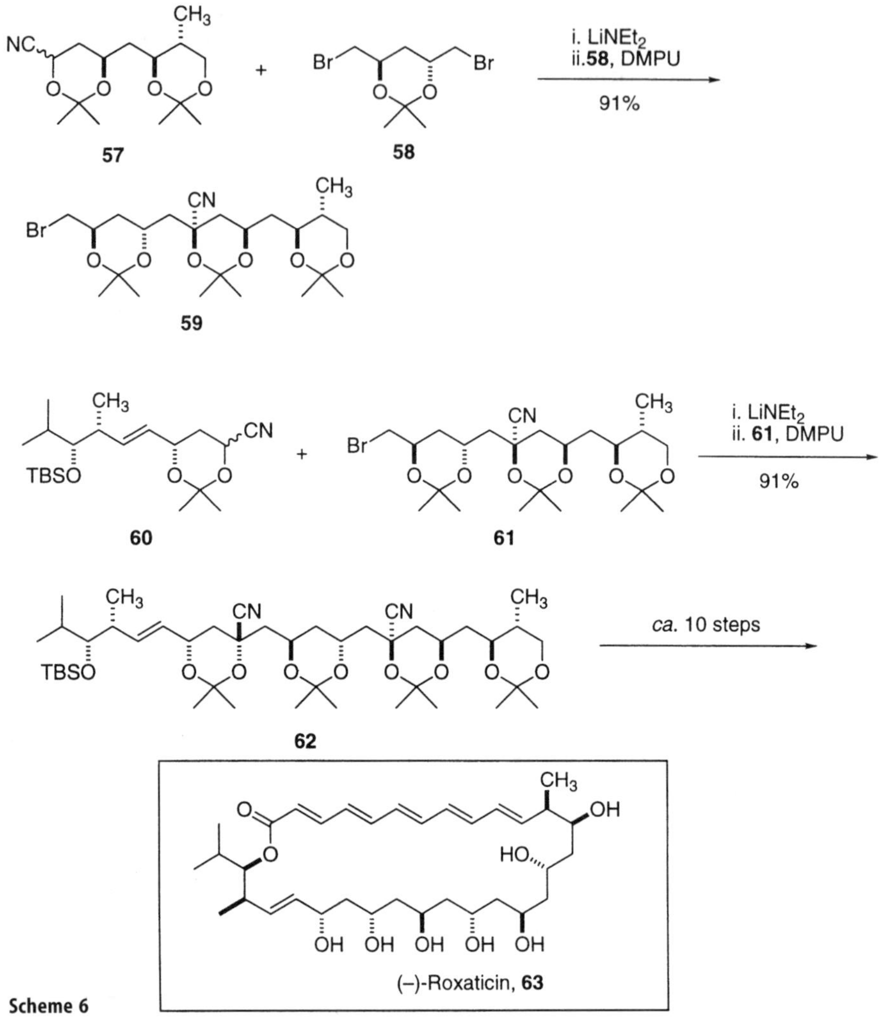

Scheme 6

nucleophilic activation with base or electrophilic activation by conversion to the more reactive iodide **65**, allowed for an approach to *syn*-polyols that is simultaneously iterative and convergent.

This strategy was applied to the synthesis [30] of the permethylated isotactic alternating all-*syn* polyols isolated from the blue-green alga *Tolypothrix conglutinata* var. *chlorata* [31] (Fig. 3). Electrophilic activation of chloride **64** by conversion to the corresponding iodide was effected under forcing conditions, namely treatment with potassium iodide and 18-crown-6 in refluxing xylenes (Scheme 7). Alkylation of **64** with this iodide then generated the homologue **68**, which possesses the same potential for orthogonal activation. Thus, this material was treated with KHMDS, and alkylated with allyl chloride to provide **69** in 89% yield.

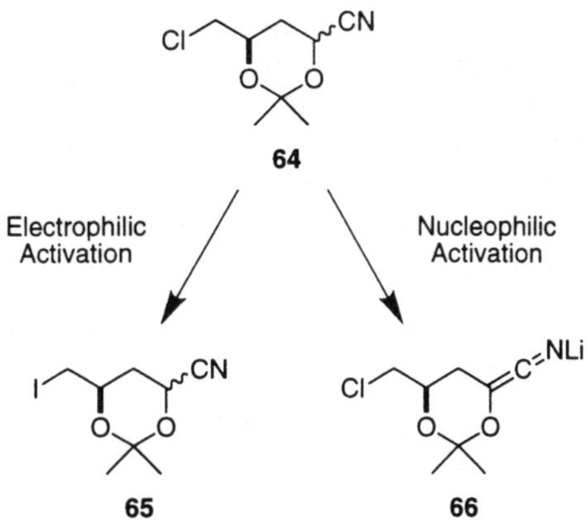

Fig. 2. Orthogonal activation of 1,3-diol synthon **64**

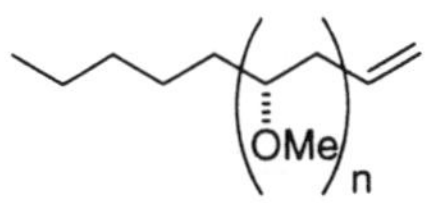

67, (n = 5,6,8,9 and 10)

Fig. 3. All-*syn* polyols from *Tolypothrix conglutinata* var. *chlorata*

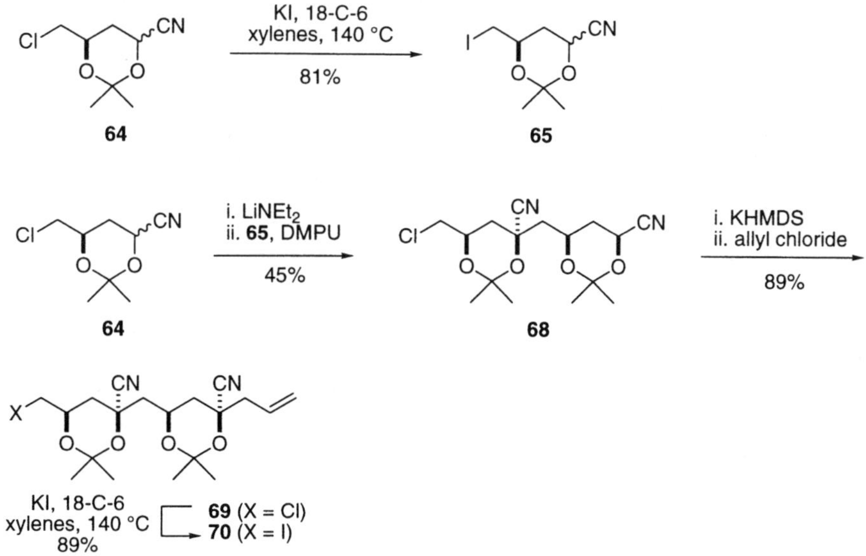

Scheme 7

Alkylation of iodide **70** with **68** generated the protected octol **71** in 87% yield (Scheme 8). A final electrophilic activation gave iodide **72**, which was then used to alkylate cyanohydrin acetonide **73**, thereby completing the assembly of the carbon backbone. The key reductive decyanation of **74**, which sets five of the target's ten stereocenters in a single step, proceeded in 85% yield. Deprotection and permethylation provided the target **67**. The strategy of orthogonal nucleophilic and electrophilic activation of a 1,3-diol synthon allowed for the total synthesis of a complex permethylated polyol in just ten steps from synthon **64**, and has since proven to be a valuable strategy for the total synthesis of polyene macrolide antibiotics (*vide infra*).

A related strategy of orthogonal nucleophilic and electrophilic activation was later employed in the synthesis of the polyene macrolide roflamycoin [32]

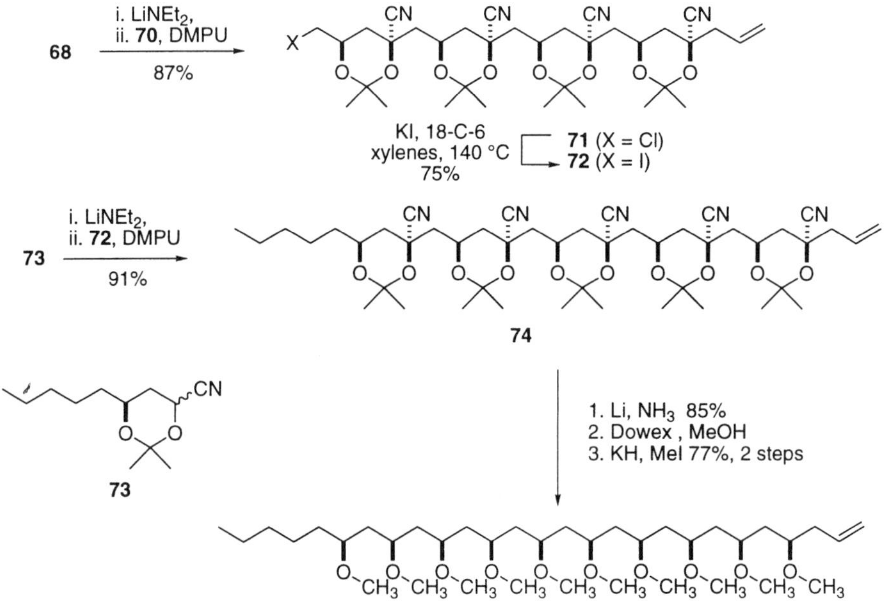

Scheme 8

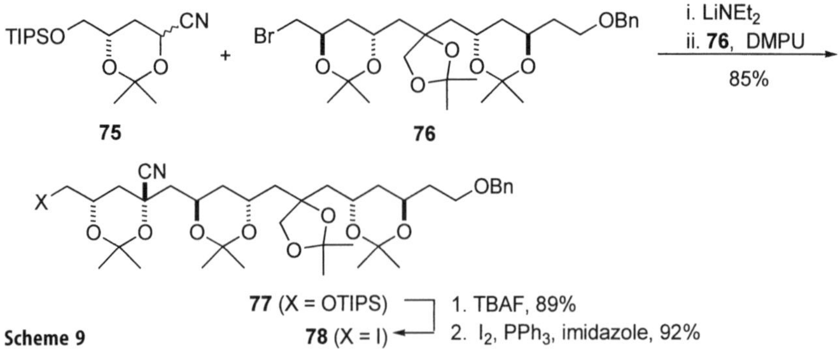

Scheme 9

(Scheme 9). Although cyanohydrin acetonide **64** could conceivably have been used, the silyl ether **75** was chosen. This compound is readily available from (L)-malic acid, and can undergo electrophilic activation under far more mild conditions than compound **64**. Alkylation of the 1,3-diol synthon **75** with bromide **76** created the C11-C26 framework of roflamycoin, in 85% yield. A two-step conversion of the terminal siloxy group to the primary iodide (**78**) proceeded in 80% overall yield.

Alkylation of cyanohydrin acetonide **79** with the iodide **78** proceeded smoothly to give pentaacetonide **80** in 70% yield (Scheme 10). This represents the entire polyol framework of roflamycoin. An eight-step sequence involving installation of the polyene, macrocyclization via Horner-Emmons reaction, and protecting group machinations, completed the first total synthesis of roflamycoin.

As part of a program aimed at elucidating the structure-activity relationships of the polyene macrolide antibiotics, the structural analogue 17-deoxyroflamycoin (**84**) was targeted [33]. This compound is completely homologous to the natural product, save the absence of the labile hemiketal at C17. Again, the orthogonal activation strategy was effectively employed. A key alkylation of **75** with the alkyl bromide **82** provided **83** in 95% yield (Scheme 11). Completion of the analogue synthesis mirrored that of the natural product. 17-Deoxyroflamycoin formed anion-selective ion channels in membrane vesicles containing cholesterol [34]. This anion selectivity is unprecedented with polyene macrolides, which have previously been shown to form cation-selective ion channels [35].

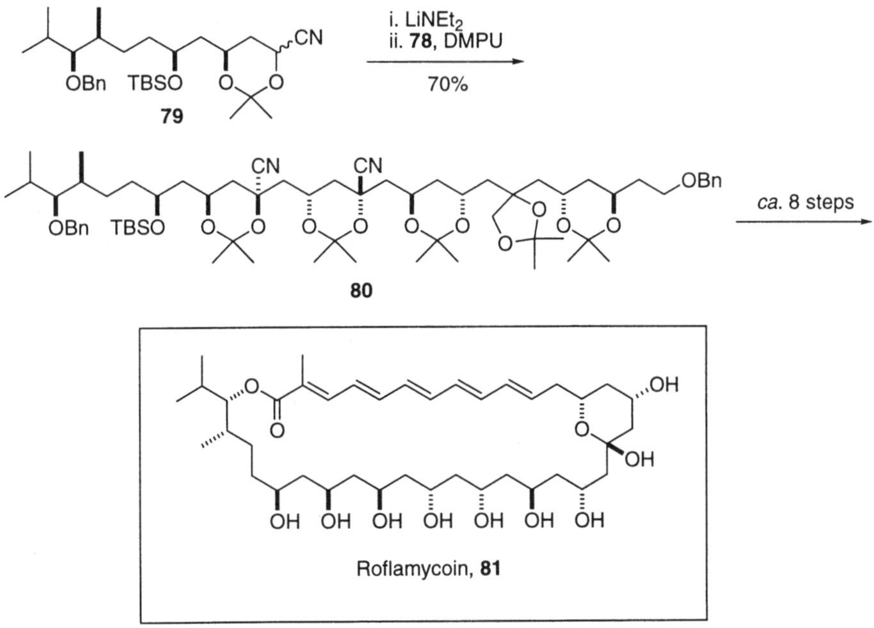

Scheme 10

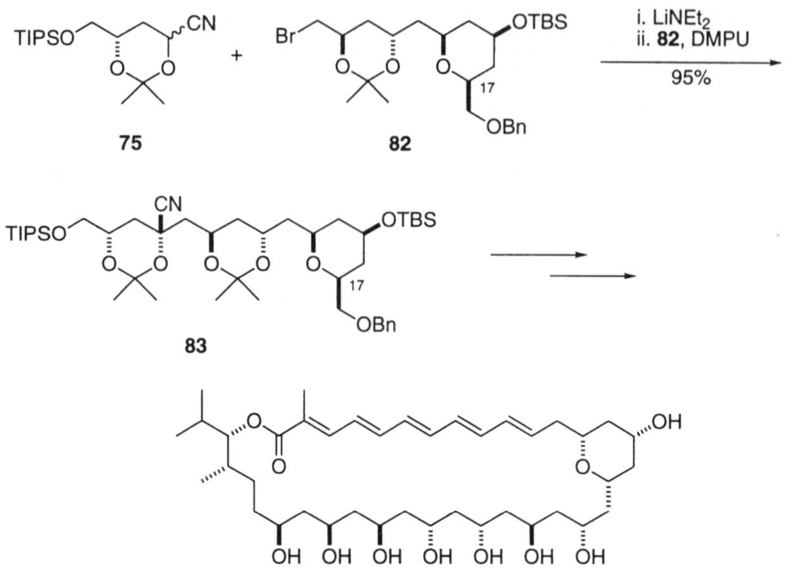

Scheme 11 17-Deoxyroflamycoin, **84**

Next we will describe the synthesis of filipin III (**114**) in greater detail, to bring to light some of the issues that arise in the total synthesis of a complex polyene macrolide [7, 8].

The polyene macrolide filipin was isolated in 1955 from the cell culture filtrates of *Sterptomyces filipinensis*, and was later shown to be a mixture of four components [36]. Although too toxic for therapeutic use, the filipin complex has found widespread use as a histochemical stain for cholesterol and has even been used to quantitate cholesterol in cell membranes [37]. The flat structure of filipin III, the major component of the filipin complex, was assigned from a series of degradation studies [38]. Rychnovsky completed the structure determination by elucidating the relative and absolute stereochemistry [39]. The total synthesis plan for filipin III relied heavily on the cyanohydrin acetonide methodology discussed above.

Based on information accrued during the stereochemical elucidation, macrolactone **85** was identified as a viable synthetic intermediate (Scheme 12). The authors were cognizant of the potential challenges that could arise. First, the required formation of a trisubstituted alkene in a projected Horner-Emmons macrocyclization was without strong precedent. Also, this strategy would necessitate a stereoselective reduction of the C15 ketone, which was predicted to be feasible based on MM2 calculations.

This convergent approach conveniently divides filipin III into a polyol segment **86** (Scheme 12), and a polyene segment **107** (*vide infra*). Application of the cyanohydrin acetonide coupling methodology would allow the polyol segment to be divided into two equally complex fragments, the C6-C15 fragment (**87**) and the C1-C5 fragment (**12**).

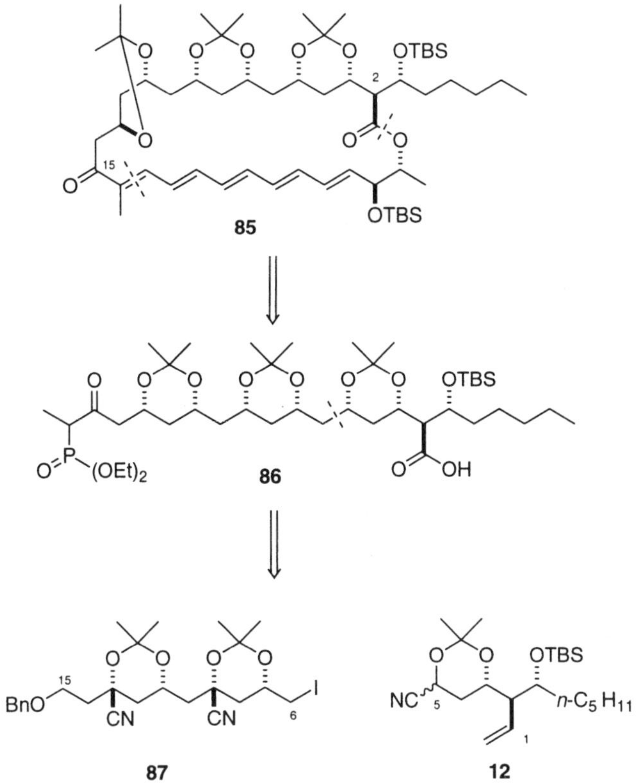

Scheme 12

The synthesis of the C1-C5 cyanohydrin acetonide fragment is shown below (Scheme 13). Allylic alcohol **89** was obtained by Red-Al reduction of 2-octyne-1-ol. An efficient Sharpless asymmetric epoxidation was followed by a copper-catalyzed opening of the resultant epoxide with vinylmagnesium bromide. The minor, undesired 1,2-diol was easily separated following periodate cleavage, allowing diol **90** to be isolated in 60% overall yield. Protecting group manipulations then gave the primary alcohol **91** in 82% overall yield. A Dess-Martin oxidation and subsequent $SnCl_2$-catalyzed homologation with ethyl diazoacetate provided β-keto ester **92** in 81% yield. Reduction of **92** with $NaBH_4$ resulted in a 4:1 mixture of the separable diastereomers **11**, in 95% total yield. Alcohol **11** was then converted to the cyanohydrin acetonide **12**.

Schemes 14 and 15 outline the polyol chain assembly. Alkylation of cyanohydrin **93** with iodide **94** provided the chlorocyanohydrin **95**, which was converted to the required iodide (**96**). A second alkylation of **93**, this time with **96**, provided bisacetonide **97** in 70–80% yield. Conversion of **97** to iodide **87** completed the synthesis of the C6-C15 fragment.

The final assembly of the polyol chain is shown below, Scheme 15. Alkylation of cyanohydrin acetonide **12** with the C6-C15 iodide **87** gave the coupled product **98**. The anion of **12** suffered extensive decomposition under standard alky-

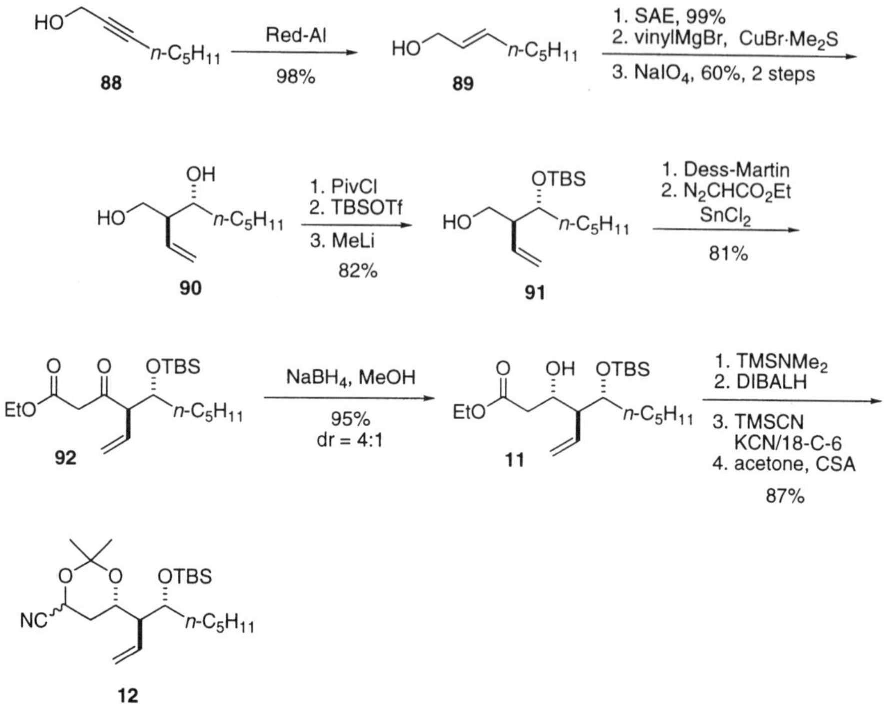

Scheme 13

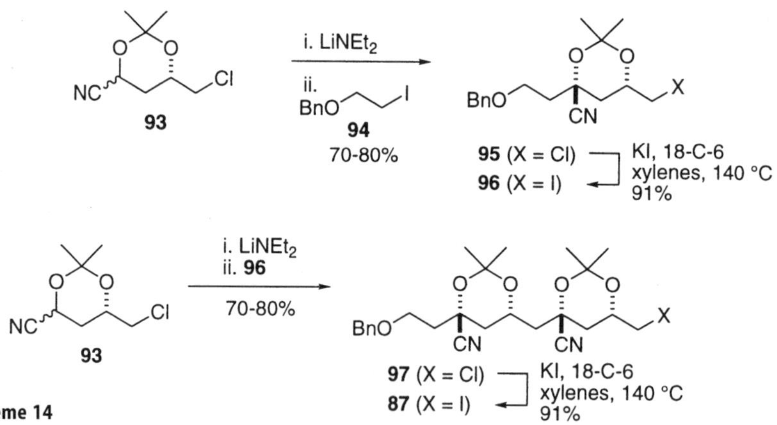

Scheme 14

lation conditions, which call for the use of DMPU. However, the anion was stable at –30 °C in THF in the absence of DMPU, and prolonged alkylation times at this temperature provided consistently high yields. Treatment of **98** with lithium in ammonia effected reductive decyanation with concomitant benzyl group removal, to give triacetonide **99** in 60–70% yield, as a single diastereomer. This compound was converted to β-ketophosphonate **100** in a 3-step sequence.

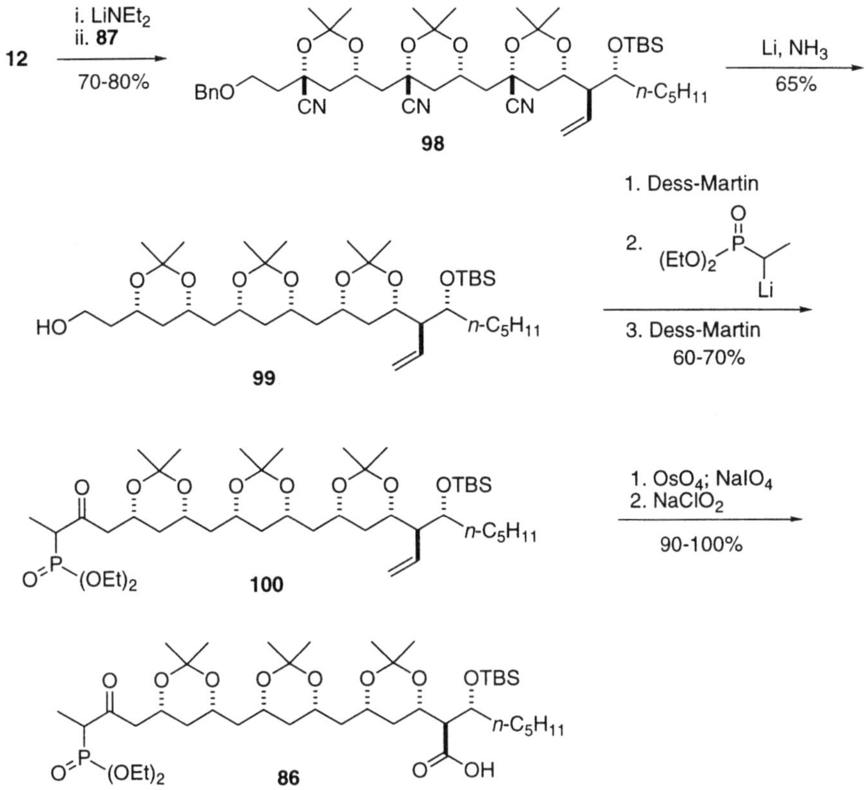

Scheme 15

Oxidation of the C1-olefin to the corresponding carboxylic acid efficiently provided the polyol subtarget **86**.

The polyene segment **107** was synthesized as described in Scheme 16. The protected triol **101** could be obtained on large scale from (L)-ascorbic acid. Acetonide cleavage, bis-silylation with TBSOTf, and selective removal of the primary silyl group gave compound **102**, which was smoothly oxidized to **103** by treatment with the Dess-Martin reagent. Aldehyde **103** was subjected to the Grignard reagent derived from Wollenberg's 1-(tributylstannyl)-4-ethoxybutadiene (**104**) [40], followed by mesylation and solvolysis of the allylic alcohol intermediate to give dienal **105**. A second application of this procedure provided the tetraenal **106** in 64% overall yield from **103**. Removal of the benzoate from **106** necessitated temporary protection of the aldehyde as a dimethyl acetal. After reductive cleavage of the benzoate, the aldehyde was unmasked to give the target fragment **107** in 70% yield from **106**.

The synthesis of the macrocycle is outlined in Scheme 17. Union of the polyol segment **86** and polyene segment **107** proved difficult, presumably due to steric hindrance about the ester linkage. After a series of standard esterification protocols failed, it was found that this coupling could be accomplished in 70% yield

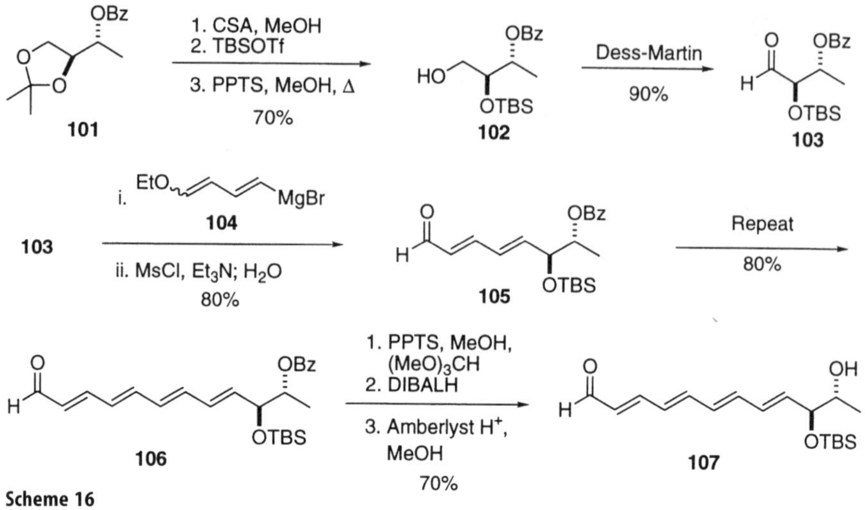

Scheme 16

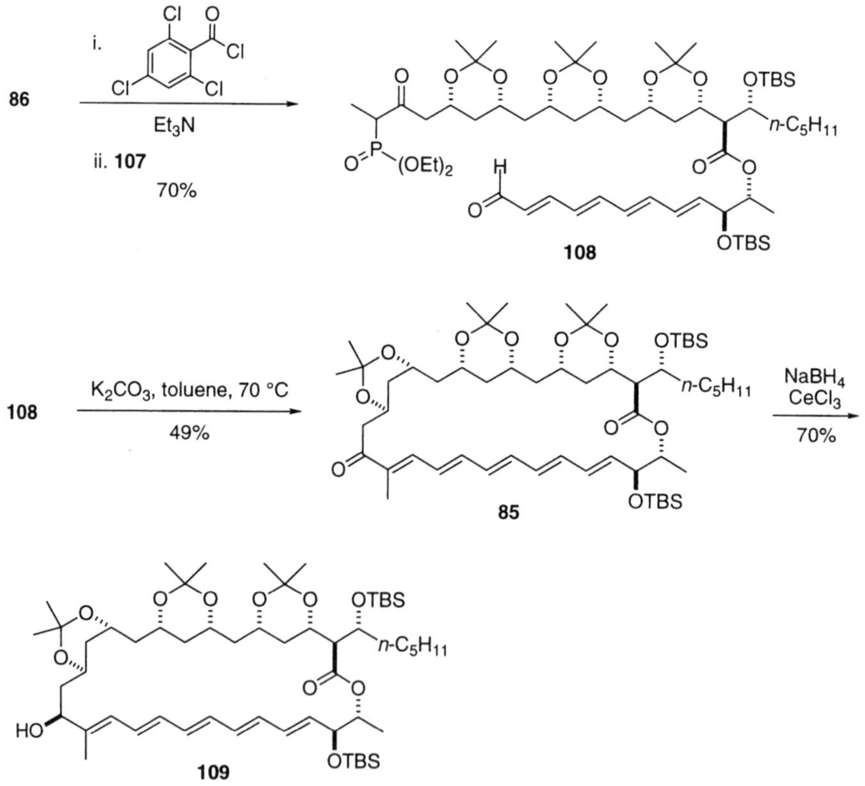

Scheme 17

using Yamaguchi's protocol. Macrocyclization was effected by treatment of **108** with potassium carbonate in warm toluene. Reduction of the resulting ketone with $NaBH_4/CeCl_3$ provided a separable 3:1 mixture of diastereomers, from which **109** could be isolated in 70% yield.

At this point, completion of the total synthesis required removal of the three acetonides and the two silyl protecting groups (Scheme 18). Removal of the silyl groups with TBAF and subsequent treatment to acidic deprotection conditions led to complete deprotection of **110**, but failed to provide filipin III. It was sus-

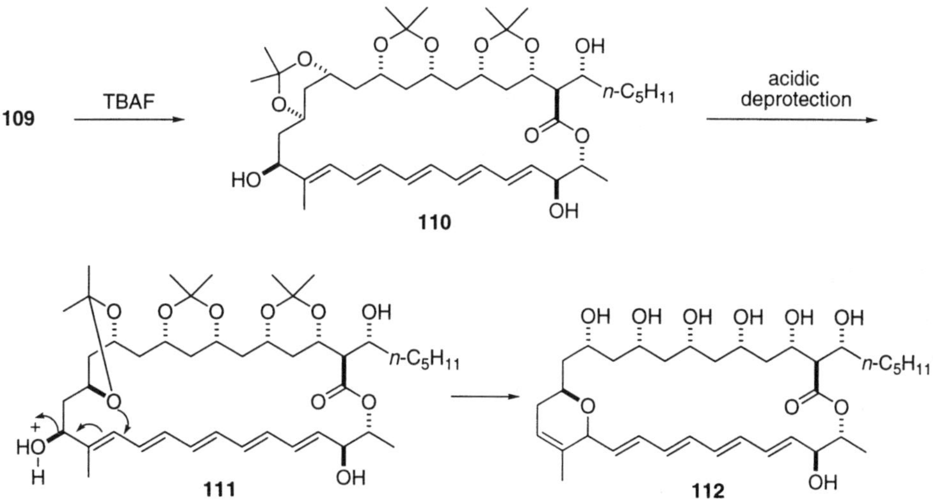

Scheme 18

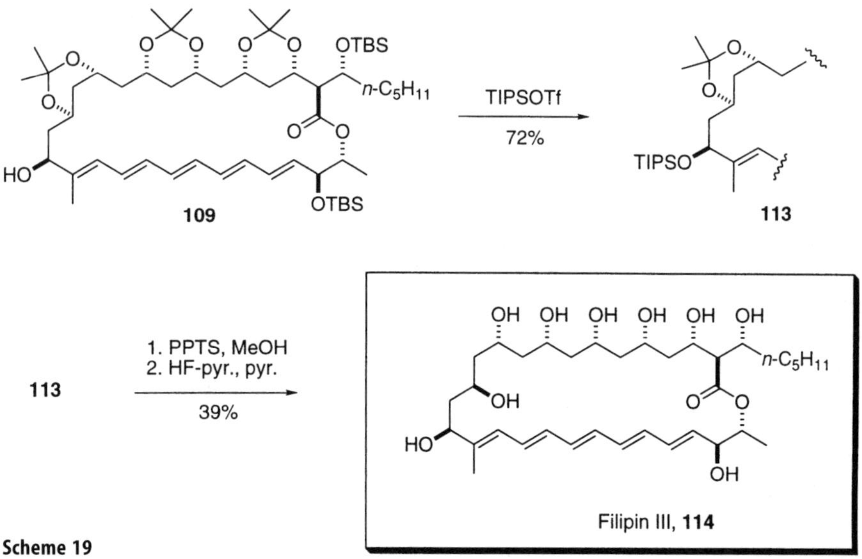

Scheme 19

pected that acetonide deprotection was complicated by solvolysis of the C15 alcohol to generate dihydropyran **112**.

In order to ameliorate the problem of solvolytic degradation, compound **109** was treated with TIPSOTf, to provide silyl ether **113** in 72% yield (Scheme 19). The acetonides were removed with PPTS in warm MeOH to provide a mixture of compounds in which the TBS groups were also partially removed. Exposure of this mixture to HF-pyridine successfully generated filipin III (**114**), in 39% overall yield from **113**.

2.5
Radical Nitrile Transfer Reactions

Our group has also reported that the alkylation products of 4-cyano-1,3-dioxanes can serve as substrates for radical atom transfer reactions [41]. One such example is shown below (Eq. 17). Slow addition of tributyltin hydride/AIBN to a refluxing solution of cyanohydrin **115** generated the radical nitrile transfer product **116**. This method, though somewhat limited in scope, can provide access to *syn*-1,3-diols which may be unstable to the vigorous Li/NH_3 reduction conditions.

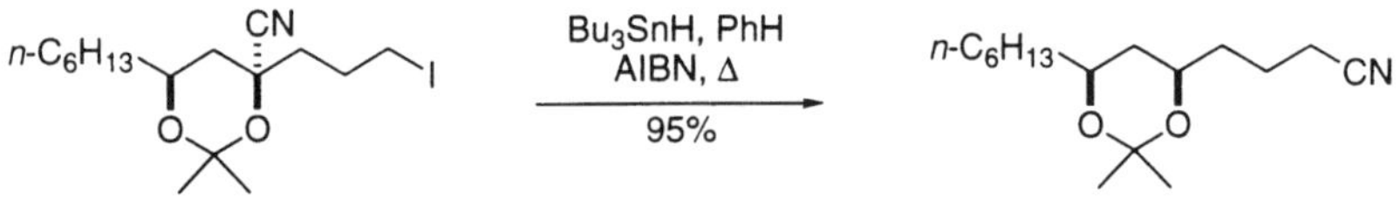

115 **116** (17)

3
4-Acetoxy-1,3-dioxanes in Synthesis

This section reviews recent developments in the chemistry of 4-acetoxy-1,3-dioxanes. Highly selective Lewis acid catalyzed nucleophilic additions are described, including their potential utility in natural product synthesis.

3.1
Background

Nucleophilic couplings with 6-membered ring oxocarbenium ions strongly favor addition from the axial direction, as a consequence of stereoelectronic effects [42, 43]. Axial addition proceeds through a chair-like transition state, while equatorial attack requires the intervention of a higher energy twist-boat transition state. Kishi et al. exploited this phenomenon in their seminal approach to *C*-glycosides [44]. Lewis acid-promoted addition of allyltrimethylsilane to *p*-nitrobenzoyl glycoside **117** provided the *α*-*C*-glycoside **118** (Eq. 18). Alternatively, the *β*-*C*-glycoside **120** could be obtained by addition of allylmagnesium bromide to lactone **119** and hydride reduction of the hemiacetal intermediate (Eq. 19). In both cases, the selectivity can be rationalized by axial attack on the corresponding oxocarbenium ion intermediate **121** (Fig. 4).

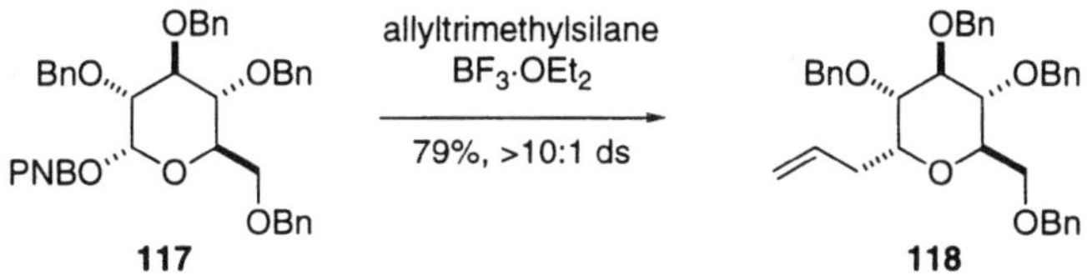

(18)

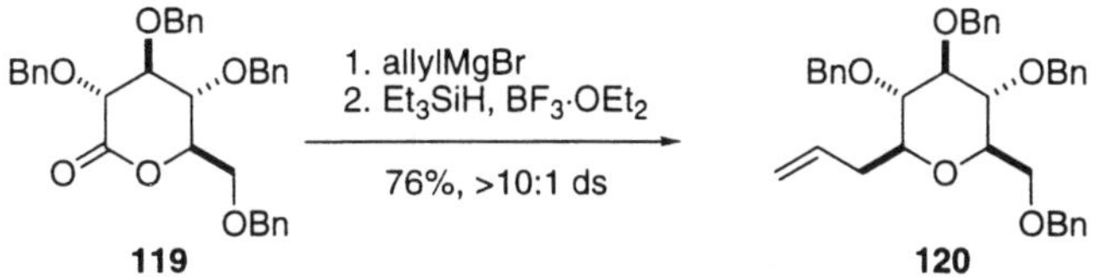

(19)

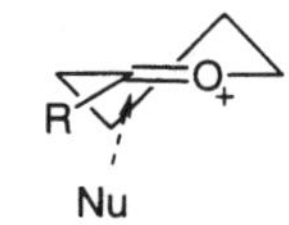

121

Fig. 4. Axial addition to oxocarbenium ion **121**

3.2 Preparation of 4-Acetoxy-1,3-dioxanes

Axial addition to oxocarbenium ions derived from 1,3-dioxanes provides protected *anti*-1,3-diols. Our group has developed 4-acetoxy-1,3-dioxanes as oxocarbenium ion precursors. This general strategy for the convergent preparation of *anti*-1,3-diols complements cyanohydrin acetonide methodology, which gives access to *syn*-1,3-diol synthons (Sect. 2).

4-Acetoxy-1,3-dioxanes can be prepared directly from the corresponding β-hydroxy aldehydes [45] (Scheme 20). β-Hydroxy aldehyde **122** exists in an

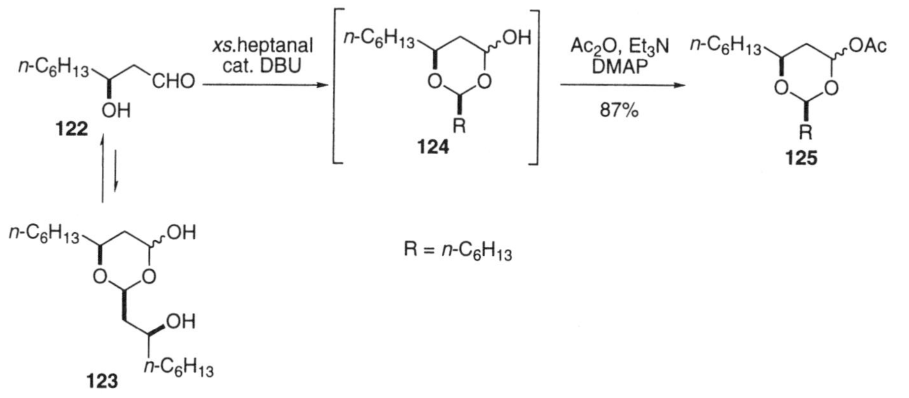

Scheme 20

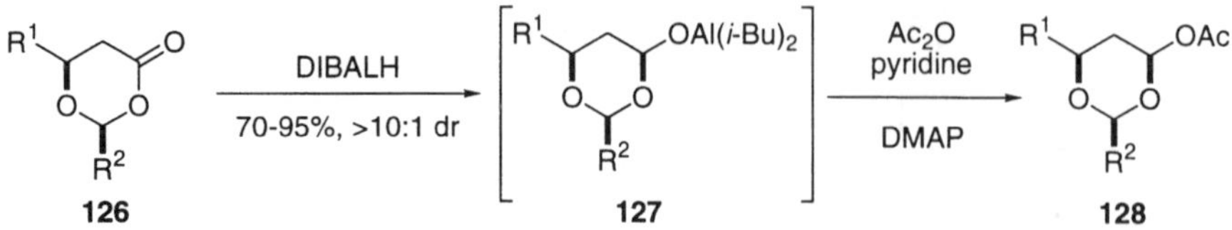

Scheme 21

equilibrium which favors the acetal dimer **123**. Treatment of the mixture with catalytic DBU and a large excess of heptaldehyde effected an exchange reaction that generated cyclic hemiacetal **124**. This acetal was then acylated *in situ* to provide 4-acetoxy-1,3-dioxane **125**; the cyclic hemiacetal **124** was prone to revert to the dimer upon standing. This method works well for simple substrates, but is limited by the large excess of aldehyde that is required to achieve good conversion. More limiting is the propensity of α-substituted-β-hydroxy aldehydes toward epimerization under the reaction conditions.

A more general route to 4-acetoxy-1,3-dioxanes utilizes the reductive acylation of 1,3-dioxane-4-ones [46] (Scheme 21). 1,3-Dioxane-4-ones **126** are prepared from the corresponding β-hydroxy carboxylic acids. Low temperature reduction with DIBALH generates a diisobutylaluminum hemiacetal (**127**) which undergoes acylation *in situ* with Ac_2O in the presence of pyridine and DMAP. This method allows for the preparation of a wide range of 4-acetoxy-1,3-dioxanes, without the problem of α-epimerization. This method also represents a general approach to acylic α-acetoxy ethers, which are themselves useful synthetic intermediates [47, 48].

Davis has described an approach to related 1,3-diol synthons [49] (Eq. 20). Silylation of the β-hydroxy ester **129** with diisopropylchlorosilane, followed by fluoride ion-catalyzed intramolecular hydrosilylation generated a 1:1 diastereomeric mixture of acetals **131**. These acetals were shown to undergo diastereoselective nucleophilic additions (*vide infra*).

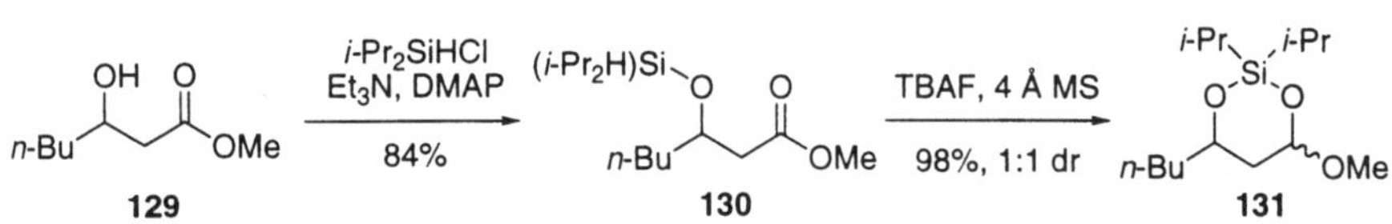

(20)

3.3 Nucleophilic Additions to 4-Acetoxy-1,3-dioxanes

3.3.1 *Allyl/Crotyl Metal Reagents*

Initial attempts to effect addition of allyltrimethylsilane with compound **132** under conditions developed in Kishi's *C*-glycoside work were plagued by epimerization of the acetal center subsequent to the coupling event, providing **133** as a 1:1 mixture of diastereomers (Eq. 21). Boons reported that acetal epimerization was suppressed when the reaction was conducted at –78 °C [50] (Eq. 22).

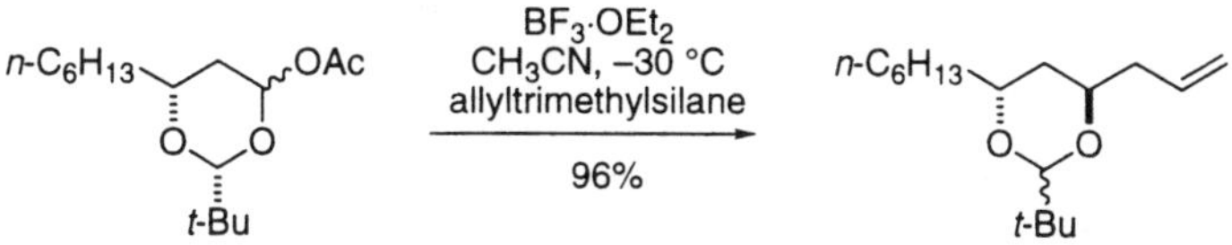

132 133 (21)

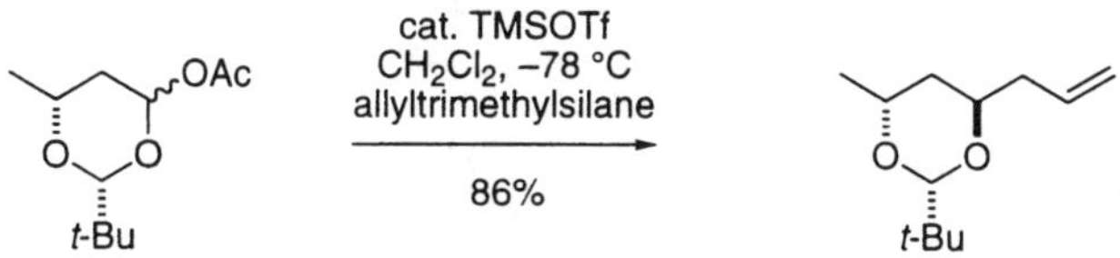

134 135 (22)

The alkoxysiladioxanes described by Davis were shown to undergo selective axial addition of allyltrimethylsilane [49]. Moderate selectivity (7–13:1) was observed when the reactions were promoted by Lewis acids (TMSOTf, $SnCl_4$), while higher selectivity was realized when a Bronsted superacid was used (Eq. 23).

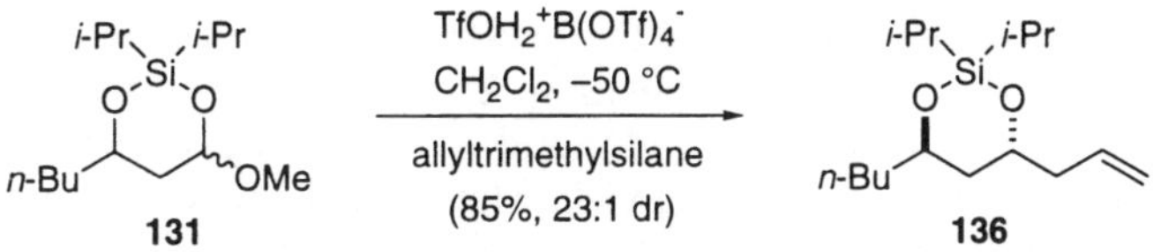

131 136 (23)

Allylsilane additions were used in a formal synthesis of roflamycoin [51] (Eq. 24). A one-pot, three-component sequential coupling of bis-allylsilane **138** with 4-acetoxy-1,3-dioxanes **137** and **139** provided the C11-C22 polyol chain (**140**) in moderate yield.

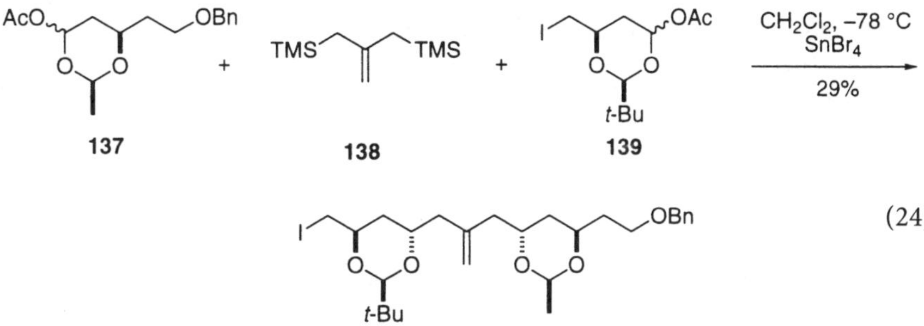

137 138 139 (24)

140

Addition of crotyl metal reagents to 4-acetoxy-1,3-dioxanes was utilized in the synthesis of dipropionate synthons [52] (Scheme 22). These reactions

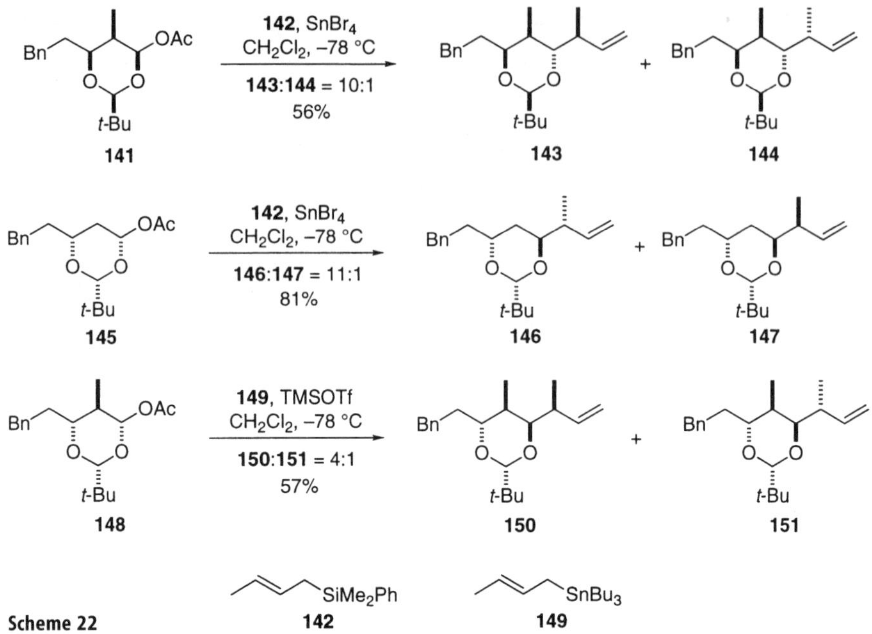

Scheme 22

showed a marked dependence on the presence and stereochemical disposition of the substituent at the 5-position. Treatment of compound **141** with (*E*)-crotyl(dimethyl)phenylsilane (**142**) and $SnBr_4$ provided a 10:1 mixture of diastereomers **143** and **144**, in 56% yield. Compound **145** underwent coupling under identical conditions to provide a similar stereochemical outcome. In contrast, 4-acetoxy-1,3-dioxane **148**, which bears an equatorial methyl group at the 5-position, gave **150** as the major isomer upon reaction with (*E*)-crotyltributylstannane (**149**) in the presence of TMSOTf.

The observed selectivities are consistent with a model introduced by Danishefsky [53], which invokes a synclinal approach of the crotylsilane with the crotyl methyl group pointed away from the ring (Fig. 5). In the case of compound **148**, such an approach results in a *syn*-pentane interaction between the crotyl methyl group and the equatorial methyl group, leading to the turnover in selectivity that was observed.

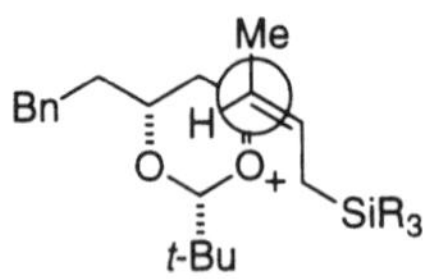

Fig. 5. Synclinal approach of crotyl silanes onto oxocarbenium ions

3.3.2
Alkyl Zinc Reagents

In an attempted preparation of protected allylic *anti*-1,3-diols by the Lewis acid-promoted addition of the vinyl alkyl zinc reagent **153** to 4-acetoxy-1,3-dioxane **152**, it was found that transfer of the alkyl group was competitive with vinyl group transfer [54] (Eq. 25). Dioxanes **154** and **155** were isolated in 50% and 35% yield, respectively, both as single diastereomers. This surprising result stands in contrast to additions of vinyl alkyl zinc reagents to aldehydes, which give exclusively allylic alcohols [55]. The use of a variety of other Lewis acids (TMSOTf, $TiCl_4$, $ZnCl_2$) resulted in almost exclusive transfer of the ethyl ligand. Given the wide range of functional groups which are compatible with dialkylzinc reagents and the host of methods for their preparation, this reaction appeared to merit further study.

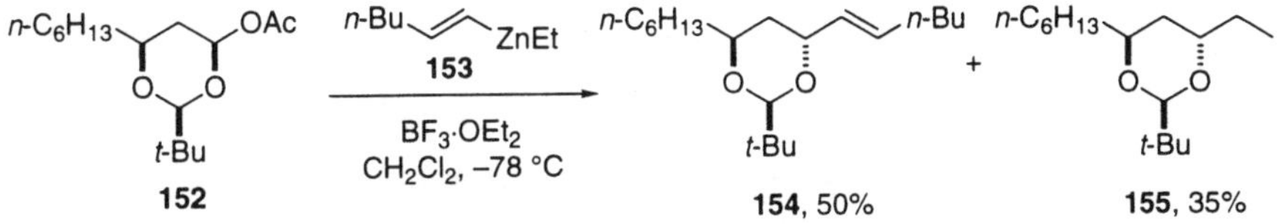

(25)

A survey of reaction conditions and Lewis acid promoters was conducted (Scheme 23). TMSOTf proved to be the most effective promoter, leading to complete conversion within minutes at –78 °C, while use of $BF_3 \cdot OEt_2$ led to sluggish reactions. Essentially complete 1,3-*anti* selectivity was observed in a variety of solvents. The minor diastereomer was not the corresponding *syn*-1,3-diol acetal, but rather the *anti*-1,3-diol acetal which was epimeric at the acetal center; this epimerization could be suppressed by carefully performing a low temperature quench. While dialkylzinc reagents typically transfer only one alkyl group, in the

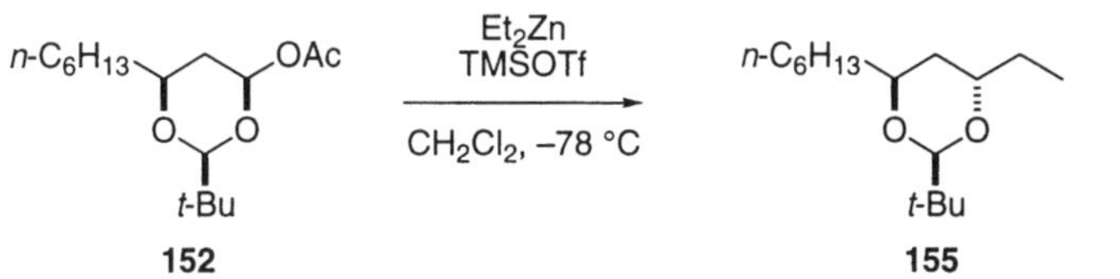

Entry	Equiv Et_2Zn	Solvent	Yield	Acetal epimer ratio
1	2	CH_2Cl_2	91%	51:1
2	0.6	CH_2Cl_2	90%	100:0
3	2	Et_2O	100%	100:0
4	2	THF	83%	21:1
5	2	$PhCH_3$	91%	15:1

Scheme 23

addition to 4-acetoxy-1,3-dioxane **152**, both ligands were transferred efficiently (entry 2). The ability to transfer both ligands would clearly aid the use of complex dialkylzinc reagents in natural product synthesis.

GC analysis revealed >290:1 selectivity (*anti*:*syn*) in the diethylzinc addition. The 4-acetoxy-1,3-dioxane **152** used in the above experiments was a 24:1 mixture of diastereomers, epimeric at the 2-position. This implies that the acetal stereocenter undergoes isomerization to the most stable oxocarbenium ion prior to reaction with Et_2Zn. Conclusive evidence for this was obtained when submission of compound **156** to the identical conditions produced **155** as the major product (Eq. 26).

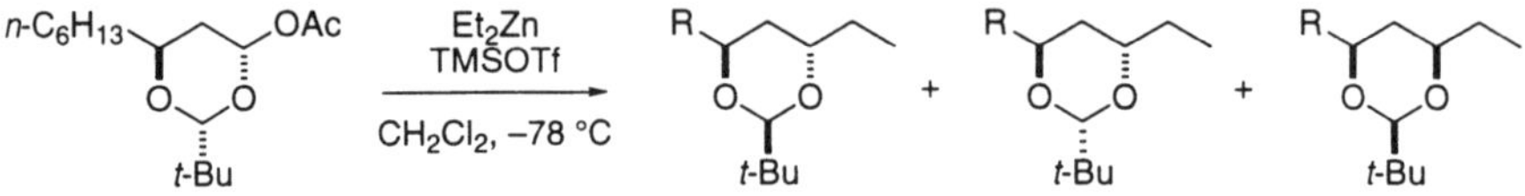

156 R = n-C_6H_{13} **155**, 93.9% **157**, 5.5% **158**, 0.5%

(26)

A survey of functionalized dialkylzincs revealed that a variety of functional groups is tolerated (Scheme 24). Esters, chlorides, and to a lesser degree sulfides, are tolerated. Also, secondary alkylzinc reagents were shown to be effective (entries 4 and 5). The more hindered dicyclohexylzinc underwent coupling to give dioxane **163** in 58% yield, but with competitive β-hydride transfer to produce **164** in 35% yield. Unless noted, the products were obtained as single diastereomers. The coupling of secondary alkylzincs is particularly intriguing, as it points to potential use of configurationally stable secondary dialkylzinc reagents in couplings with 4-acetoxy-1,3-dioxanes.

An unexpected result was obtained in the coupling of bis(3-alkoxypropyl)zincs with **152** [56] (Scheme 25). A substantial erosion in 1,3-*anti* selectivity was observed. The lower diastereoselectivity observed in these couplings may be a consequence of an equilibrium between the open-chain form (**167**) and the intramolecular chelated form (**168**) of the dialkylzinc, which increases the steric bulk about the C-Zn bond [57, 58] (Fig. 6). As the steric bulk of the chelated form increases, addition from the less hindered equatorial trajectory begins to become competitive, and substantial amounts of the *syn*-1,3-diol synthon are generated.

The *tert*-butyl acetal protecting group employed in many of the model systems described above is particularly robust, and may not be well suited for deprotection in a complex system. Thus, a series of less sterically demanding acetal protecting groups was explored (Scheme 26). Switching from the *tert*-butyl acetal protecting group to the more readily deprotected methyl acetal led to only a minor erosion in selectivity. The *n*-benzyloxybutanal (BOB) acetal protecting group allows for the use of alternative deprotection conditions. This protecting group can be removed in 91% yield by hydrogenolysis of the benzyl ether and concomitant transacetalization of the resultant free alcohol (Eq. 27).

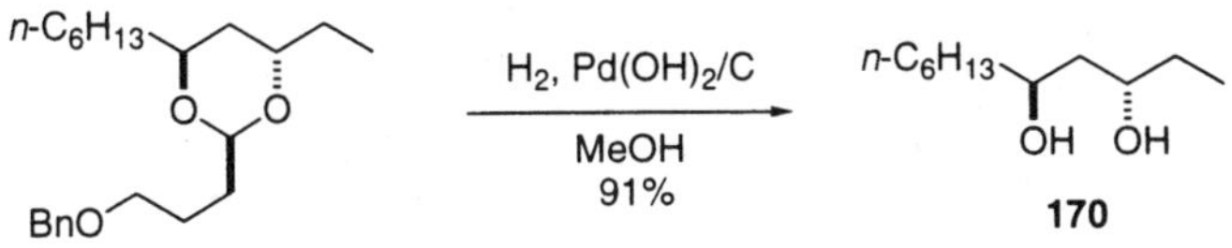

169 170 (27)

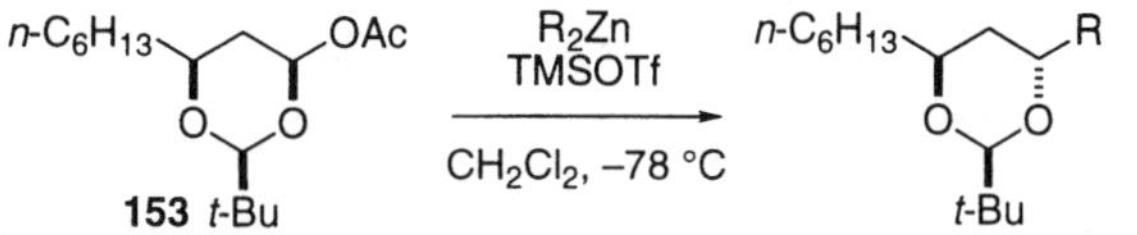

153

Entry	R	Product	Yield
1	$Cl(CH_2)_4$	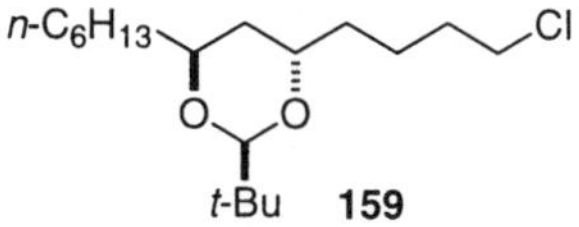 **159**	82%
2	$EtO_2C(CH_2)_3$	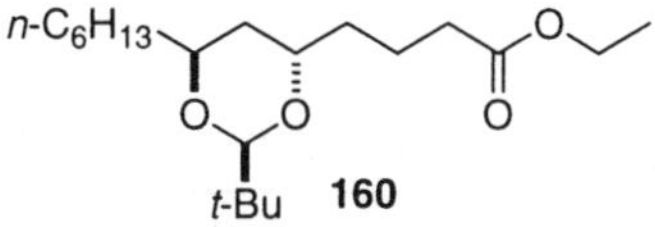 **160**	77%
3	$PhS(CH_2)_3$	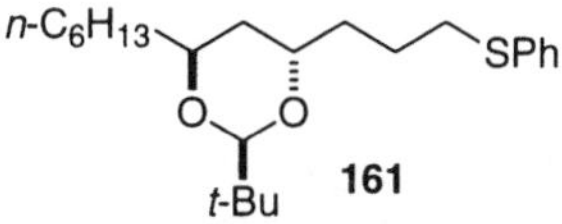 **161**	36% (8:1 mixture of acetal epimers)
4	*i*-Pr	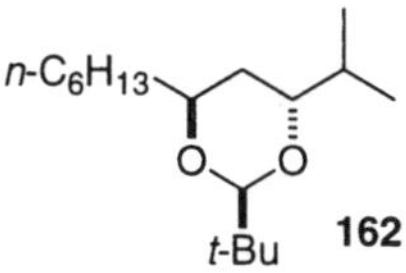 **162**	90%
5	C_6H_{11}	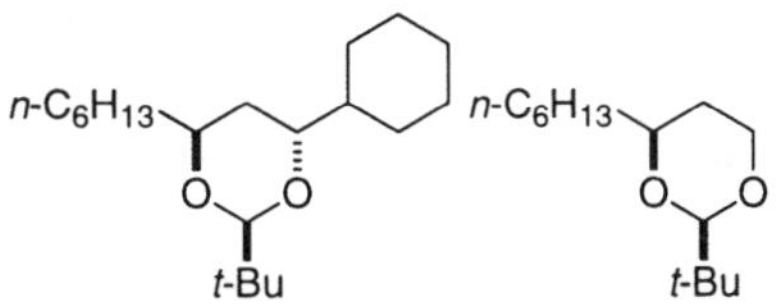 **163** **164**	**163**, 58% **164**, 35%

Scheme 24

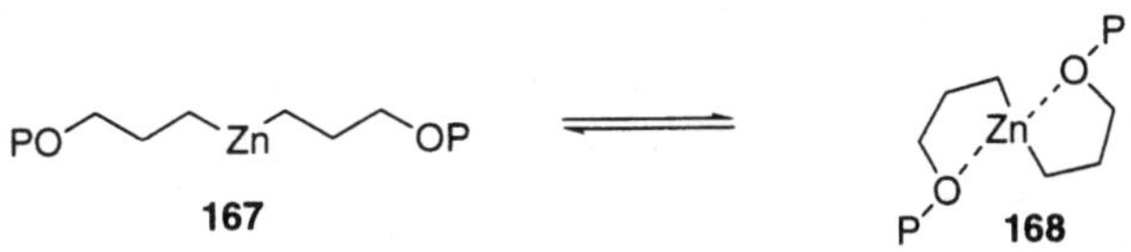

167 168

Fig. 6. Acyclic/chelated equilibrium for bis(3-alkoxypropyl)zinc reagents

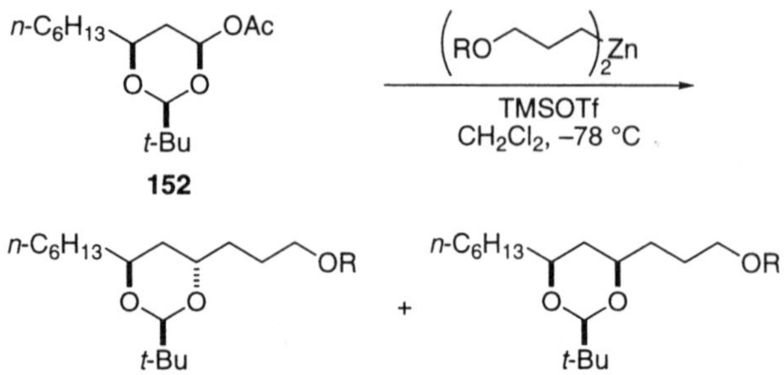

152

165 166

Entry	R	Yield	*Anti*:*Syn* Ratio (**165**:**166**)
1	TBS	98%	6.3:1
2	TIPS	60%	9.5:1
3	Bn	93%	33:1
4	Piv	36%	100:0

Scheme 25

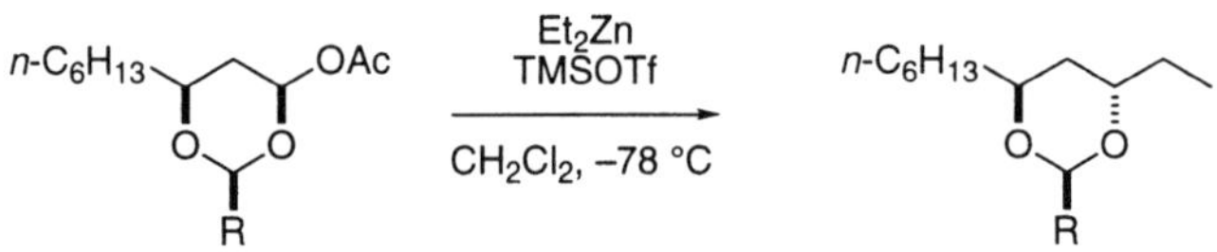

152

Entry	R	Yield	*Anti*:*Syn* ratio
1	$C(CH_3)_3$	91%	290:1
2	$CH(CH_3)_2$	86%	287:1
3	CH_3	63%	200:1
4	H	66%	120:1

Scheme 26

3.3.3 *Alkynyl Metal Reagents*

The Lewis acid promoted coupling of alkynyl organometallics with 4-acetoxy-1,3-dioxanes is a highly selective route to acetal protected propargylic *anti*-1,3-diols [59] (Scheme 27). Alkynyl diethylalanes in which the acetylene is substituted with a silyl (entry 1) or sp^2-centered group (entries 2 and 3) couple with **152** in excellent yield and with very high (>200:1) diastereoselectivity. Nucleophilic additions of alkynyl diethylalanes in which the acetylene is substituted with an sp^3-centered group, however, were plagued by competitive transfer of the ethyl ligand. The use of alkyl substituted alkynyl tributylstannanes solved the problem; these nucleophiles readily coupled with **152** with very high diastereoselectivity (entries 4 and 5).

The masked propargylic *anti*-1,3-diols obtained in these reactions are useful precursors to more functionalized systems. Lindlar reduction of alkyne **171** generated the (*Z*)-allylic diol **172**, which underwent diastereoselective osmium tetraoxide-catalyzed dihydroxylation to provide the partially protected tetraol **173** (Scheme 28). The propargylic *anti*-1,3-dioxane **175**, obtained in 88% yield from

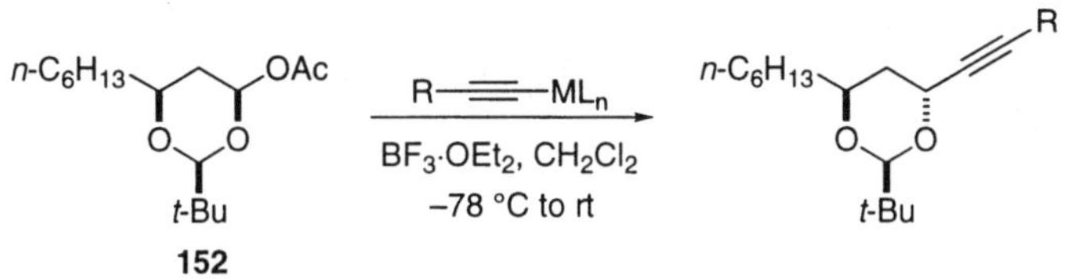

Entry	R	ML_n	Yield
1	TMS	$AlEt_2$	85%
		$SnBu_3$	71%
2	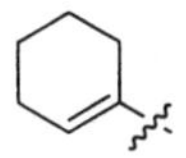	$AlEt_2$	90%
		$SnBu_3$	84%
3	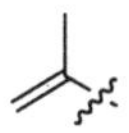	$AlEt_2$	69%
4	Bu	$SnBu_3$	94%
5	$TBSO(CH_2)_2$	$SnBu_3$	87%

Scheme 27

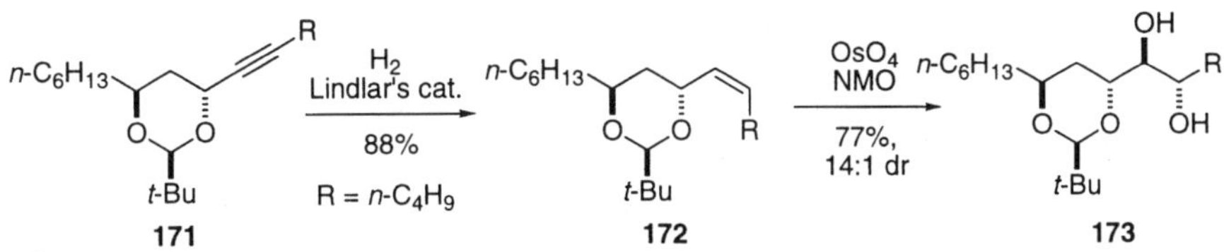

Scheme 28

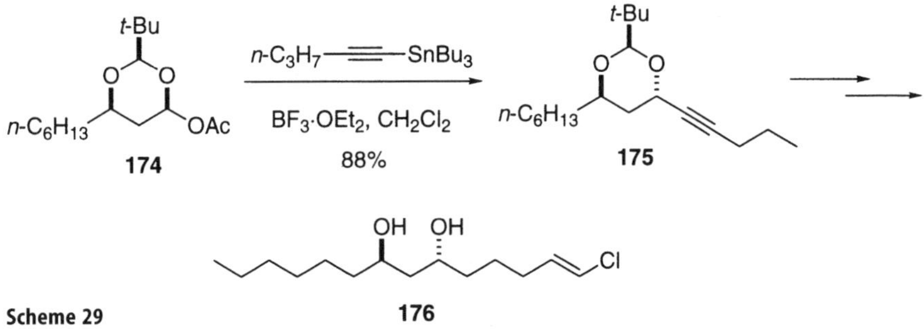

Scheme 29

coupling with the corresponding alkynyltributylstannane, was a key intermediate in the synthesis [59] of the antibiotic (–)-1-chlorotridec-1-ene-6,8-diol (**176**) [60] (Scheme 29).

4
4-Lithio-1,3-dioxanes in Synthesis

4-Lithio-1,3-dioxanes are prepared by reductive lithiation of 4-(phenylthio)-1,3-dioxanes and by transmetallation of the corresponding 4-(tributylstannyl)-1,3-dioxanes. This section describes the use of 4-lithio-1,3-dioxanes in the synthesis of *syn*- and *anti*-1,3-diols.

4.1
Background

Since Still's seminal report, chiral α-alkoxylithium reagents have been of interest to synthetic chemists [61]. Still and McGarvey [62] prepared these reagents by transmetallation of diastereomerically or enantiomerically pure alkylstannanes, and demonstrated their configurational stability at low temperature. Reductive lithiation of alkyl phenyl sulfides is an attractive alternative preparation of α-alkoxylithium reagents [14, 63]. Cohen first demonstrated this approach in the synthesis of 2-lithiotetrahydropyrans and related intermediates [64]. Single electron reduction of **177** with lithium (dimethylamino)naphthalenide (**178**) generated an α-alkoxy radical intermediate. Anomeric stabilization of the intermediate radical provided the axial alkyllithium **180** as the kinetic product, which could undergo alkylation at low temperature with retention of configuration (Scheme 30). The axial alkyllithium was shown to equilibrate to the more stable equatorial isomer upon warming. These equilibrations stand in contrast to acyclic α-alkoxylithiums, which decompose when allowed to warm above -30 °C [61].

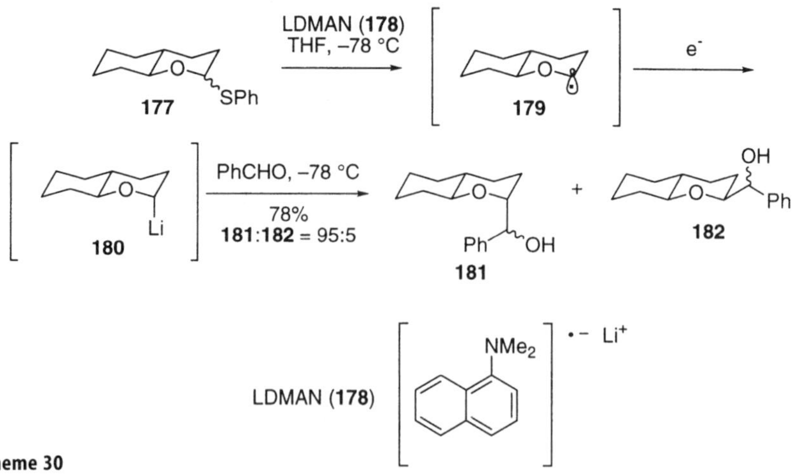

Scheme 30

4.2
Kinetic and Thermodynamic Stability of 4-Lithio-1,3-dioxanes

4.2.1
Experimental Studies

Our group has exploited 4-phenylthio-1,3-dioxanes as convenient precursors to 4-lithio-1,3-dioxanes [45, 65–69]. 4-Phenylthio-1,3-dioxanes **184** were originally prepared from β-silyloxy aldehydes **183** [65] (Eq. 28). Lewis acid-promoted addition of phenylthiotrimethylsilane gave an unstable thioacetal intermediate, which could be converted *in situ* to the corresponding 1,3-dioxane. Yields for this process are variable, as the product is unstable under the conditions of its formation. The reaction slowly evolves to a mixture of the desired product, the phenylthio acetal of **183**, the phenylthio acetal of acetone, and a variety of other unidentified products.

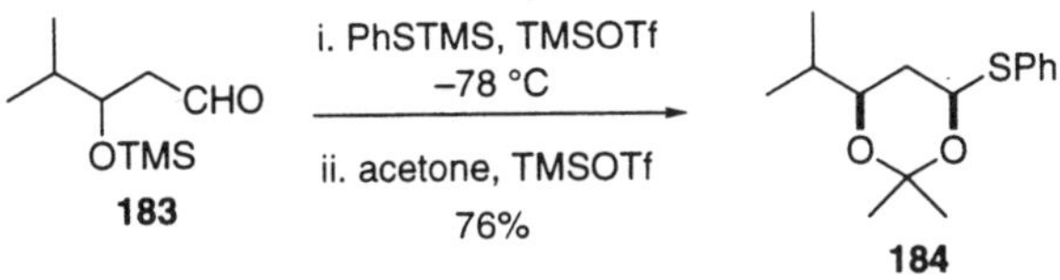

(28)

Phenylthio acetals are much more easily prepared from the corresponding 4-acetoxy-1,3-dioxanes [45] (Eq. 29). The dioxane **125** gave the phenyl sulfides **185** in essentially quantitative yield on treatment with $BF_3 \cdot OEt_2$ and thiophenol at –78 °C. This method has been used to prepare compound **185** on a 40 g scale.

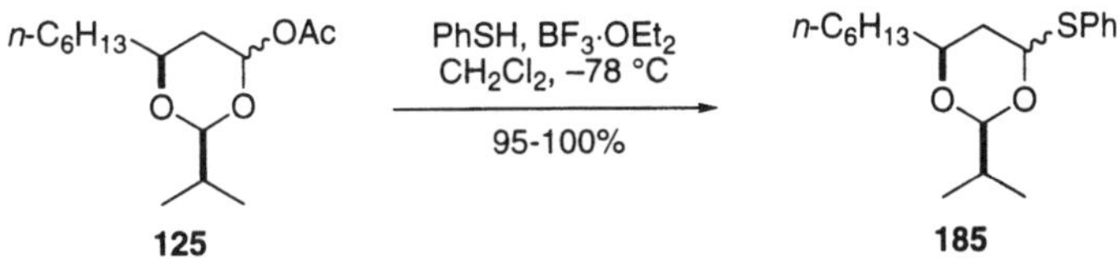

(29)

Early studies involving the phenyl sulfide **184** demonstrated the potential of these reagents as *syn*- or *anti*-1,3-diol synthons [45] (Scheme 31). Reductive lithiation at low temperature with lithium (di-*tert*-butyl)biphenylide (LiDBB) generated an axial alkyllithium, which could be trapped with retention of configuration to provide *anti*-1,3-diols. Alternatively, *syn*-1,3-diols could be accessed by briefly warming to –30 °C to effect equilibration to the more stable equatorial lithium species.

Later work examined substituent effects on kinetically controlled alkylations [68, 69] (Scheme 32). Substitution at the 5-position is well tolerated in these reactions. Reductive lithiation of a series of 4-phenylthio-1,3-dioxanes and quenching of the axial alkyllithium intermediate with dimethyl sulfate provided the *anti*-1,3-diols in good yield, with essentially complete selectivity.

Efforts to obtain the corresponding *syn*-1,3-diols by equilibration of the same intermediates gave mixed results (Scheme 33). The C5-unsubstituted acetal

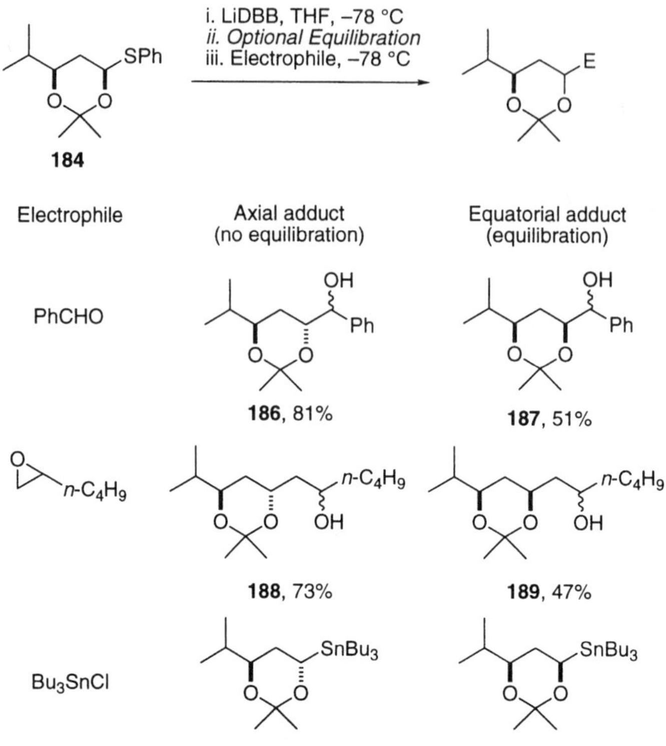

Scheme 31

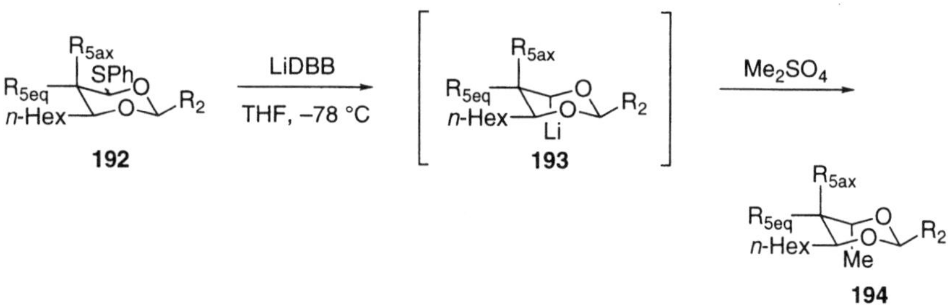

Entry	R_2	R_{5ax}	R_{5eq}	Protonated (%)	Alkylated[a] (%)	Ratio (ax/eq)[b]
a	*i*-Pr	H	H	*c*	79	99:1
b	Me	H	Me	8	77	99:1
c	Me	Me	H	17	75	99:1
d	Me	Me	Me	9	83	99:1

Scheme 32 [a] Isolated yields [b] Ratios by GC analysis [c] Not measured

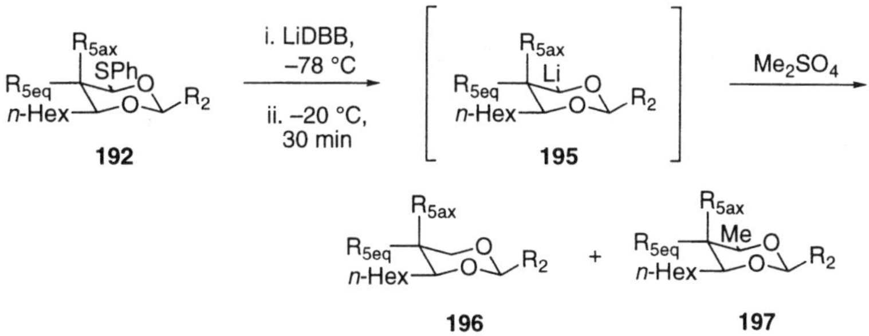

Entry	R_2	R_{5ax}	R_{5eq}	Protonated[a] (**196**) (%)	Alkylated[a] (**197**) (%)	Ratio (eq/ax)[b]
a	*i*-Pr	H	H	*c*	79	>99:1
b	Me	H	Me	45	47	35:1
c	Me	Me	H	27	54	1:2
d	Me	Me	Me	49	31	1.5:1

Scheme 33 [a] Isolated yields [b] Ratios by GC analysis [c] Not measured

(entry a) equilibrated well, giving the desired equatorial adduct with >99:1 selectivity and in good yield. The $C5_{eq}$-methyldioxane (entry b) also gave the equatorial product with good selectivity, but in lower yield. Neither the $C5_{ax}$-methyldioxane nor the 5,5-dimethyldioxane (entries c and d) equilibrated effectively. The very strong basicity of the 4-lithio-1,3-dioxane intermediates led to substantial amounts of product **196**, which arose from proton transfer with solvent.

To examine whether the breakdown in equilibration arose from a slow rate of equilibration or if the equilibrium does not favor the equatorial alkyllithium, the time course of equilibration was investigated (Scheme 34). The diastereomerically pure α-alkoxystannanes **198** were prepared in analogy to Linderman's procedures [70], and their equilibrations were examined. Transmetallation and alkylation under kinetic control (entries 1 and 2) provided the expected equatorial and axial adducts, respectively, with no detectable minor isomer. Attempted equilibrations of the axial stannanes (entries 4 and 6) led to mixtures of axial and equatorial products. The equatorial stannane gave only the equatorial adduct under both equilibration conditions (entries 3 and 5). From these data, the authors concluded that the equilibration does take place under these conditions, and that the equilibrium lies essentially completely toward the equatorial alkyllithium. The observed product ratios in entries 4 and 6 can be attributed to a slow rate of equilibration.

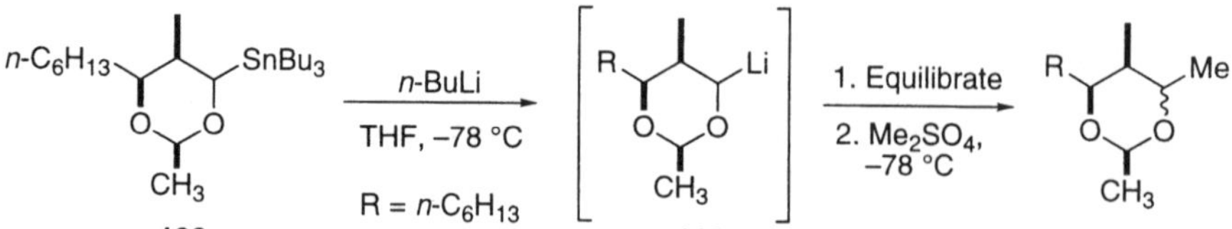

198 **199** **200**

Entry	Stannane	Equilibration temp. (time)	Protonated[a] (%)	Alkylated[a] (%)	Ratio (ax:eq)[b]
1	eq	none	3	97	0:100
2	ax	none	18	82	100:0
3	eq	–30 °C (15 min)	12	88	0:100
4	ax	–30 °C (15 min)	37	63	89:11
5	eq	–20 °C (30 min)	63	37	0:100
6	ax	–20 °C (30 min)	74	26	58:42

Scheme 34 [a]Isolated yields [b]Ratios by GC analysis

4.2.2
Theoretical Studies

The equilibration of alkyllithium reagents was investigated by *ab initio* methods [68]. It was found that the unsubstituted 4-lithio-1,3-dioxane preferred a chair conformation with an equatorial lithium substituent. Using the B3LYP/6–31+G(d)//B3LYP/6-31+G(d) method, the equatorial alkyllithium was favored over the chair axial alkyllithium by 5.3 kcal/mol. Non-chair conformations were also investigated, and the lowest energy "axial lithium" was found to have a 3,6-twist boat structure that was 3.5 kcal/mol higher in energy than the equatorial alkyllithium. This overwhelming preference for the equatorial lithium geometry behaves like an *anti*-anomeric effect, but it may reflect a more effective coordination between the lithium atom and the adjacent oxygen atom. Is the equatorial alkyllithium preference general? Cohen's examples show that there is a significant preference for equatorial alkyllithium geometry in 2-lithio tetrahydropyrans [64, 71]. To shed light on the mixed equilibration results discussed above, our group calculated the energy preference for equatorial versus axial lithium geometries with substituents at the 5-position. Dioxanes with no C5 substituent and with either an axial or equatorial C5 methyl group were investigated. In each case the equatorial alkyllithium was preferred over the axial alkyllithium by more than 5 kcal/mol at B3LYP/6-31+G(d)//HF/3-21G, Fig. 7. The authors conclude that there is no significant difference in the thermodynamic driving force for equilibration of the different 4-lithio-1,3-dioxanes. The different efficiency of equilibration is a manifestation of the rate of equilibration rather than the thermodynamic

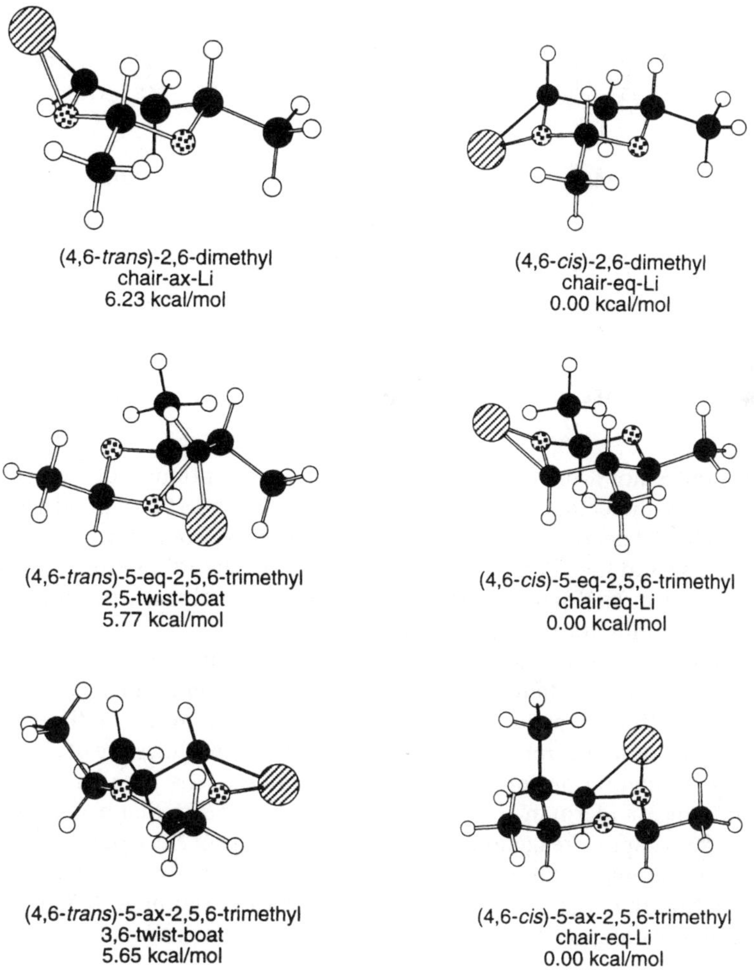

Fig. 7. Conformations of substituted 4-lithio-1,3-dioxanes calculated at B3LYP/6-31+G(d)//HF/3-21G. Only the lowest energy 4,6-*cis* and 4,6-*trans* conformations are listed, along with the relative energy for each pair. The (4,6-*cis*)-5-axial conformation is 1.81 kcal/mol higher in energy than the (4,6-*cis*)-5-equatorial conformation

driving force, and this conclusion is consistent with both experimental and computational results.

4.3 Synthetic Applications

The use of excess lithium LiDBB in reductive lithiations is a drawback for preparative-scale reactions. A modification of Yus' procedure [72, 73] allowed for the generation of α-alkoxylithium reagents under catalytic conditions [45] (Scheme 35). Slow addition of the phenyl sulfide **185** to a suspension of lithium

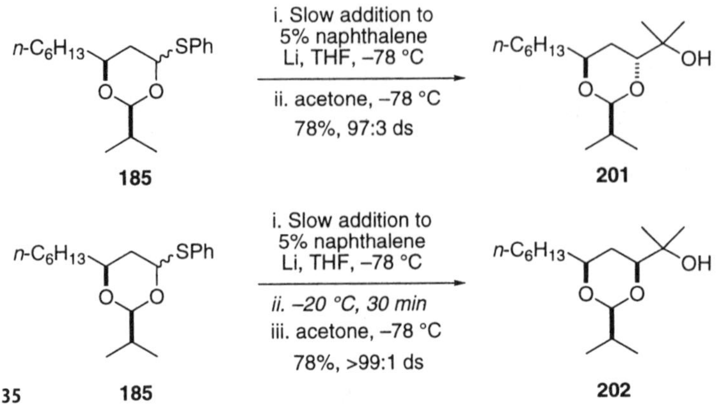

Scheme 35

powder and 5 mol % naphthalene in THF, followed by optional equilibration and quenching with acetone, provided the adducts **201** and **202** in good yield and with excellent selectivity. The slow addition prevents proton transfer reactions between **185** and the alkyllithium intermediate that are otherwise problematic. While the use of LiDBB presents little problem on a small scale, the catalytic procedure is recommended for preparative scale.

A limited range of electrophiles reacts efficiently with α-alkoxylithium reagents. Dimethyl sulfate [61, 74], CO_2 [75], and some aldehydes [71, 76], ketones [76, 77] and epoxides [65, 66] have been shown to react efficiently with retention of configuration. Reactions with other electrophiles are often complicated by competing proton transfer reactions, due to the high basicity of the alkyllithiums. Linderman [78] and Fuchs [79] have shown that α-alkoxycopper and α-alkoxycuprate reagents react with a wider range of electrophiles, although the configurations of the reacting centers are not always retained. Linderman reported that the organocopper species derived from stannane **203** underwent conjugate addition to ethyl propiolate with complete retention of configuration

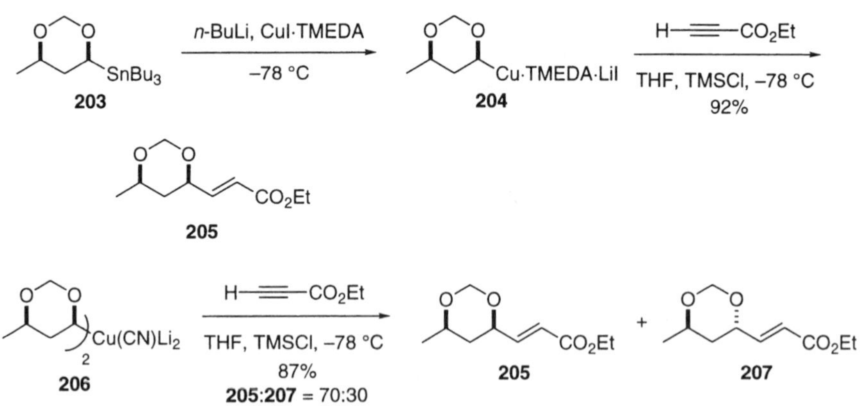

Scheme 36

[80] (Scheme 36). Interestingly, the higher order cuprate **206** underwent conjugate addition with only moderate selectivity. This is likely due to the intervention of an electron transfer pathway. Competing electron transfer reactions involving α-alkoxymetal reagents of this type have also been reported by Cohen [81].

Our group has also investigated transmetallation of α-alkoxylithiums to copper [68] (Scheme 37). These α-alkoxycuprates smoothly undergo conjugate additions under kinetic conditions, as shown in the generation of *anti*-1,3-dioxane **208**. The exocyclic 1,2-*syn* relationship is consistent with related work reported by Linderman [82]. The corresponding *syn*-1,3-dioxane **209** could also be prepared by equilibration prior to transmetallation, but a lower yield and reduced selectivity were observed. The exocyclic configuration could not be determined in this case. Transmetallation to copper facilitates additions to acid chlorides as well; the ketone **210** was cleanly prepared from the corresponding cuprate. The same reaction proceeded poorly in the absence of copper.

While there are noteworthy examples of the use of 2-lithiotetrahydropyrans in complex molecule synthesis [77], 4-lithio-1,3-dioxanes have seen relatively limited use in natural product synthesis. The reductive lithiation approach allowed for a convergent synthesis of the hexaol **216** from a common intermediate [65] (Scheme 38). Hexaol **216** corresponds to a skipped polyol segment of the antibacterial and antifungal agent lienomycin. While the alkyllithium fragment coupling with epoxides could be performed directly from 4-phenylthio-1,3-dioxanes, the 1 equiv. of lithium thiophenoxide generated under these condi-

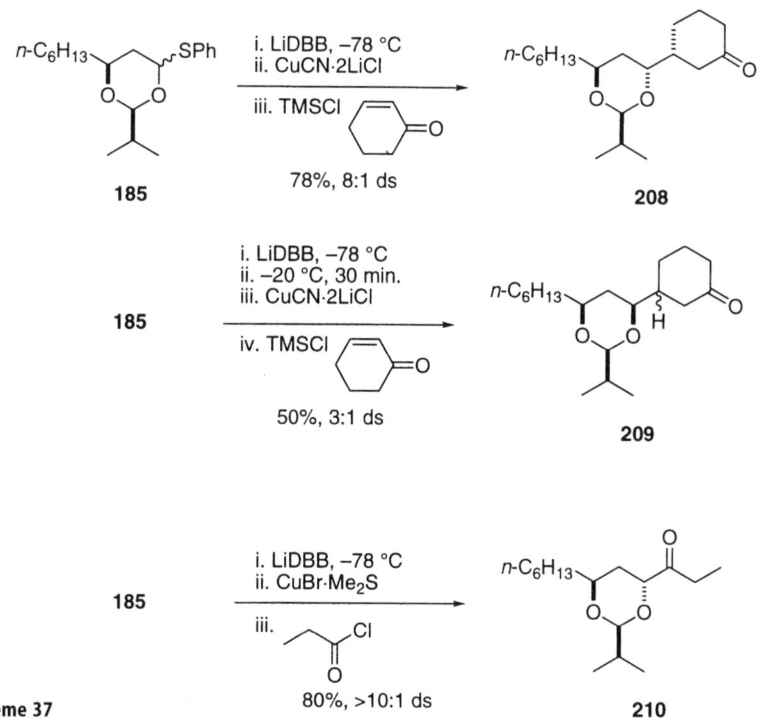

Scheme 37

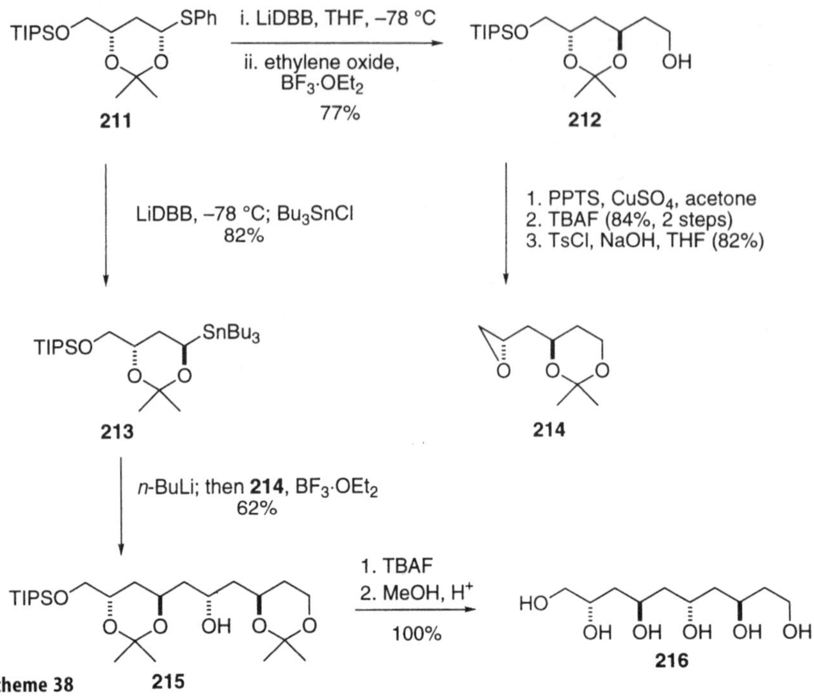

Scheme 38

tions also consumes the (potentially precious) electrophile. Transmetallation of the corresponding stannane **213** avoided this problem.

5 Alternative Approaches to 1,3-Diol Synthesis

Numerous other approaches to 1,3-polyol synthesis exist; however, most fall outside the scope of this review. Some of these approaches have been reviewed [83]. Particularly noteworthy advances in the direct preparation of protected 1,3-diols have recently been reported by Smith [84], Leighton [85], and Evans [86].

6 Conclusions

New synthetic methods are the lifeblood of organic chemistry. Synthetic efforts toward natural products often provide the impetus for the development of novel methodology. Reactive synthons derived from 1,3-dioxanes have proven to be valuable intermediates for both *syn*- and *anti*-1,3-diols found in many complex natural products. Coupling reactions at the 4-position of 1,3-dioxanes exploit anomeric effects to generate *syn*-1,3-diols (cyanohydrin acetonides), *anti*-1,3-diols (4-acetoxy-1,3-dioxanes), and either *syn*- or *anti*-1,3-diols (4-lithio-1,3-dioxanes). In the future, as biologically active polyol-containing natural products continue to be discovered, the methods described above should see much use.

Acknowledgments. We thank the NIH and NSF for their support of our program. We acknowledge the many talented coworkers that have contributed to the design and execution of this research program; their work is cited in the references. C. J. S. is grateful to the Department of Education and Pharmacia and Upjohn for fellowship support. We thank Alex Buckmelter and Claudia Roberson for helpful comments on this manuscript.

7 References

1. Beau J-M, Sinäy P (1985) Tetrahedron Lett 26:6185
2. Stork G, Maldonado L (1971) J Am Chem Soc 93:5286
3. For recent reviews, see: (a) Gregory RJH (1999) Chem Rev 99:3649; (b) Albright JD (1983) Tetrahedron 39:3203
4. Stork G, Manabe K, Liu L (1998) J Am Chem Soc 120:1337
5. Rychnovsky SD, Swenson SS (1997) J Org Chem 62:1333
6. Evans DA, Truesdale LK, Carroll GL (1973) J Chem Soc Chem Commun 55
7. Richardson TI, Rychnovsky SD (1997) J Am Chem Soc 119:12360
8. Richardson TI, Rychnovsky SD (1999) Tetrahedron 55:8977
9. Hu Y (1998) PhD thesis, University of California, Irvine
10. Packard G, Rychnovsky SD, unpublished results, University of California, Irvine
11. Hu Y, Skalitzky DJ, Rychnovsky SD (1996) Tetrahedron Lett 37:8679
12. Brzezinski LJ, Levy DD, Leahy JW (1994) Tetrahedron Lett 35:7601
13. Rychnovsky SD, Powers JP, Lepage TJ (1992) J Am Chem Soc 114:8375
14. Cohen T, Bhupathy M (1989) Acc Chem Res 22:152
15. Yue CW, Royer J, Husson HP (1990) J Org Chem 55:1140
16. Husson H-P, Royer J (1999) Chem Soc Rev 28:383
17. Rychnovsky SD, Skalitzky DJ (1990) Tetrahedron Lett 31:945
18. Rychnovsky SD, Rogers BN, Richardson TI (1998) Acc Chem Res 31:9
19. Evans DA, Rieger DL, Gage JR (1990) Tetrahedron Lett 31:7099
20. Rychnovsky SD, Rogers B, Yang G (1993) J Org Chem 58:3511
21. Rychnovsky SD, Zeller S, Skalitzky DJ, Griesgraber G (1990) J Org Chem 55:5550
22. Takahashi T, Nemoto H, Tsuji J (1983) Tetrahedron Lett 24:2005
23. Stork G, Depezay J-C, d'Angelo J (1975) Tetrahedron Lett 389
24. Matthews CF, Rychnovsky SD, unpublished results, University of California, Irvine
25. Ojika M, Nagoya T, Yamada K (1995) Tetrahedron Lett 36:7491
26. Rychnovsky SD, Griesgraber G (1993) J Chem Soc Chem Commun:291
27. Powers JP (1995) PhD Thesis, University of Minnesota
28. Lee J (1997) PhD Thesis, University of Minnesota
29. Rychnovsky SD, Hoye RC (1994) J Am Chem Soc 116:1753
30. Rychnovsky SD, Griesgraber G (1992) J Org Chem 57:1559
31. Mynderse JS, Moore RE (1979) Phytochemistry 18:1181
32. Rychnovsky SD, Khire UR, Yang G (1997) J Am Chem Soc 119:2058
33. Rychnovsky SD, Yang G, Hu YQ, Khire UR (1997) J Org Chem 62:3022
34. Ruckwardt, T, Scott, JS, Rychnovsky, SD, Yang, G, Hartsel, SC (1997) 17th Intl Congress of Biochemistry and Molecular Biology. San Francisco, CA
35. Bolard J (1986) Biochim. Biophys. Acta 864:257
36. Whitfield GB, Brock TD, Ammann A, Gottlieb D, Carter HE (1955) J Am Chem Soc 77:4799
37. Behnke O, Tranum-Jensen J, van Deurs B (1984) Eur Cell Biol 35:189
38. (a) Berkoz B, Djerassi C (1959) Proc Chem Soc 316; (b) Djerassi C, Ishikawa M, Budzikiewicz H (1961); Tetrahedron Lett 383; (c) Golding BT, Rickards RW (1964) Tetrahedron Lett 2615; (d) Ceder O, Ryhage R (1964) Acta Chem Scand 18:558
39. (a) Rychnovsky SD, Richardson TI (1995) Angew Chem Int Ed Engl 34:1227; (b) Richardson TI, Rychnovsky SD (1996) J Org Chem 61:4219
40. Wollenberg RH (1978) Tetrahedron Lett 717

41. Rychnovsky SD, Swenson SS (1997) Tetrahedron 53:16489
42. Deslongschamps P (1993) In: Thatcher GJ (ed), The Anomeric Effect and Associated Stereoelectronic Effects. American Chemical Society, Washington, DC, p 26
43. Recent reviews on *C*-glycoside synthesis: (a) Du Y, Linhardt RJ, Vlahov IR (1998) Tetrahedron 54:9913; (b) Levy DE, Tang C (1995) The Chemistry of C-Glycosides. Elsevier Science, Oxford
44. Lewis MD, Cha JK, Kishi Y (1982) J Am Chem Soc 104:4976
45. Rychnovsky SD, Skalitzky DJ (1992) J Org Chem 57:4336
46. Dahanukar VH, Rychnovsky SD (1996) J Org Chem 61:8317
47. Kopecky DJ, Rychnovsky SD (2000) J Org Chem 65:191
48. Rychnovsky SD, Hu YQ, Ellsworth B (1998) Tetrahedron Lett 39:7271
49. Davis AP, Hegarty SC (1992) J Am Chem Soc 114:2745
50. Boons G-J, Eveson R, Smith S, Stauch T (1996) Synlett 536
51. Rychnovsky SD, Fryszman O, Khire UR (1999) Tetrahedron Lett 40:41
52. Rychnovsky SD, Sinz CJ (1998) Tetrahedron Lett 39:6811
53. Danishefsky SJ, DeNinno S, Lartey P (1987) J Am Chem Soc 109:2082
54. Rychnovsky SD, Powell NA (1997) J Org Chem 62:6460
55. Wipf P, Jahn H (1996) Tetrahedron 52:12853
56. Powell NA, Rychnovsky SD (1999) J Org Chem 64:2026
57. Langer F, Schwink L, Devasagayaraj A, Chavant P-Y, Knochel P (1996) J Org Chem 61:8229
58. Berninger J, Koert U, Eisenberg-Hohl C, Knochel P (1995) Chem Ber 128:1021
59. Powell NA, Rychnovsky SD (1998) Tetrahedron Lett 39:3103
60. Mynderse JS, Moore RE (1978) Phytochemistry 17:1325
61. Still WC (1978) J Am Chem Soc 100:1481
62. McGarvey GJ, Kimura M (1982) J Org Chem 47:5420
63. Screttas CG, Micha-Screttas M (1978) J Org Chem 43:1064
64. Cohen T, Matz JR (1980) Synth. Commun 10:311
65. Rychnovsky SD (1989) J Org Chem 54:4982
66. Rychnovsky SD, Mickus DE (1989) Tetrahedron Lett 30:3011
67. Rychnovsky SD, Plzak K, Pickering D (1994) Tetrahedron Lett 35:6799
68. Rychnovsky SD, Buckmelter AJ, Dahanukar VH, Skalitzky DJ (1999) J Org Chem 64:6849
69. Rychnovsky SD, Buckmelter AJ, Dahanukar VH, Skalitzky DJ (1999) J Org Chem 64:7678
70. Linderman RJ, Ghannam A (1990) J Am Chem Soc 112:2392
71. Cohen T, Lin MT (1984) J Am Chem Soc 106:1130
72. Ramon DJ, Yus M (1992) Tetrahedron Lett 33:2217
73. Yus M (1996) Chem Soc Rev 25:155
74. Chan PC-M, Chong JM (1988) J Org Chem 53:5584
75. Chan PC-M, Chong JM (1990) Tetrahedron Lett 31:1985
76. Sawyer JS, Kucerovy A, Macdonald TL, McGarvey GJ (1988) J Am Chem Soc 110:842
77. Boeckman RK, Enholm EJ, Demko DM, Charette AB (1986) J Org Chem 51:4743
78. Linderman RJ, Godfrey A (1986) Tetrahedron Lett 27:4553
79. Hutchinson DK, Fuchs PL (1987) J Am Chem Soc 109:4930
80. Linderman RJ, Griedel BD (1991) J Org Chem 56:5491
81. Lin M-T (1984) PhD Thesis, University of Pittsburgh
82. Linderman RJ, McKenzie JR (1988) Tetrahedron Lett 29:3911
83. Oishi T, Nakata T (1990) Synthesis 635
84. Multicomponent linchpin couplings of silyl dithianes: (a) Smith AB III, Boldi AM (1997) J Am Chem Soc 119:6925; (b) Smith AB III, Pitram SM (1999) Org Lett 1:2001
85. Hydroformylation of 4-methylene-1,3-dioxanes: (a) Leighton JL, O'Neil DN (1997) J Am Chem Soc 119:11118; (b) Sarraf ST, Leighton JL (1998) Tetrahedron Lett 39:6423
86. Elaboration of homoallylic alcohols to protected 1,3-diols: (a) Evans DA, Gauchet-Prunet JA (1993) J Org Chem 58:2446; (b) Sarraf ST, Leighton JL (2000) Org Lett 2:403; (c) Duan JJ-W, Smith AB III (1993) J Org Chem 58:3703

The Synthetic Potential of Three-Membered Ring Aza-Heterocycles

Binne Zwanenburg · Peter ten Holte

Department of Organic Chemistry, NSR Institute for Molecular Structure, Design and Synthesis, University of Nijmegen, Toernooiveld, 6525 ED Nijmegen, The Netherlands.
E-mail: zwanenb@sci.kun.nl

Functionalized aziridines, especially aziridine-2-carboxylic esters, are highly valuable small-ring heterocycles for the synthesis of a large variety of anomalous amino acids, new types of ligands for catalytic purposes, new synthons, and four- and five-membered ring heterocycles through ring expansion reactions. This review highlights the aziridine chemistry with the focus on the utility of aziridine esters. The synthesis and chemistry of the highly strained azirine carboxylic esters is also briefly reviewed.

Keywords: Aziridine-2-carboxylic esters, Ring expansion reactions, Azirine carboxylic esters, Aziridine carbinols, Anomalous amino acids

1 Introduction . . . 94

2 Synthesis . . . 94

2.1 Functionalized Aziridines from Oxiranes . . . 94
2.2 Aziridines from Vicinal Diols . . . 97
2.3 Aziridines from Amino Epoxides by a Modified Payne Rearrangement . . . 97
2.4 Aziridines from β-Hydroxy-α-amino Acids . . . 97
2.5 Azirines from Aziridines . . . 100
2.6 Azirine Synthesis by a Modified Neber Reaction . . . 103

3 Reactions . . . 105

3.1 Anomalous Amino Acids from Aziridine-2-carboxylic Esters . . . 105
3.2 Aziridine Ring Expansions . . . 109
3.3 Aziridine Carbinols as Chiral Ligands . . . 115
3.4 *N*-Tritylaziridine-2-carboxaldehyde as Synthon . . . 117
3.5 Supramolecular Chemistry . . . 120
3.6 Reactions of 2*H*-Azirines . . . 121

4 Concluding Remarks . . . 122

5 References . . . 122

1 Introduction

The chemistry of three-membered ring heterocycles, especially epoxides and aziridines, has attracted the attention of synthetic chemists for more than a century. This is primarily due to the intrinsically high reactivity of these small-ring heterocycles, which makes them very versatile species in organic synthesis. In this chapter the main focus will be on functionalized, nitrogen-containing three-membered rings. The presence of an additional functionality makes these heterocycles especially valuable, for instance, aziridine-2-carboxylic acids can be considered as α- as well as β-amino acids and therefore they are attractive for synthetic elaboration. The four-membered ring counterparts, namely azetidines, are much less popular, which may be attributed to their reduced reactivity and to the fact that they are less readily accessible. Ring expansion of aziridines to five-membered ring heterocycles is an attractive method for the preparation of a variety of these aza-heterocycles. The focus will be again on expansion of functionalized three-membered rings.

During our research in this field of small-ring heterocycles we found that functionalized aziridines are attractive chiral catalysts, e.g., in the diethylzinc addition to aldehydes. Aspects of such uses of aziridines will be discussed as well. This overview does not pretend to be an exhaustive coverage of all existing literature on small-ring aza-heterocycles as that would require a separate monograph. Instead, emphasis is put on functionalized three-membered aza-heterocycles, that were investigated in the author's laboratory [1], and relevant related literature. The older literature on these heterocycles is adequately summarized in some extensive reviews [2]. Chiral aziridines have been reviewed recently by Tanner [3], by Osborn and Sweeney [4], and by McCoull and Davis [5].

2 Synthesis

2.1 Functionalized Aziridines from Oxiranes

Aziridine-2-carboxylic acids represent a special type of amino acid and therefore their synthesis in enantiopure form is important. An attractive method for the preparation of aziridine-2-carboxylic esters of high enantiopurity from the corresponding oxirane-2-carboxylic esters was developed by us in the late 1980s [6]. In this approach effective use is made of the Sharpless asymmetric epoxidation of allylic alcohols [7] to install a defined chirality. The initially obtained epoxy alcohols are converted into the oxirane-2-carboxylic esters **1** by subsequent oxidation and esterification. The key step in this sequence leading to aziridine esters involves treatment of oxirane esters with sodium azide in the presence of ammonium chloride as the buffer, followed by reaction with triphenylphosphine. The first step leads to a mixture of regioisomeric azido alcohols **2a** and **2b**, which do not need to be separated. The subsequent reaction with triphenylphosphine produces oxazaphospholidines **3**, again as a mixture of

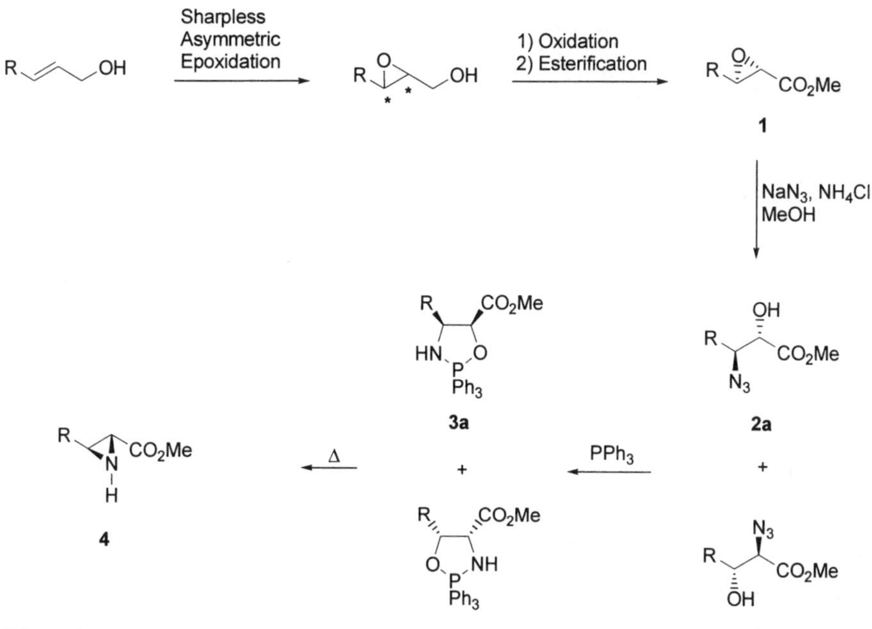

Scheme 1

regioisomers (**3a** and **3b**). Heating then results in the elimination of triphenylphosphine oxide and the concurrent formation of the desired aziridine-2-carboxylic esters **4**. It should be noted that both regiomers of **3** lead to the same aziridine, which explains why no separation of the regioisomeric intermediates is required. In this sequence, which is outlined in Scheme 1, an inversion of configuration at both stereogenic centers takes place, the first one during the treatment with azide ions and the second one during the thermal extrusion of triphenylphosphine oxide.

The stereochemistry of the first step was ascertained by an X-ray analysis [8] of an isolated oxazaphospholidine **3** (R = Ph). The overall sequence from oxirane to aziridine takes place with an excellent retention of chiral integrity. As the stereochemistry of the oxirane esters is determined by the chiral inductor during the Sharpless epoxidation, both enantiomers of aziridine esters can be readily obtained by choosing the desired antipodal tartrate inductor during the epoxidation reaction. It is relevant to note that the required starting allylic alcohols are conveniently prepared by chain elongation of propargyl alcohol as a C_3 synthon followed by an appropriate reduction of the triple bond, e.g., with lithium aluminum hydride [6b].

The oxirane to aziridine conversion method was also utilized for the successful synthesis of naturally occurring (+)-(2*S*,3*S*)-aziridine-2,3-dicarboxylic acid **5** as is shown in Scheme 2 [9].

Tanner [3, 10] used essentially the same methodology for the conversion of a monobenzyl derivative of a C_2-symmetric epoxy bis-alcohol into the corresponding derivative **6** of a C_2-symmetric aziridine bis-alcohol (Scheme 3).

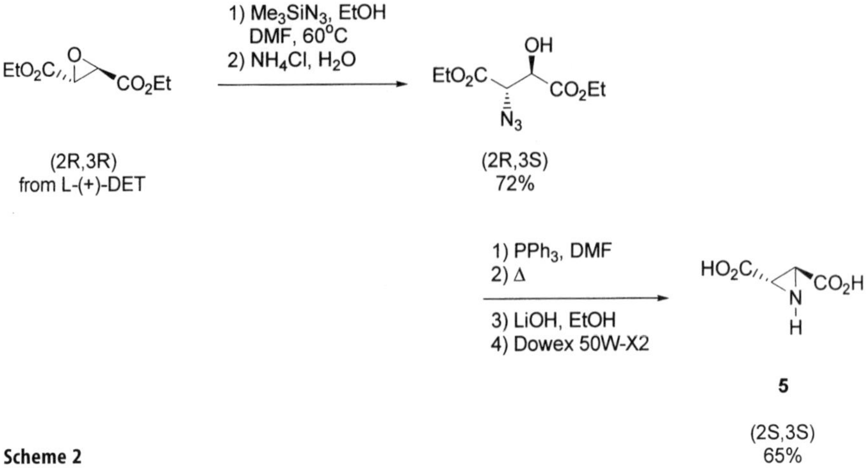

Scheme 2

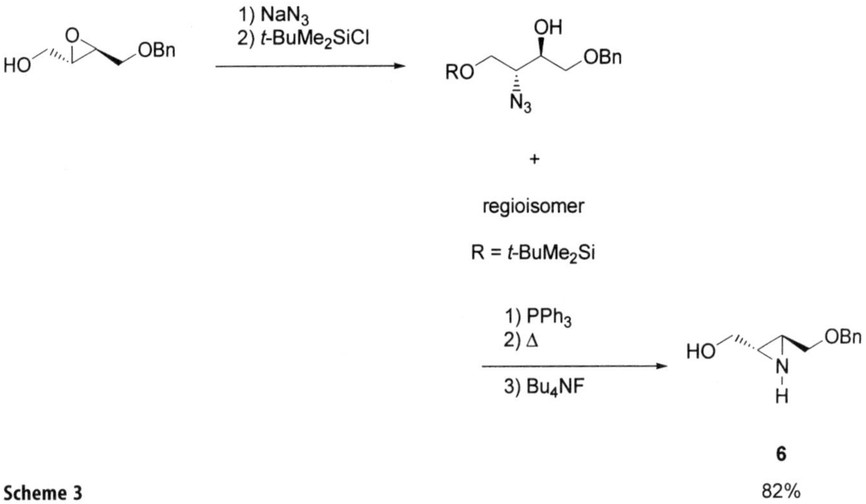

Scheme 3

Glycidyl butyrate **7** was similarly conveniently converted in the corresponding aziridinylmethyl butyrate **8** (Scheme 4) [11].

It may be concluded that the conversion of functionalized oxiranes into the corresponding aziridines by an azide ring opening followed by a Staudinger ring closure with triphenylphosphine constitutes a general method for the preparation of aziridines with high enantiopurity.

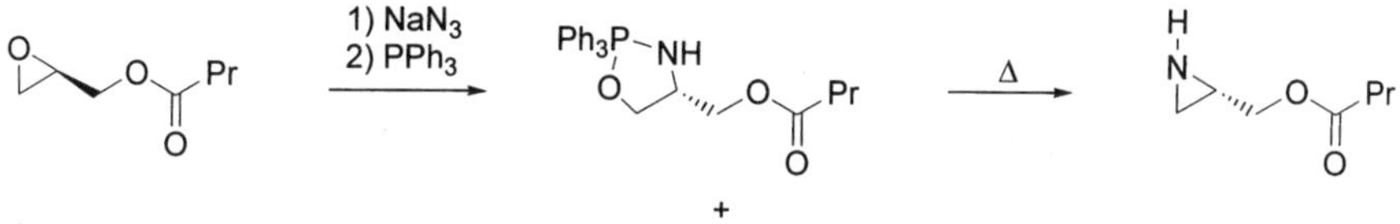

7 (R) **8** (S)

Scheme 4

64% overall

2.2
Aziridines from Vicinal Diols

Although the Sharpless asymmetric epoxidation is an elegant method to introduce a specific defined chirality in epoxy alcohols and thus, in functionalized aziridines (see Sect. 2.1), it is restricted to the use of allylic alcohols as the starting materials. To overcome this limitation, cyclic sulfites and sulfates derived from enantiopure *vic*-diols can be used as synthetic equivalents of epoxides (Scheme 5) [12, 13].

The required *vic*-diols can readily be prepared from olefins by the Sharpless asymmetric dihydroxylation method [14]. Treatment of the cyclic sulfites with lithium azides regioselectively leads to hydroxy azides, which, as shown in Sect. 2.1, can be ring-closed by reaction with triphenylphosphine to give aziridines **9** [12]. In the case of the cyclic sulfates, ring opening with lithium azide gives an azido sulfate, which upon reduction with lithium aluminum hydride and subsequent base treatment results in the desired aziridines **9** in high yields. Alternatively, the cyclic sulfates can be treated with a primary amine, which gives a zwitterionic β-amino sulfate **10.** In a subsequent reaction with either butyl lithium or lithium aluminum hydride, ring closure to *N*-substituted aziridines **11** can be accomplished in good overall yields [13]. The scope of the methods described above is very attractive.

2.3
Aziridines from Amino Epoxides by a Modified Payne Rearrangement

A modified Payne rearrangement of amino epoxides catalyzed by Lewis acid or induced by base, represents an interesting but a limited method for the synthesis of functionalized aziridines of high enantiopurity. The limitations are primarily due to the accessibility of the starting materials (Scheme 6) [15].

2.4
Aziridines from β-Hydroxy-α-amino Acids

An attractive and useful method for the preparation of aziridine-2-carboxylic esters makes use of the readily available amino acids serine and threonine. Essentially, this synthesis involves the ring closure of 1,2-amino alcohols,

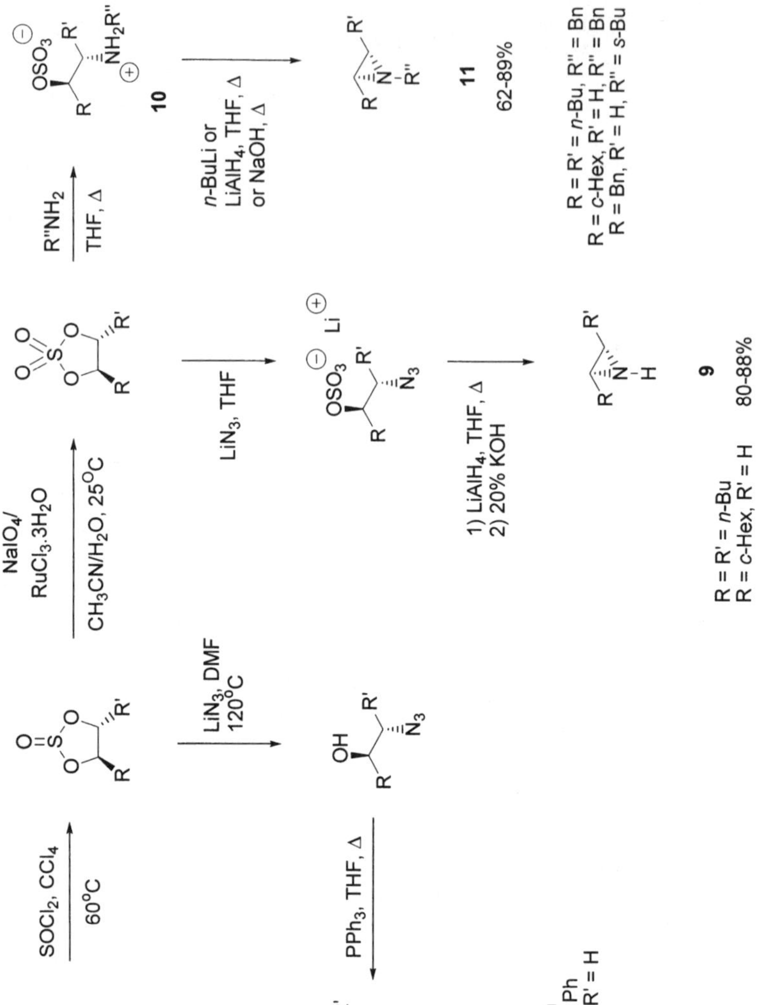
SOCl$_2$, CCl$_4$
60°C
NaIO$_4$/
RuCl$_3$.3H$_2$O
CH$_3$CN/H$_2$O, 25°C
R''NH$_2$
THF, Δ
LiN$_3$, DMF
120°C
LiN$_3$, THF
n-BuLi or
LiAlH$_4$, THF, Δ
or NaOH, Δ
PPh$_3$, THF, Δ
1) LiAlH$_4$, THF, Δ
2) 20% KOH
10
11
62-89%
9
49-53%
9
80-88%
R = R' = Ph
R = H, R' = Ph
R = c-Hex, R' = H
R = R' = n-Bu
R = c-Hex, R' = H
R = R' = n-Bu, R'' = Bn
R = c-Hex, R' = H, R'' = Bn
R = Bn, R' = H, R'' = s-Bu

Scheme 5

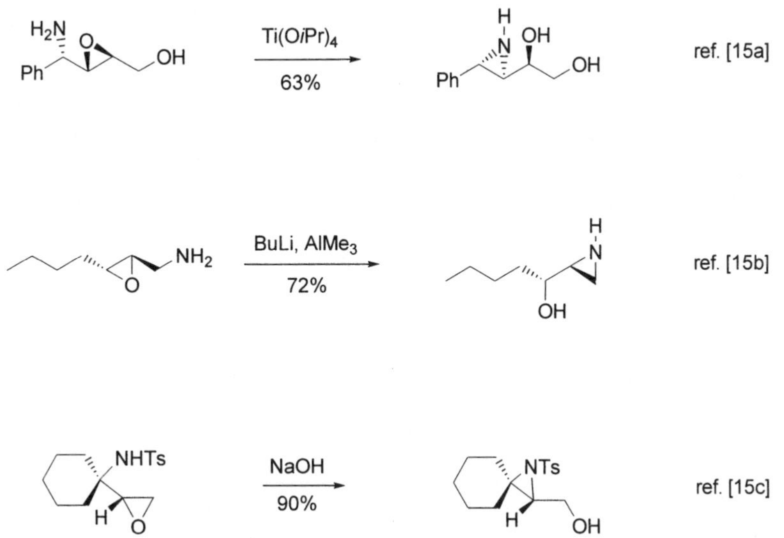

Scheme 6

whereby the nitrogen is protected and the hydroxy function converted into a good leaving group. In the present case, the amino acids are *N*-tritylated and subsequently treated with *p*-toluenesulfonyl chloride [16] or with methanesulfonyl chloride [17]. Van Boom [18] reported an improved "one-pot, one-step" procedure using sulfuryl chloride and an excess of triethylamine for the ring closure. On a multigram scale this procedure gave also a by-product in ca. 30% yield, arising from an aziridine ring-opening reaction by chloride ions. We reported a modified and convenient experimental procedure whereby two equivalents of methanesulfonyl chloride were used in a one-pot operation [19] (Scheme 7). In this manner multigram quantities of products **12** and **13** could be prepared routinely. Usually, the protecting trityl group is removed during a subsequent reaction with these aziridine esters.

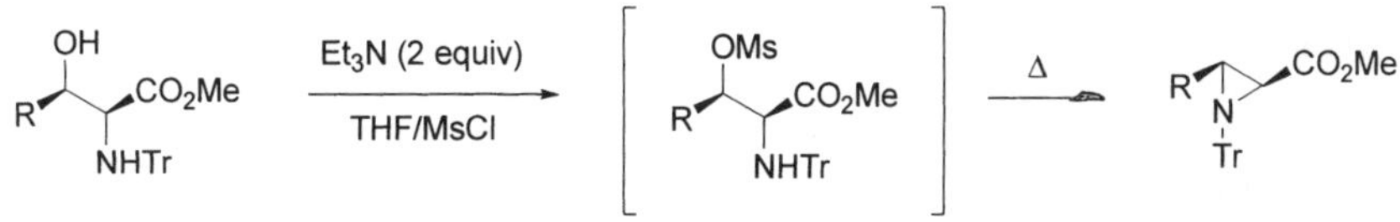

12 (R = H)
13 (R = Me)

R = H, from L-serine
R = Me, from L-threonine

cy quantitative
purity 95% (GLC)

Scheme 7 cy = chemical yield

2.5 Azirines from Aziridines

Azirines are the most strained three-membered aza-heterocycles. Remarkably, these highly strained types of molecules occur in nature, namely azirinomycin **14** and dysidazirine **15** (Fig. 1).

The first compound is an antibiotic isolated from *Streptomyces aureus* [20], while the second compound is a cytotoxic antibiotic isolated from *Dysidea fragilis*, a marine sponge [21]. A logical approach to the synthesis of azirines would be an elimination reaction of a suitably *N*-substituted aziridine. Thus, *N*-chlorination of aziridine-2-carboxylic esters was carried out using *tert*-butyl hypochlorite (Scheme 8).

Interestingly, both invertomers of the obtained *N*-chloroaziridines **16** were clearly observable in the ^{1}H-NMR spectrum and they could even be separated by chromatography. The dehydrochlorination was investigated with a variety of bases; however, the resulting yields were disappointingly low. Only for R = Ph, a yield of 39% of azirine **17** was obtained using DBU as the base, in all other cases the yields were lower [22]. Davis et al. [23] successfully applied the β-elimination of the sulfinyl group in chiral non-racemic *N*-sulfinylaziridines (Scheme 9), whereby the eliminated sulfenate was trapped by an excess of methyl iodide, which facilitated the isolation of the desired product (**18**).

Fig. 1 **14** **15**

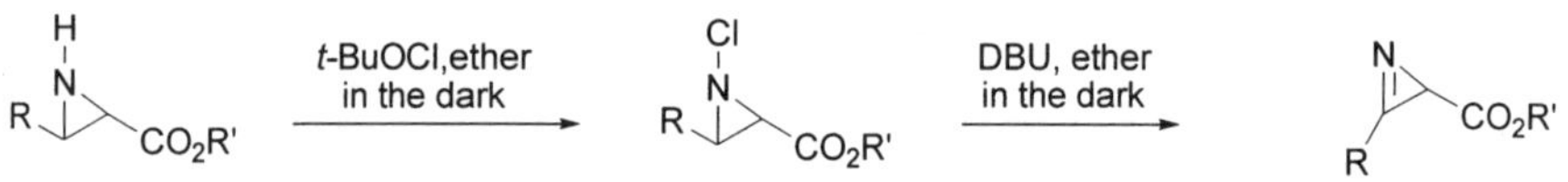

16

92-97%
2 isomers

17

R = Ph, 39%
R = alkyl, 10-20%
(aziridine ester)

Scheme 8

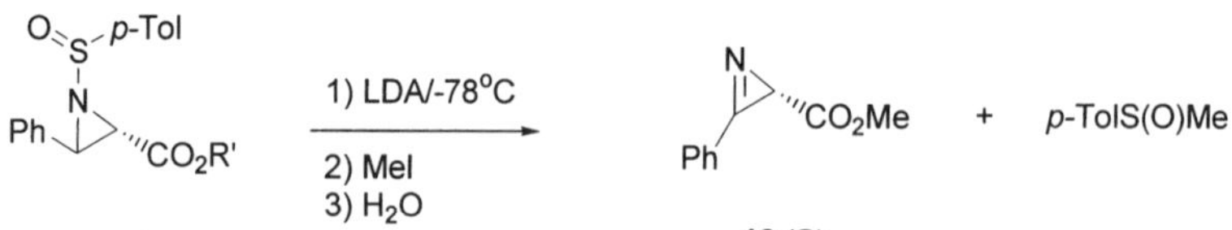

(enantiopure)

18 (S)

47-52%

Scheme 9

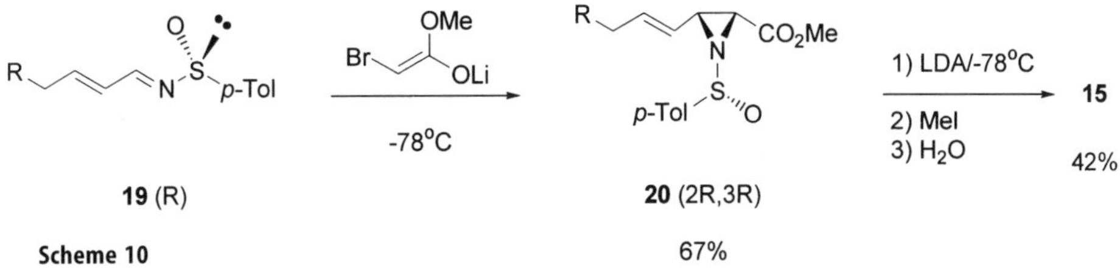

Scheme 10

The same elimination strategy was used for the synthesis of the natural product (*R*)-(−)-dysidazirine **15** as is shown in Scheme 10 [23]. The requisite aziridine ester was prepared by treatment of sulfimine **19** with the lithium enolate of methyl bromoacetate. This reaction is a Darzens-type condensation leading to *cis*-*N*-sulfinylaziridine ester **20**. The elimination of sulfenate was accomplished in the same manner as mentioned above (see Scheme 9). The natural product **15** (see Fig. 1) was obtained in 42 % yield. Attempts to prepare azirinomycin **14** in a similar fashion all failed [23].

A highly remarkable and entirely unexpected conversion of aziridine esters **21** into azirine esters **22** was accomplished by subjecting the aziridine to a Swern oxidation (Scheme 11).

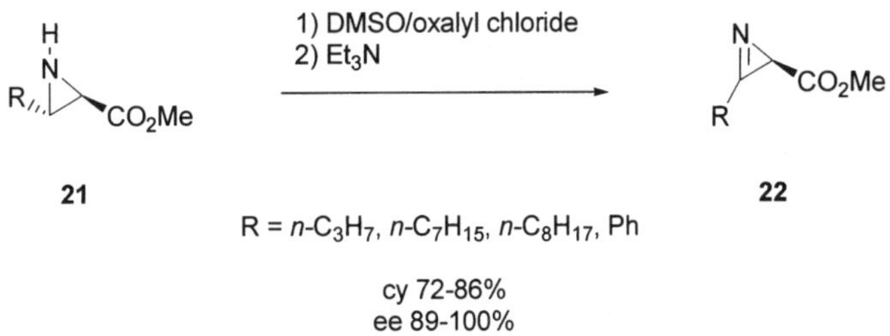

Scheme 11

This simple procedure resulted in 2*H*-azirines exclusively in high yields and with retention of chirality at C-2 [24]. Mechanistically, this oxidative synthesis of azirines can be rationalized by invoking the intermediacy of chloro-dimethylsulfonium chloride ($Me_2SCl^{+\cdot}$ Cl^-), which then reacts with the aziridine as indicated in Scheme 12.

Subsequent cyclo-elimination induced by base then results in the formation of the azirine product. To explain the exclusive formation of the 2*H*-azirine ester, it is conceivable that the invertomer of intermediate **23** with the Me_2S^+ group *syn* to the ester function is (strongly) preferred. Considering this proposed course of the elimination one would expect that for *cis*-aziridine esters **24** the intermediate sulfonium unit **25** has a preferred conformation in which the Me_2S^+ group is positioned *anti* to both the R substituent and the ester function and *syn* to both abstractable hydrogens (Scheme 13). Accordingly, formation of 3*H*-azirine ester with the imine bond in conjugation with the ester moiety would be expected, at least in part. Surprisingly however, Swern oxidation of the *cis*-substrates again gave exclusive formation of the 2*H*-azirine ester **22**.

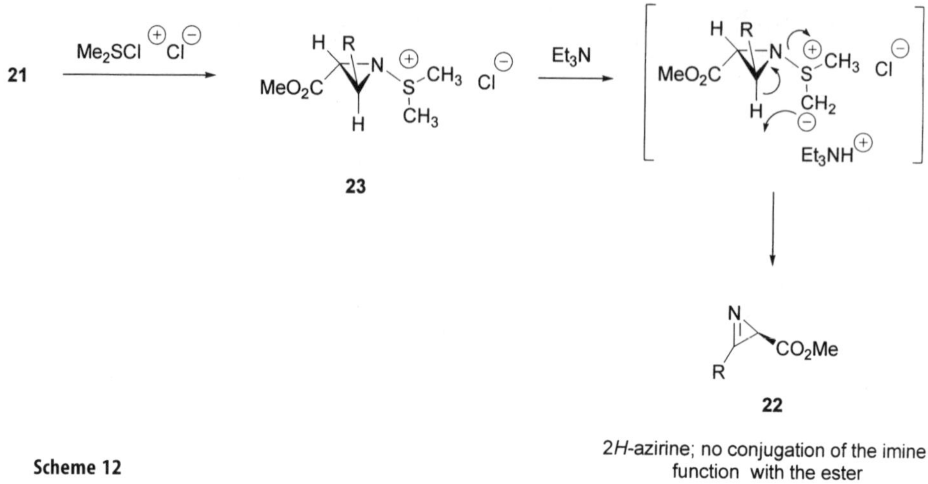

Scheme 12

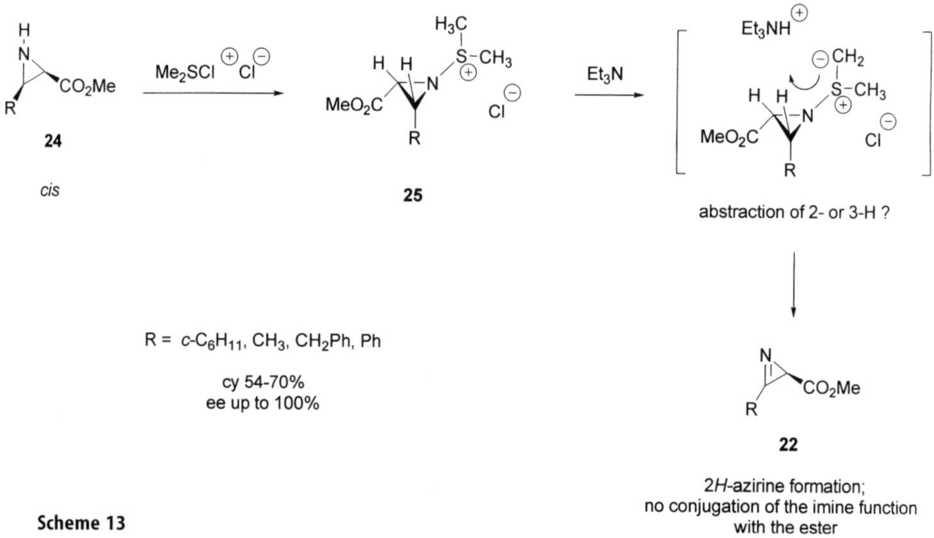

Scheme 13

The regiochemistry of this elimination reaction resembles that observed by Davis et al. (see Scheme 9) [23]. The special nature of the bonds in three-membered rings is probably responsible for this exclusive regiochemistry. It is of interest to note that 3,3-dimethylaziridine-2-carboxylic ester indeed leads to the corresponding 3*H*-azirine ester upon Swern oxidation; here there is, of course, no choice.

The Swern oxidation was also employed by Davis and McCoull [25] for the synthesis of 2*H*-azirinephosphonates **27** and **28** from the corresponding aziridines **26** (Scheme 14). Interestingly, in this case a mixture of the regioisomeric azirines is obtained, whereby the proton abstraction adjacent to the phos-

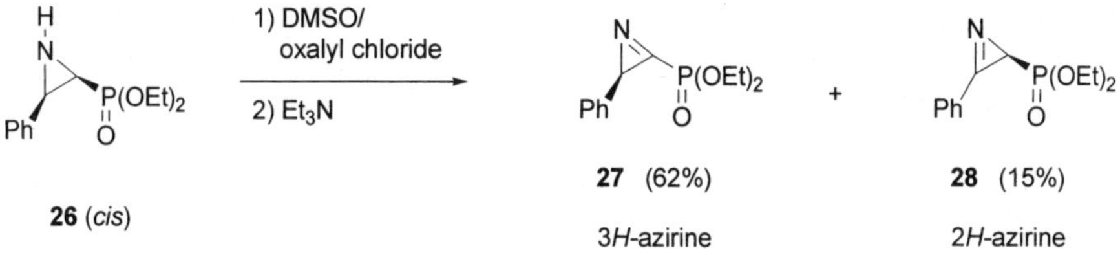

Scheme 14

phonate function leads to the predominant product **27**. Here the phosphonate stabilizes the pyramidal α-carbanion on the aziridine ring, which is much less favorable in the case of an adjacent carboxylic ester function [24]. The starting aziridinephosphonates were prepared by a modified Darzens-type condensation of enantiopure sulfinimines with diethyl chloromethylphosphonate and subsequent removal of the *N*-sulfinyl auxiliary under acidic conditions [25] (see also Scheme 10).

2.6 Azirine Synthesis by a Modified Neber Reaction

An entirely different approach to the synthesis of 2*H*-azirinecarboxylic esters involves a modified Neber reaction of oxime tosylates derived from β-keto esters (Scheme 15) [26].

Interestingly, certain chiral tertiary bases, viz., the *Cinchona* alkaloids, result in an asymmetric 1,3-elimination to give enantiomerically enriched azirine esters **29** (Scheme 15). The best results were obtained with quinidine in toluene as the solvent at a rather high dilution (2 mg mL^{-1}) at 0 °C. In an alcoholic solvent no asymmetric conversion was observed. It is of importance to note that the pseudoenantiomers of the alkaloid bases gave opposite antipodes of the azirine ester, whereby quinidine leads to the predominant formation of the (*R*)-enantiomer (ee = ~80%). To explain this asymmetric Neber reaction, it is suggested

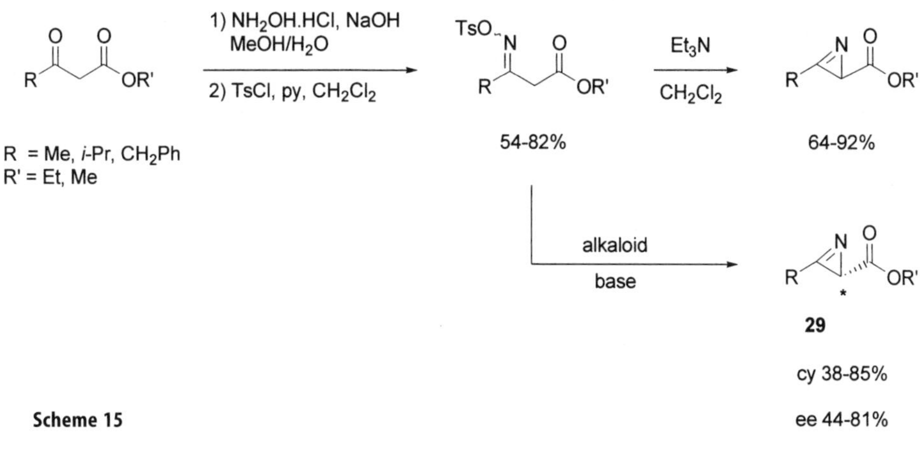

Scheme 15

that the alkaloid bases form a tightly bound complex with the ketoxime tosylate. As it was found that chiral bases lacking an alcohol function, e.g., sparteine and strychnine, do not give an asymmetric 1,3-elimination, it may be concluded that hydrogen bonding of the alkaloid base and the substrate through this requisite hydroxy group is governing the enantiodifferentiation during the abstraction of the methylene protons. This conclusion is substantiated by the observation that hydroxylic solvents, additives like LiCl and water lower the optical yield considerably. In the procedures discussed so far, a stoichiometric amount of alkaloid base is used. A catalytic version of this asymmetric Neber reaction could be achieved by regenerating the alkaloid base *in situ* by adding 10–20 equiv of potassium carbonate to the reaction mixture. Under these conditions only 10 mol % of quinidine was needed, however, the ee dropped to 70 % (Scheme 16).

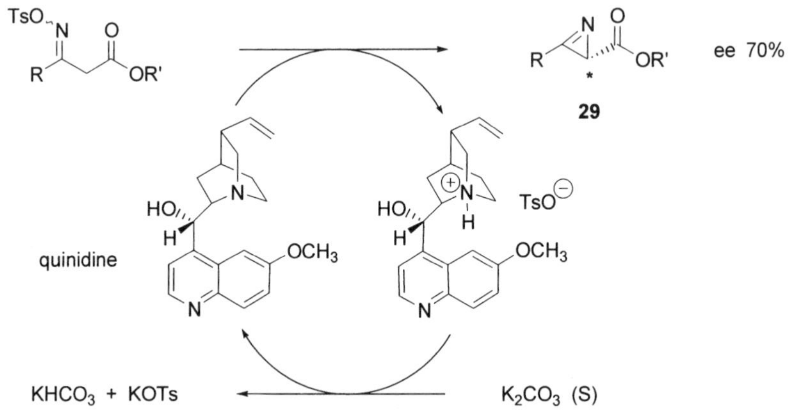

Scheme 16

A possible extension of the modified Neber reaction would be the synthesis of sulfonyl-substituted 2*H*-azirines following the chemistry shown in Scheme 17 [27]. Unexpectedly, the oxime derived from *β*-keto sulfones could not be converted into the oxime tosylate. Therefore, a different route to these requisite starting materials was designed, viz., via the corresponding sulfides **30** which were then oxidized with peracid to the sulfones **31**.

1,3-Elimination gave also here the azirines **32**. The use of chiral base in these cases did not result in a chirality transfer. Attempts were also made to prepare the corresponding 2-sulfinyl-2*H*-azirines. It turned out that these compounds

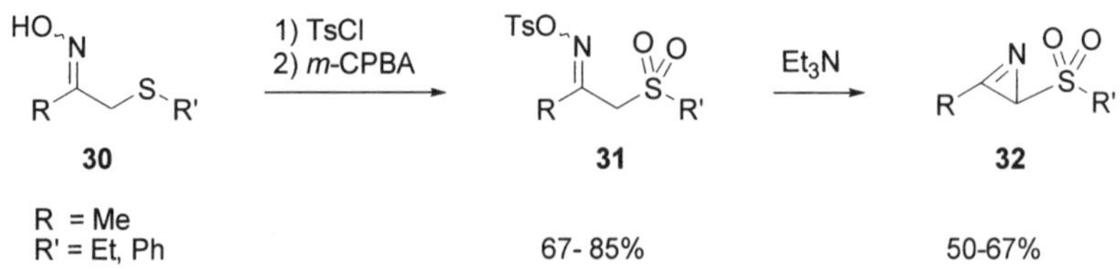

Scheme 17

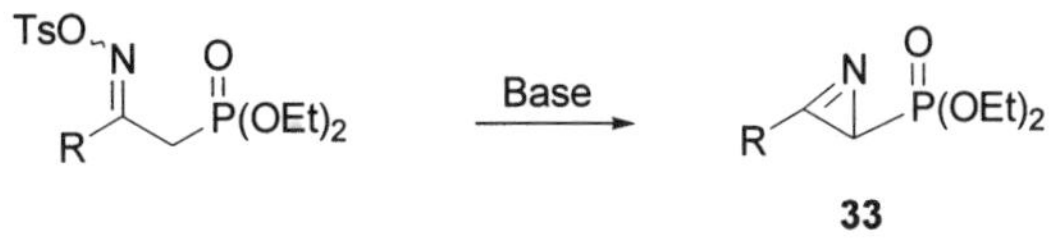

Scheme 18 R = Me, Et, Ph 69-95%

are rather unstable and escaped isolation, although their formation could be ascertained by means of ^{1}H-NMR spectroscopy.

Palacios et al. utilized the modified Neber reaction for the preparation of 2*H*-azirine-2-phosphonates **33** as shown in Scheme 18 [28a]. The use of quinidine and dihydroquinidine as the chiral base resulted in moderate chirality transfer (20–52% ee). Similarly, 2-phosphinoyl-2*H*-azirines could be obtained by the Neber 1,3-elimination reaction [28b].

3 Reactions

3.1 Anomalous Amino Acids from Aziridine-2-carboxylic Esters

Aziridine esters are α- and β-amino acid derivatives at the same time. A characteristic reaction of α-amino acids is their reaction with triethylboron to give boroxazolidines. We showed that aziridinecarboxylic acids exhibit the expected behavior in their reaction with triethylboron, viz., that they form stable boroxazolidines **34** (Scheme 19) [29]. These boron heterocycles can be reconverted into the free amino acids by treatment with 8-hydroxyquinoline.

The majority of reactions of aziridines deal with acid-catalyzed ring opening with various nucleophiles. In this review only reactions with aziridine-2-carboxylic esters are summarized (for other types of aziridines, see ref. [3]).

Whether the nucleophilic attack takes place at C-2 or C-3 strongly depends on the reaction conditions and the substituents at the three-membered ring. Ring-opening reactions of 3-arylaziridine-2-carboxylic esters with HCl in ether gave mixtures of regioisomeric chloro-amino acids with a preference for halide attack at C-3. When the aryl group is capable of extra stabilizing a cationic center at C-3, as is the case for Ar = *p*-methoxyphenyl, then the ring opening is not

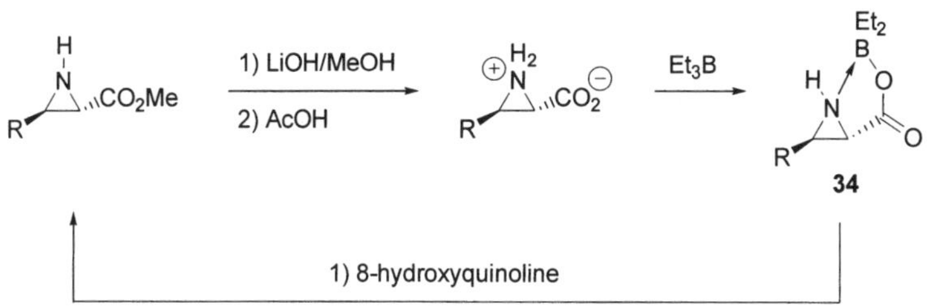

R = Aryl, alkyl

Scheme 19

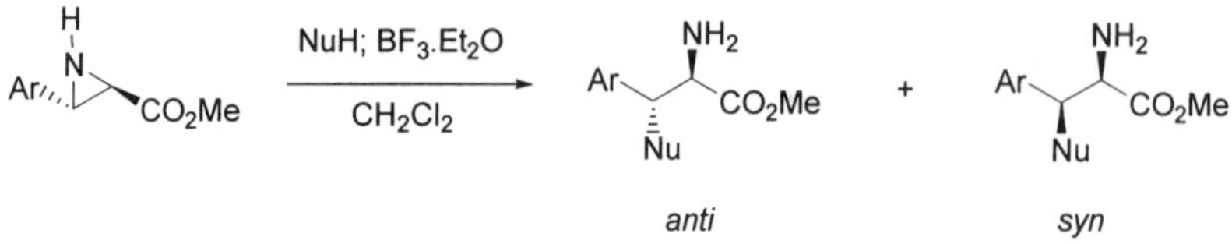

Ar	NuH	*anti:syn*	*cy*
p-$MeOC_6H_4$	PhSH	3:2	65%
Ph	PhSH	*anti* only	80%
p-$NO_2C_6H_4$	PhSH	*anti* only	65%
p-$MeOC_6H_4$	indole	55:45	68%
Ph	indole	*anti* only	53%
p-$NO_2C_6H_4$	indole	*anti* only	60%

Scheme 20

stereospecific as is evident from the formation of *anti*- and *syn*-3-chloro-2-amino acid in the ratio of 7:3. Such a *syn*-product is not observed for the ring opening of 3-(*p*-nitrophenyl)aziridine carboxylic ester. Similarly, the ring opening of 3-(*p*-methoxyphenyl)aziridinecarboxylic ester with nucleophiles, such as benzenethiol and indole, under BF_3 catalysis, gives a mixture of *anti*- and *syn*-products arising from the reaction with the nucleophile at C-3.

Completely stereospecific and regiospecific ring-opening reactions were observed with the just mentioned nucleophiles when the 3-aryl substituent is a phenyl or a *p*-nitrophenyl group (Scheme 20) [30].

No products arising from a nucleophilic reaction at C-2 were observed. The reaction with acetic acid also takes place regioselectively at C-3, but the initially formed *anti*-3-acetoxy product undergoes an acyl migration to give the *anti*-3-hydroxy-2-acetamino product (see Scheme 21).

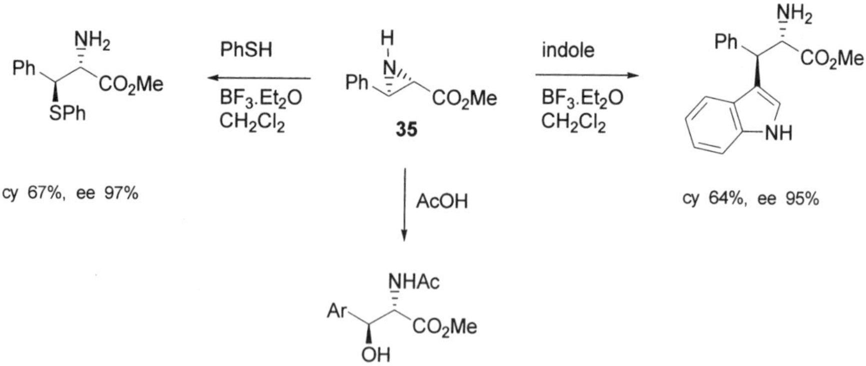

Scheme 21

cy 69%, ee 99%

The results described above clearly demonstrate that for 3-phenylaziridine-2-carboxylic esters the ring opening is completely regio- and stereospecific. This conclusion was further substantiated by using (almost) enantiopure aziridine ester **35** as the substrate. The results shown in Scheme 21 speak for themselves [30].

Ring-opening reactions with 3-alkylaziridine esters **36** take a similar course. The reactions are in practically all cases regio- and stereospecific with attack at C-3. An important difference is that the aziridine ring needs to be activated by an electron-withdrawing substituent, such as a tosyl or a benzyloxycarbonyl group. In addition, for benzenethiol, indole, and DMF, catalysis with BF_3 was necessary (Scheme 22) [31].

The behavior described above towards nucleophiles under acid catalysis can be considered as typical. Additional examples are collected in the reviews by Tanner [3] and Davis [5].

Ring opening by halide in a fully stereocontrolled manner can be accomplished with $MgBr_2$ (exclusive opening at C-3) (Scheme 23) [32]. However, with NaBr a predominant C-2 opening was observed with a considerable dependence of the regiochemistry on the substituent at C-3. The ring opening with organometallics, such as higher order cuprates, is only regiospecific for C-3 unsubstituted aziridine esters [33].

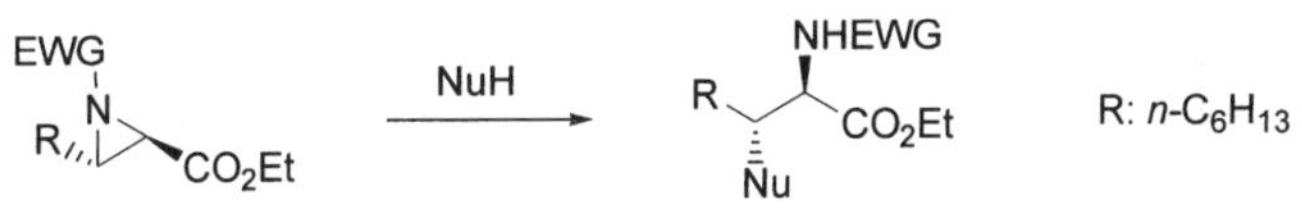

36

EWG: Ac, BnOC(O), Ts, Ms, $Me_3SiCH_2CH_2CO_2$
NuH: $PhSH/BF_3.Et_2O$; indole/$BF_3.Et_2O$; HCO_2H (neat); DMF/$BF_3.Et_2O$; AcOH

Scheme 22

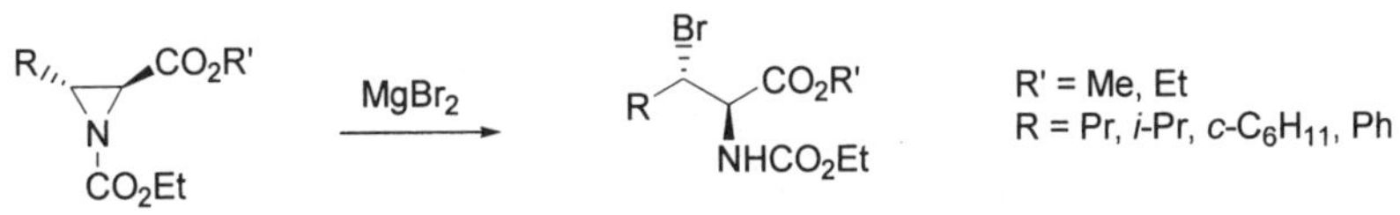

Scheme 23

quant.

An indirect nucleophilic opening is depicted in Scheme 24. The functionalized vinyl aziridine **37** undergoes a *M*ichael-*i*nitiated *r*ing *c*losure (MIRC) reaction upon treatment with suitable nucleophiles to give cyclopropanes with concomitant opening of the aziridine ring [34].

The optimum results were obtained with Grignard reagents in the presence of 10 mol % of Cu(I)CN. The stereochemical course of this MIRC reaction can be explained by adopting Yamamoto's model for conjugate addition of cyanocuprates to γ-alkoxy-α,β-unsaturated esters (Fig. 2) [35]. In this model, it is proposed that the larger substituent (L), in our case the tosyl group, will adopt the

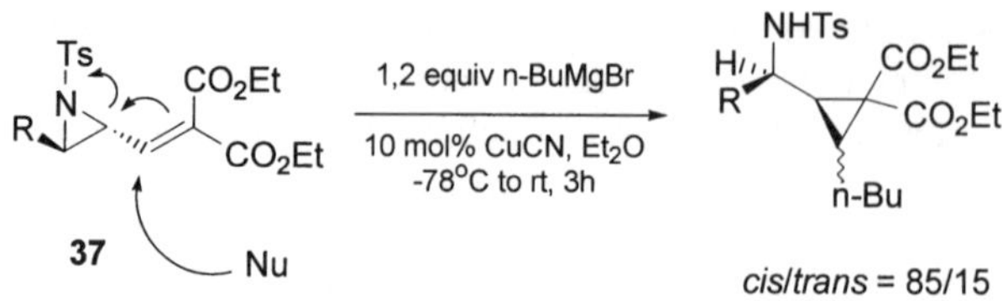

Scheme 24

R = *n*-hexyl, Me

anti-position (see Newman projection). This orientation of the tosyl group causes shielding of the upper face of the double bond from π complexation and as a consequence also from nucleophilic attack. As a result, complexation and reaction will preferably take place from the lower side (Fig. 2), which leads to the predominant formation of the *cis*-isomer.

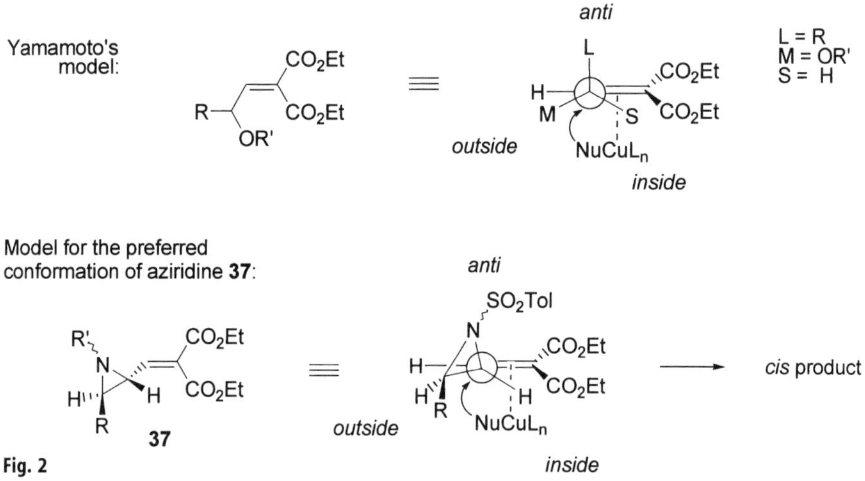

Fig. 2

The *cis*-aziridine substrate shows, as expected on the basis of this model, predominant formation of the *trans*-cyclopropane product. The starting materials for this MIRC reaction can readily be obtained from the aziridine esters by reduction to the corresponding aldehyde and a subsequent Knoevenagel reaction with malonate ester (Scheme 25) [34].

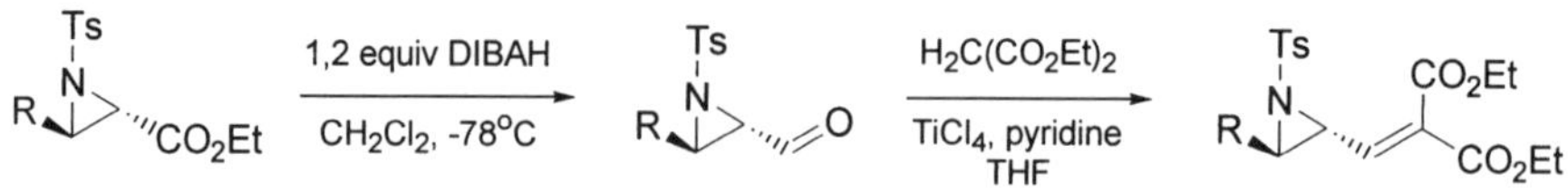

Scheme 25

37

Special mention must be made of the control of the regioselectivity of the ring opening of *N*-acylaziridines **38** at an organic-aqueous interface (Scheme 26) [36]. The fatty acid chains and the phenoxy substituent will orient the substrate such that the unsubstituted aziridine carbon atom points to the aqueous layer

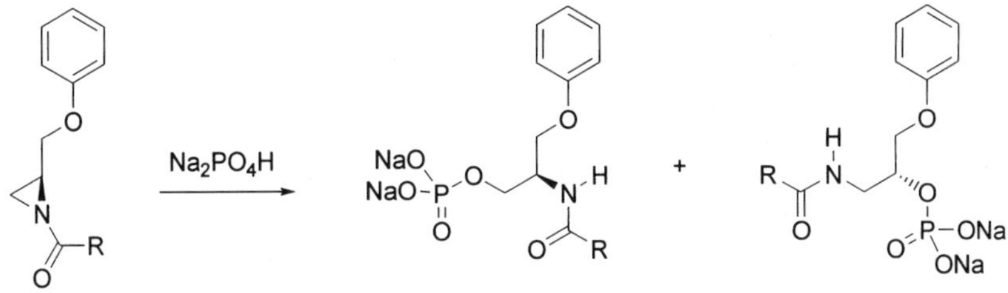

38

39a

39b

R = $C_{12}H_{23}$ or $C_{17}H_{35}$

Scheme 26

making this carbon accessible for nucleophiles present in the aqueous phase. This interface orientation leads to an exclusive nucleophilic reaction of phosphate ions at this unsubstituted carbon atom. In contrast, in organic medium both regioisomers **39a** and **39b** are obtained in equal amounts (see also Scheme 47).

Reductive ring-opening reaction of aziridine ester constitutes a convenient access to amino acids; a typical example is given in Scheme 27 [37].

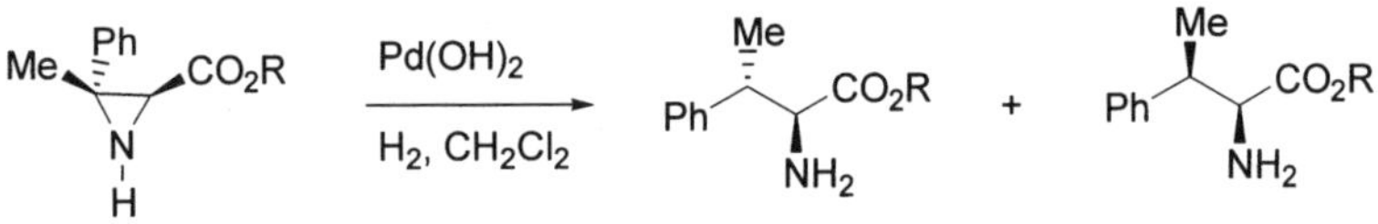

R = Me, 88% (86:14)
R = *t*-Bu, 84% (96:4)

Scheme 27

3.2 Aziridine Ring Expansions

Treatment of aziridinecarboxylic esters having an electron-withdrawing substituent at nitrogen with acetonitrile under BF_3 catalysis leads to a smooth ring expansion reaction as depicted in Scheme 28 [31].

In this modified Ritter reaction inversion at C-3 takes place, implying that the five-membered rings **40** have a *cis*-relationship between the alkyl substituent

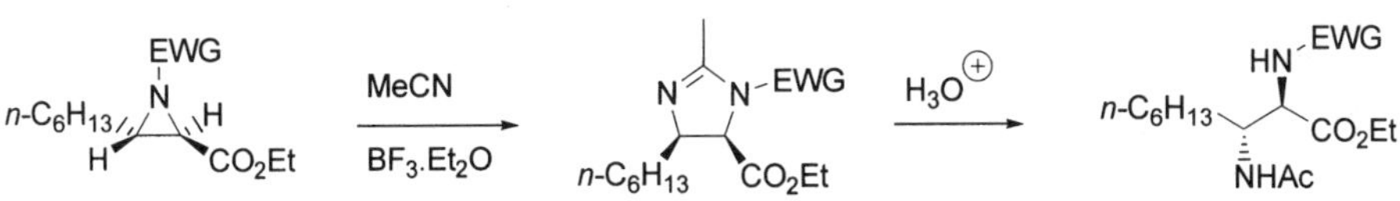

EWG = Z, Ms, Ts

40

65-92%

Scheme 28

and the ester function. The thus obtained imidazolines **40** can readily be hydrolyzed to α,β-diaminocarboxylic acid derivatives. It should be noted that both nitrogen atoms carry a different protecting function, which can be of value during further synthetic elaboration of these products. A deviant behavior was observed for *N*-acetylaziridine esters as ring expansion afforded oxazolines **41** (Scheme 29) instead of imidazolines [30]. This result can be rationalized by assuming an initial ring opening of the three-membered ring by the nitrile, followed by a ring closure via an S_N2 displacement of the nitrile unit as shown in Scheme 29.

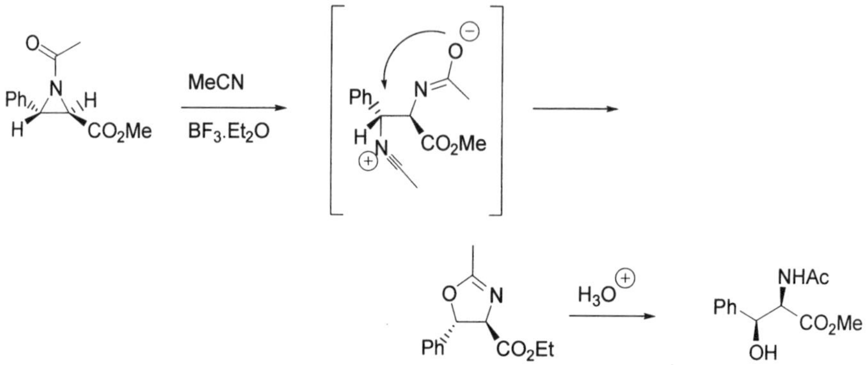

Scheme 29 **41**

This explanation resembles the intramolecular ring expansion of *N*-acylaziridines catalyzed by iodide, the so-called Heine reaction (see also Scheme 31) [1, 38]. The stereochemistry of the obtained five-membered ring is in full accordance with this explanation (double inversion which equals net retention). In this context it is relevant to mention a similar ring expansion of oxiranecarboxylic esters **42** with nitrile leading to oxazolines (Scheme 30) [39]. Note that in this case the oxazolines have a *cis*-relationship between the R substituents and the ester group, when the starting ester has the *trans*-configuration. This is due

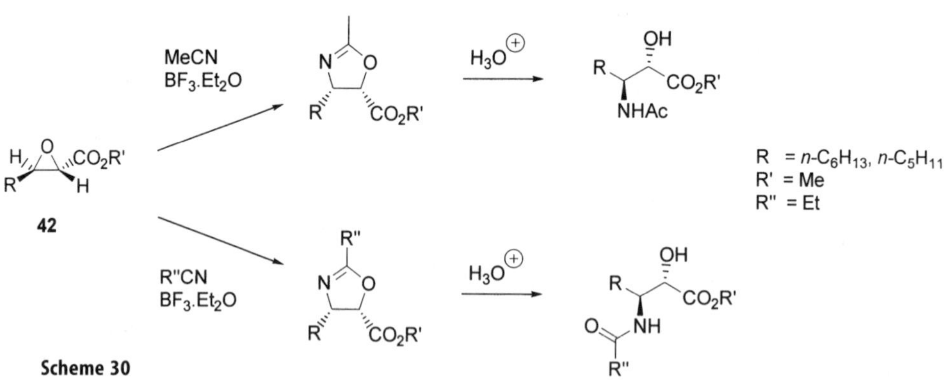

Scheme 30

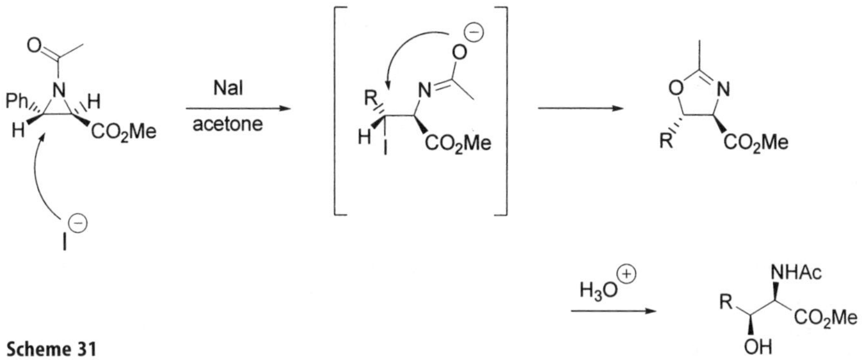

Scheme 31

to the fact that the reaction with acetonitrile at C-3 takes place with inversion. Hydrolysis of these five-membered ring products constitutes a convenient synthesis of α-hydroxy-β-amino acid derivatives. This sequence was recently re-discovered by Garcia Ruano et al. [40].

An elegant example of an intramolecular ring expansion is the formation of oxazolines from *N*-acylaziridinecarboxylic esters upon treatment with sodium iodide (Scheme 31) [1, 38].

In this aforementioned Heine reaction the initial ring opening takes place by iodide ions. Subsequent ring closure by S_N2 displacement of iodide by reaction with the negative oxygen center then leads to the products. This process proceeds with double inversion at the same carbon atom, thus with net retention. Hydrolysis of these oxazolines gives β-hydroxy-α-amino acids (Scheme 31) [1, 38]. The stereochemical course of ring expansion is the same as that observed in Scheme 29.

It should be noted that this sequence shown in Scheme 31 is complementary to that depicted in Scheme 30 in which α-hydroxy-β-amino acids are prepared. This chemistry of aziridine- and oxiranecarboxylic esters and the corresponding ring expansion reactions has been elaborated to a general protocol for the synthesis of β-hydroxy-α-amino and α-hydroxy-β-amino acids, respectively [1, 41].

Thermal ring expansion of *N*-acylaziridines by simple heating in chloroform also gives oxazolines in excellent yields, despite the fact that the aziridine ring has a rather sophisticated substituent (Scheme 32) [42].

An *N*-Boc-protected aziridine ester ring-expands upon treatment with $BF_3 \cdot Et_2O$ as shown in Scheme 33 [43]. Thus, care should be taken by choosing

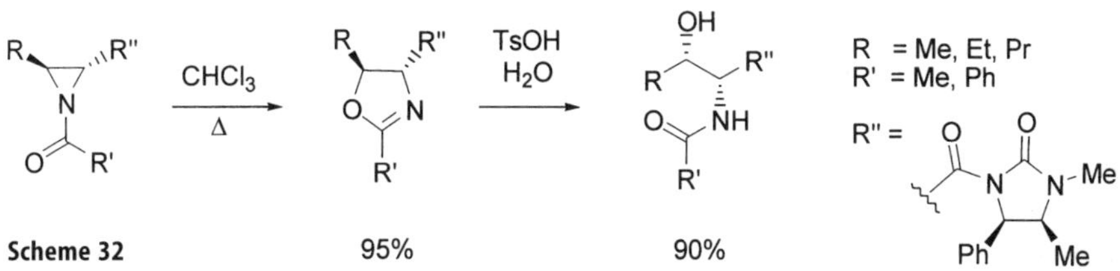

Scheme 32

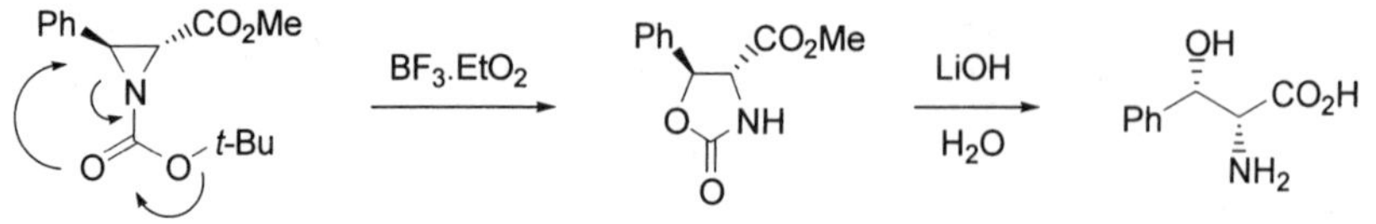

Scheme 33

an *N*-protecting function as they may become involved in an undesired reaction later on in a synthetic sequence (see also Scheme 36).

Weinreb amides of aziridinecarboxylic acids readily react with *ortho*-lithiated *O*-methoxymethyl phenols. The thus produced benzoylaziridine **43** undergoes an intramolecular ring expansion upon treatment with acid in ethanol. Base treatment leads to benzofuranones as shown in Scheme 34 [44].

An intramolecular ring expansion of aziridine esters can be accomplished by installing an appropriate nucleophilic entity in these substrates. Conversion of the ester moiety into carboxamides derived from aminomalonate ester gives compounds **44** containing the requisite nucleophilic site in the malonate moiety (Scheme 35).

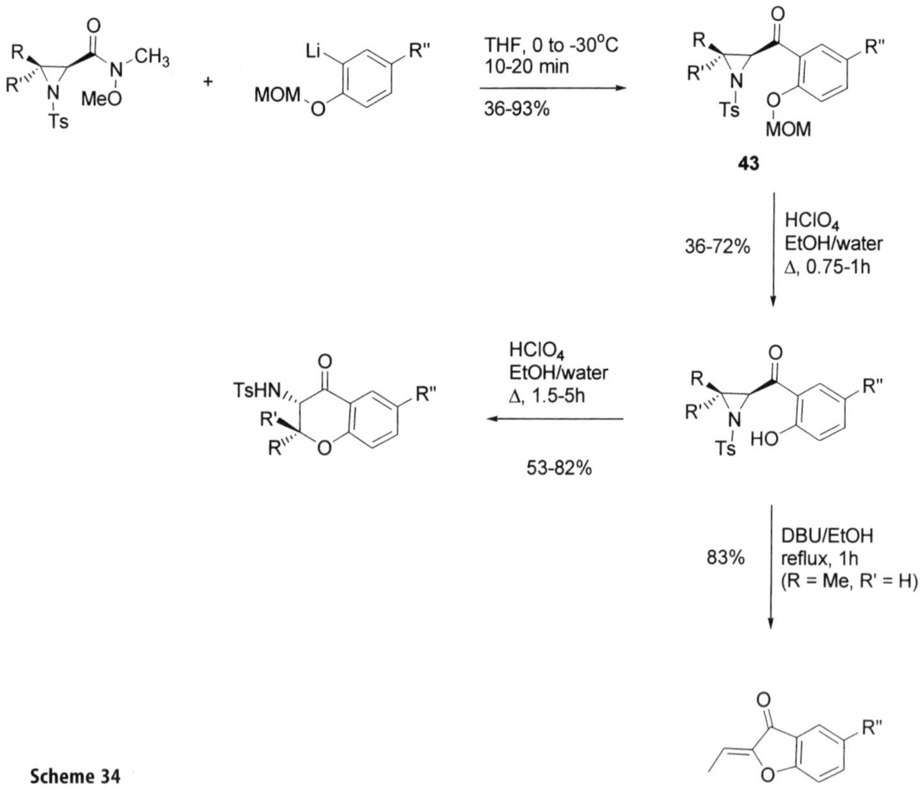

Scheme 34

Treatment of compounds **44** with a catalytic amount of sodium ethoxide in ethanol led to a smooth intramolecular reaction, which resulted in the formation of γ-lactams **45** in good yields [45] (Scheme 35). The stereochemical course of this ring expansion was unambiguously established by an X-ray analysis of product **45** [45]. Clearly an intramolecular S_N2 reaction has taken place.

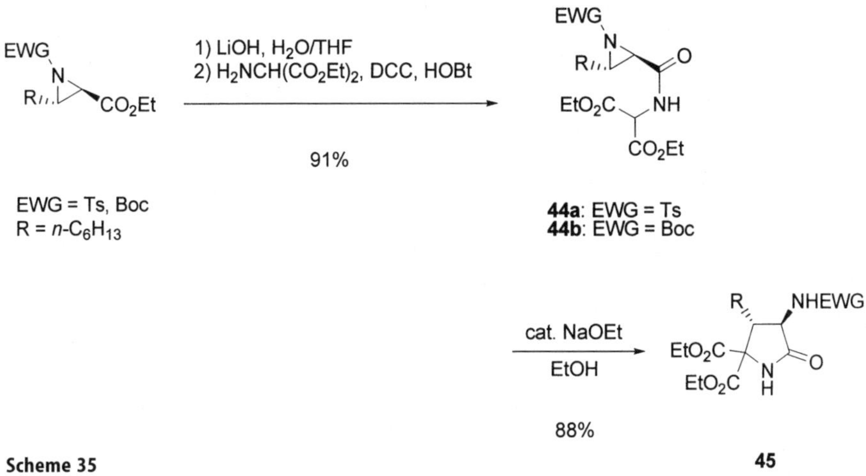

Scheme 35

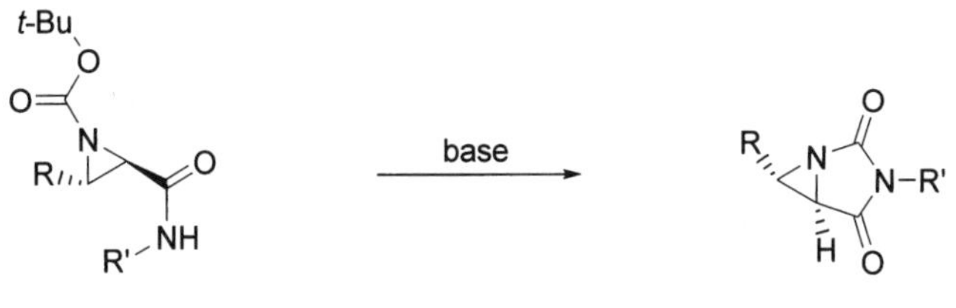

44b-f **46**

substrate	substituents		reaction conditions	cy
44b	R = C_6H_{13}	R' = $CH(CO_2Et)_2$	2.2 equiv LDA/THF -78°C, 0.5h, 0°C, 2.5h	73%
44c	C_4H_9	$CH_2C(O)Ph$	cat. *t*-BuOK/THF rt, 6h	99%
44d	C_6H_{13}	CH_2CO_2Me	cat. NaOMe/MeOH rt, overnight	52%
44e	C_6H_{13}	CH_2Ph	cat. *t*-BuOK/THF rt, overnight	65%
44f	CH_2Ph	$CH_2C(CO)Ph$	cat. *t*-BuOK/THF 0°C, 3.5h, rt, 2.5h	87%

Scheme 36

A deviant reaction was observed when the *N*-Boc-aziridinecarboxamide **44b** was treated with LDA in THF as the base. Under these kinetically controlled conditions an intramolecular reaction of the amide nitrogen with the Boc group takes place leading to the bicyclic product **46** in which the aziridine ring is retained (Scheme 36) [45].

This reaction to bicyclic compounds containing the aziridine group was also observed for other amides, viz., **44c–f**, when treated with a catalytic amount of *t*-BuOK in THF or MeONa in methanol. LDA treatment of the tosyl-activated substrate **44a** gave the five-membered ring product albeit in a low yield (31%). Remarkably, the carboxamide derived from the *cis*-aziridine failed to react with base, probably due to steric hindrance.

When the aziridinecarboxamide **44a** contains an anisyl or better a benzyl group at nitrogen instead of hydrogen, base treatment leads also to the formation of *β*-lactams **47**, in the latter case even exclusively (Scheme 37) [45, 46].

The aza-[2,3]-Wittig rearrangement [47] and the related aza-[3,3]-Claisen rearrangement [48] of vinylaziridines are elegant examples of expansion of the aziridine ring in a stereocontrolled fashion (Scheme 38).

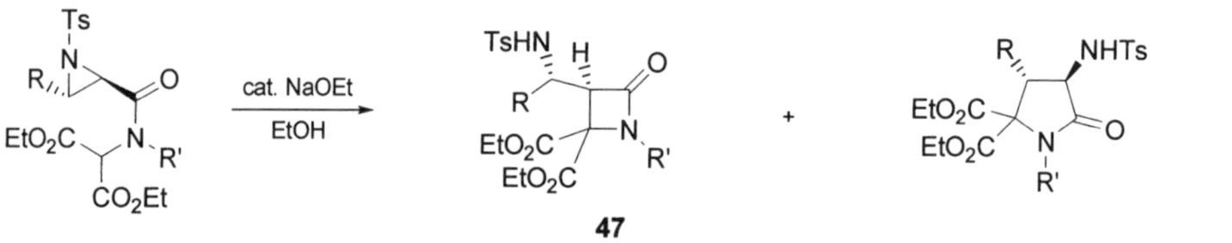

47

R = n-C_6H_{13}, R' = p-$MeOC_6H_4$, 26%
R = CH_3, R' = $C_6H_5CH_2$, 86%

R = n-C_6H_{13}, R' = p-$MeOC_6H_4$, 69%

Scheme 37

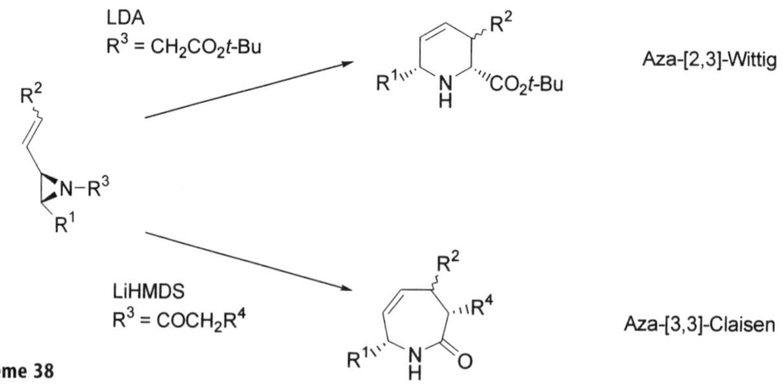

Scheme 38

3.3 Aziridine Carbinols as Chiral Ligands

Essentially, aziridine carbinols are β-amino alcohols and therefore they are potential ligands for the enantioselective addition of organometallics to the carbonyl group. These aziridine carbinol ligands are readily accessible from aziridine-2-carboxylic esters by either reduction or by reaction with Grignard reagents. An additional structural variable is the substituent at nitrogen. Tanner et al. [49] prepared a large series of aziridine carbinols, which all were tested as chiral catalysts in the diethylzinc addition to benzaldehyde. This diethylzinc reaction has been extensively studied with a large variety of β-amino alcohols as chiral ligands [50]. The best results are always obtained with aromatic aldehydes, and therefore the real challenge is to achieve a good enantioselectivity for the diethylzinc addition to aliphatic aldehydes. The best ligand from the Tanner series with aromatic aldehydes is **48** with 97% ee at best. Remarkably, the ligands having only one carbinol unit performed less well than the C_2-symmetric ligands, the best ones being **49** and **50** (90% ee in the reaction with benzaldehyde, Fig. 3).

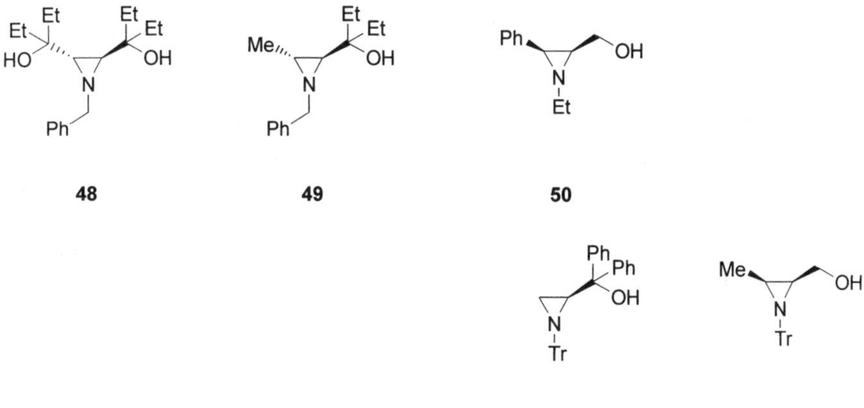

51 52

Fig. 3

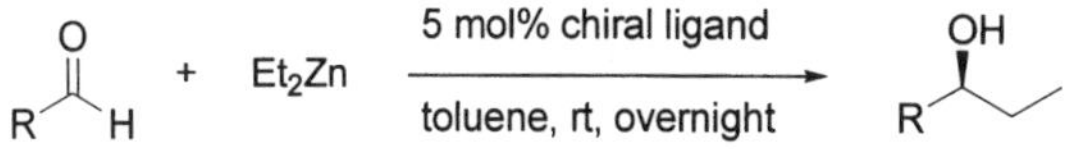

For ligand **51**: R = aryl, ee 99%
R = alkyl, ee 80-99%

Scheme 39

In the author's laboratory, ligand **51** (Fig. 3) was investigated and gratifyingly very high ee's were obtained in the diethylzinc addition to both aromatic and aliphatic aldehydes (Scheme 39) [51].

This result is in strong contrast with that obtained by Tanner et al. [49] for ligand **52** (ee of 2% for the addition of diethylzinc to benzaldehyde, Fig. 3). It was argued by Tanner that the *N*-trityl group would be too large to allow efficient chelate formation with diethylzinc [49], which is however clearly not the case

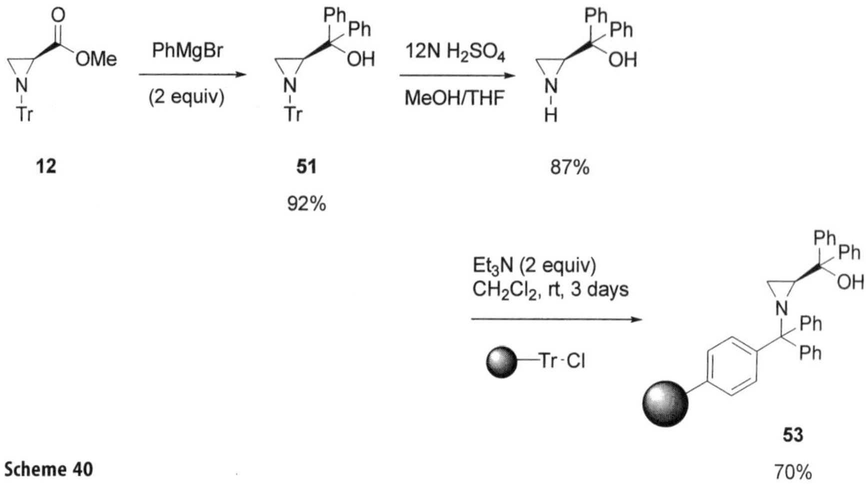

Scheme 40

[51]. Comparison of our results with those of Tanner et al. demonstrate that fine-tuning of the chiral ligands leads to remarkable improvements in the enantioselectivity.

A polymer-supported version of our optimal ligand was also developed [52]. Its preparation involves attachment of aziridine carbinols to polymer-bound triphenylchloromethane (Scheme 40). This polymer-bound ligand **53** was almost equally effective in the enantioselective addition of diethylzinc to aromatic and aliphatic aldehydes with ee's ranging from 77–97% for the latter type of substrate [52]. It is of practical interest that this polymer-supported ligand could be reused without losing much of its efficiency.

It is of interest to note that our chiral ligands contain the so-called "magic" diarylhydroxymethyl group, which also in other types of ligand showed a remarkable efficiency [53].

Tanner et al. also used an aziridine carbinol (viz. **54**) as chiral ligand in asymmetric addition of diethylzinc to *N*-(diphenylphosphinoyl)imines (Scheme 41) [54].

The aziridine carbinols are also effective ligands in the preparation of oxazaborolidine catalysts for the asymmetric ketone reduction with borane (Fig. 4) [55].

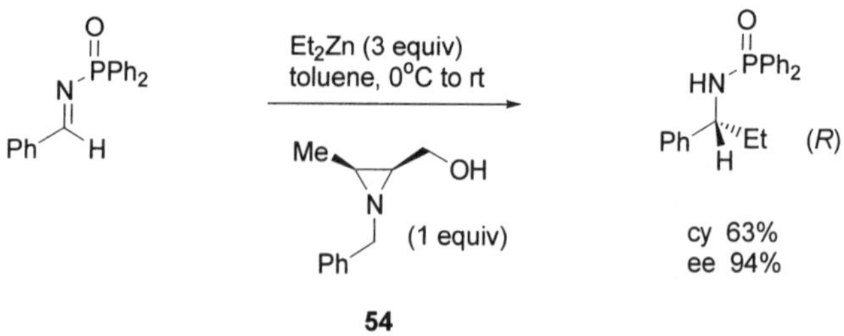

Scheme 41

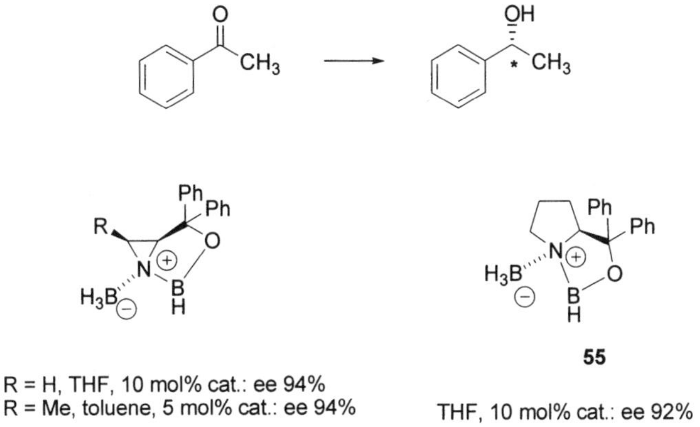

Fig. 4

These ligands can readily be obtained by a Grignard reaction of aziridine esters, followed by an acidic detritylation (see Scheme 40) [19, 55]. These aziridine carbinol-derived catalysts are equally efficient as the Corey ligand **55** derived from proline carbinols (Fig. 4) [55, 56].

3.4 *N*-Tritylaziridine-2-Carboxaldehyde as Synthon

Enantiopure α-amino aldehydes are valuable synthons in natural product synthesis [57]. However, problems are often encountered with their configurational instability [58]. Aziridine-2-carboxaldehydes are also α-amino aldehydes and accordingly have a potential synthetic value. We found that *N*-tritylaziridine-2-carboxaldehyde **56** is a perfectly stable compound and therefore comparable to Garner's aldehyde (*tert*-butyl 2,2-dimethyl-4-(*S*)-formyl-oxazolidine-3-carboxylate). Aldehyde **56** can readily be prepared from aziridine-2-carboxylic ester **12** by the sequence shown in Scheme 42 [59].

The aziridine aldehyde **56** undergoes a facile Baylis-Hillman reaction with methyl or ethyl acrylate, acrylonitrile, methyl vinyl ketone, and vinyl sulfone [60]. The adducts **57** were obtained as mixtures of *syn*- and *anti*-diastereomers. The synthetic utility of the Baylis-Hillman adducts was also investigated. With acetic anhydride in pyridine an S_N2'-type substitution of the initially formed allylic acetate by an acetoxy group takes place to give product **58**. Nucleophilic reactions of this product with, e.g., morpholine, thiol/Et_3N, or sodium azide in DMSO resulted in an apparent displacement of the acetoxy group. Tentatively, this result may be explained by invoking the initial formation of an ionic intermediate **59**, which is then followed by the reaction with the nucleophile as shown in Scheme 43.

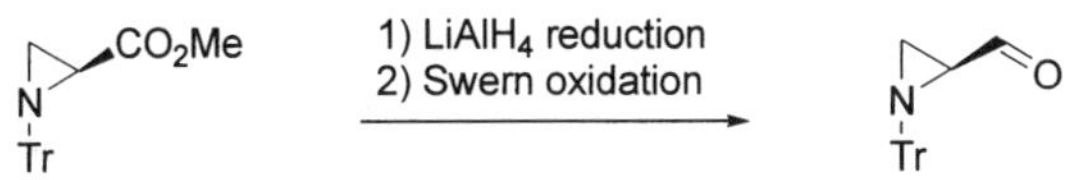

Scheme 42 **12** **56**

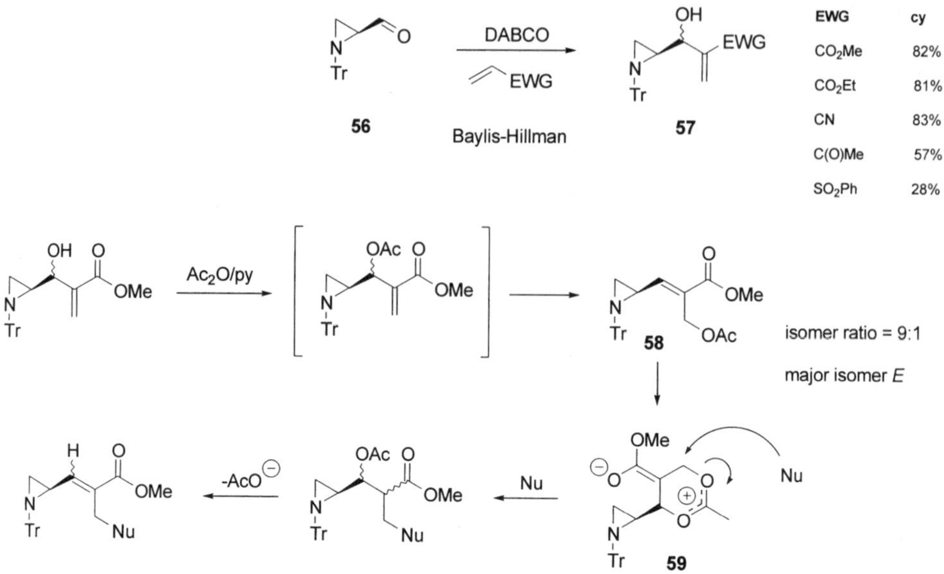

Nucleophiles, conditions and results: Morpholine, $E/Z = 4{:}1$, 92%; RSH/Et_3N/EtOH, R = Ph, only *Z*, 91%, R = Bn, $E/Z = 9{:}5$, 89%; NaN_3/DMSO, $E/Z = 7{:}3$, 92%.

Scheme 43

It should be noted that Baylis-Hillman reaction of Garner's aldehyde with methyl acrylate and DABCO results in racemization of the stereocenter of the amino aldehyde [61]. In the case of substrate **56** such racemization is seriously hampered due to the large inversion barrier in three-membered ring compounds [62].

Another successful reaction of aldehyde **56** is the diastereoselective carbonyl-ene reaction, pictured in Scheme 44, leading to products **60** [63].

Of the various Lewis acid catalysts tested, $SnCl_4$ gave the highest diastereoselective product formation with predominance for the *anti*-diastereoisomer. This *anti*-selectivity can be rationalized by invoking the Cram chelation model.

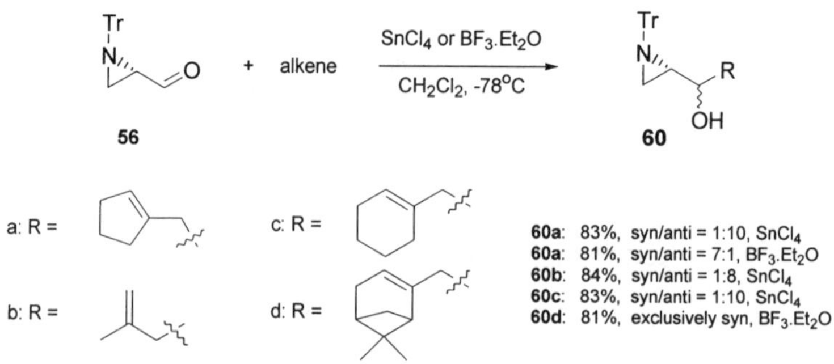

Scheme 44

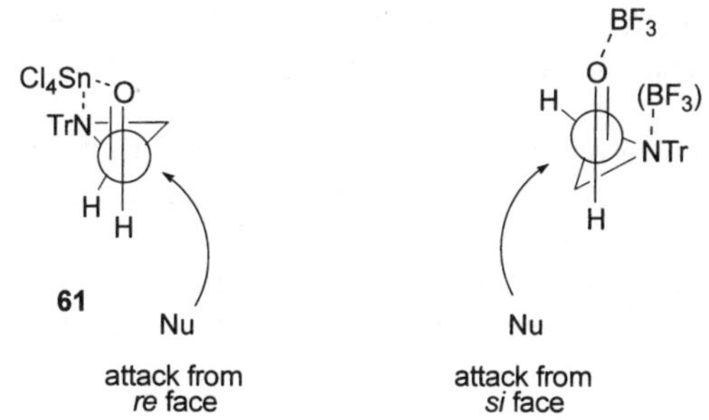

Fig. 5

In this chelate complex **61**, the bulky trityl group on nitrogen is sterically encumbering the *si* face of the carbonyl function and thus attack occurs preferentially from the unhindered *re* face leading to the observed *anti*-diastereoisomer (Fig. 5). Catalysis with $BF_3 \cdot OEt_2$, which is incapable of chelation, leads to attack from the *si* face to give the *syn*-adduct according to the Felkin-Anh model (Fig. 5). It should however be noted that only a few enes could be brought into reacton with $BF_3 \cdot OEt_2$ as the catalyst.

The aziridine-2-carboxaldehyde **56** can also serve as synthon for the synthesis of sphingosines, which are important biomembrane constituents [64]. One possible route involves the addition of an alanate to the aldehyde. In a later stage of this synthetic plan the aziridine can be opened, either via the intermediacy of an oxazoline or directly with dilute acid. Unfortunately, the reaction of aldehyde **56** with a vinylalanate has a poor diastereoselectivity of 3:2. Therefore, an alternative approach was considered, namely one involving the addition of a vinylzinc reagent to the aldehyde thereby employing our *N*-tritylaziridinediphenylmethanol **51** as the chiral catalyst. Gratifyingly, only one diastereomer was obtained. Reductive removal of the trityl function, acetylation of the hydroxy

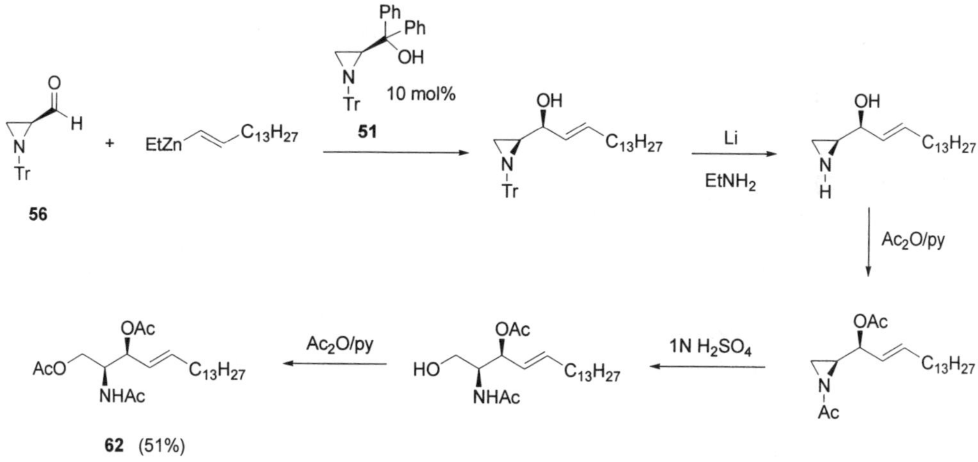

Scheme 45

group, and subsequent hydrolytic opening of the three-membered ring afforded sphingosine triacetate **62** after an extra acetylation step with acetic anhydride (Scheme 45) [65].

The elegant feature of this synthetic sequence is that an aziridine-derived chiral catalyst is used to achieve the desired diastereoselectivity.

Sphingosine precursors were prepared by Davis et al. [66]. The vinylaziridine **20**, which is a precursor for the azirine-containing natural product dysidazirine (see Sect. 2.5, Scheme 10), also can serve as a precursor for L-*threo*-sphingosines (Scheme 46).

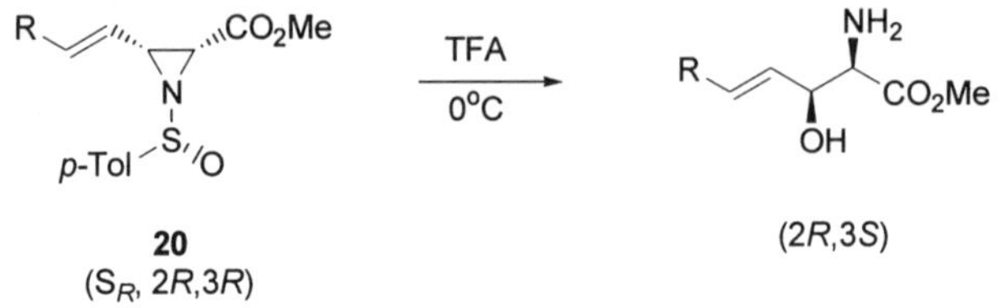

Scheme 46

3.5 Supramolecular Chemistry

Chiral amide-containing surfactants can readily be obtained by ring opening of an aziridine carrying a fatty acid-derived acyl group at nitrogen. Treatment of *N*-acylaziridine **63** with dibenzyl phosphate gave in contrast to the expectation a 1:1 mixture of regioisomeric products **64a** and **64b**, which after separation were hydrogenolytically debenzylated and converted into the corresponding disodium salts (Scheme 47) [11].

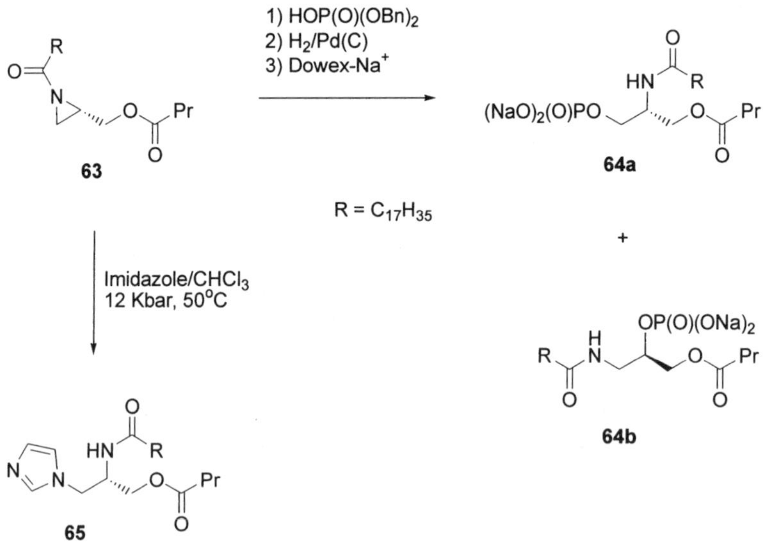

Scheme 47

The thus obtained amphiphilic species showed an entirely different aggregation behavior in water [67]. The expected regioisomer **64a** arising from aziridine ring opening at the unsubstituted carbon atom only gave plate-like structures, whereas the other regioisomer **64b** gave left-handed helical strands, which coagulated to rope-like structures [67]. This different behavior of **64a** and **64b** can be attributed to intermolecular hydrogen bonding for compound **64b** through which alignment of the molecules in an organized aggregate is made possible, whilst in **64a** intramolecular hydrogen bonding precludes such an organization. This expression of molecular chirality on a supramolecular level opens interesting possibilities for chiral catalysis by embedding catalytic species in such chiral supramolecular structures. Opening of *N*-acylaziridine **63** by reaction with imidazole under high pressure (12 Kbar) leads to compound **65** which on complexation with Cu(II) ions produces nice helical aggregates clearly through the coordination of imidazolyl groups by Cu(II) ions [68].

3.6 Reactions of 2*H*-Azirines

The highly strained and reactive 2*H*-azirines have been extensively studied for various synthetic purposes, such as ring expansion reactions, cycloaddition reactions, preparation of functionalized amines and substituted aziridines. The older literature on azirines in synthesis has extensively been reviewed [69]. Concerning azirines with defined chirality only scarce information is available. Practically all reactions of azirines take place at the activated imine bond. Reduction with sodium borohydride leads to *cis*-substituted aziridines as is shown in Scheme 48 [26, 28].

Grignard reagents react with the imine bond as expected to give substituted aziridines in 60–90% yield (Scheme 49) [23b].

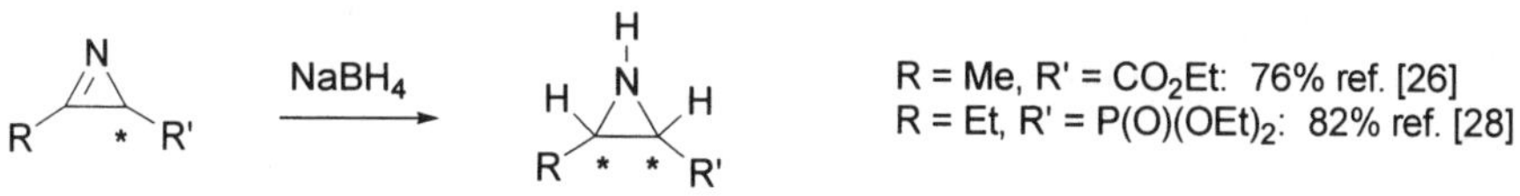

Scheme 48 *cis*

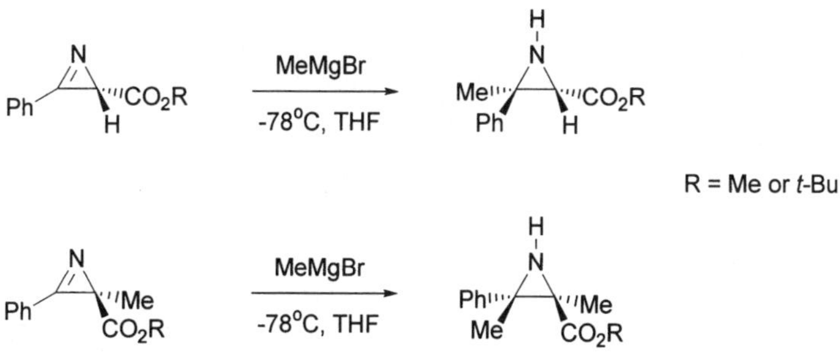

Scheme 49

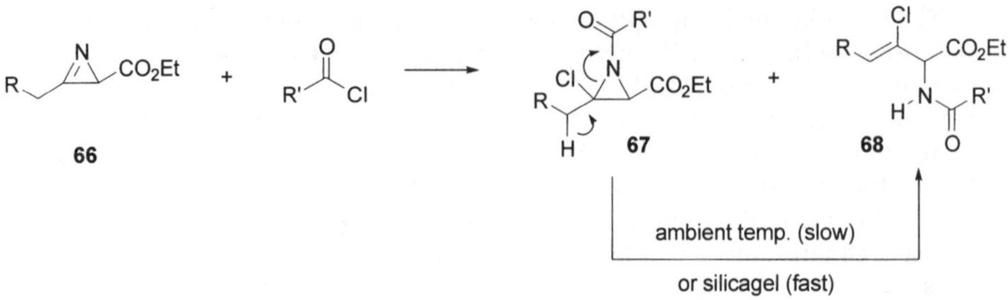

R	R'	67	68
H	Me	77%	-
H	Ph	-	63%
Ph	Ph	58%	31%

Scheme 50

Reaction of azirine ester **66** with acyl halides led essentially to two types of products, viz., the addition products **67** and halovinylamino esters **68** as depicted in Scheme 50.

The formation of product **68** from **67** can readily be explained as depicted in Scheme 50. For R = H, R′ = Ph, this reaction was also performed with enantiomerically enriched azirine ester; in product **68** this chirality was retained [27b, 70].

4 Concluding Remarks

The chemistry of functionalized aziridines is still in full swing. There is no doubt that many new applications will be found, especially when optimal use is made of the stereogenic centers in these small-ring heterocycles. In addition, the chemistry of aziridinium ions [71, 72] will add much to the flavor of this area of chemistry.

5 References

1. (a) Zwanenburg B (1999) Pure Appl Chem 71:423; (b) Zwanenburg B, Thijs L (1996) Pure Appl Chem 68:735
2. (a) Dermer OC, Ham GE (1969) Ethylenimine and other aziridines: Chemistry and applications. Academic Press, New York; (b) Deyrup JA (1983) Aziridines. In: Hassner A (ed), Small-ring heterocycles part 1. Wiley, New York, Vol 42, chap 1
3. Tanner D (1994) Angew Chem Int Ed Engl 33:599
4. Osborn HMI, Sweeney J (1997) Tetrahedron Asymmetry 8:1693
5. McCoull W, Davis FA (2000) Synthesis 1347
6. (a) Legters J, Thijs L, Zwanenburg B (1989) Tetrahedron Lett 30:4881, (b) Legters J, Thijs L, Zwanenburg B (1992) Recl Trav Chim Pays-Bas 111:1
7. Gao Y, Hanson RM, Klunder JM, Ko SY, Masamune H, Sharpless KB (1987) J Am Chem Soc 109:5765

8. Smits JMM, Beurskens PT, Thijs L, van Loon AAWM, Zwanenburg B (1988) J Cryst Spectr Res 18:625
9. Legters J, Thijs L, Zwanenburg B (1991) Tetrahedron 47:5287
10. Tanner D, Somfai P (1987) Tetrahedron Lett 28:1211
11. Sommerdijk NAJM, Buynsters PJJA, Akdemir H, Geurts DG, Nolte RJM, Zwanenburg B (1997) J Org Chem 62:4955, 9388
12. (a) Lohray BB, Ahuja JR (1991) J Chem Soc Chem Commun 95; (b) for a review on cyclic sulfites and cyclic sulfates, see: Lohray BB (1992) Synthesis 1035
13. Lohray BB, Gao Y, Sharpless KB (1989) Tetrahedron Lett 30:2623
14. Kolb HC, VanNieuwenhze MS, Sharpless KB (1994) Chem Rev 94:2483
15. (a) Urabe H, Aoyama Y, Sato F (1992) Tetrahedron 48:5639; (b) Najime R, Pilard S, Vaultier M (1992) Tetrahedron Lett 33:5351; (c) Moulines J, Charpentier P, Bats JP, Nuhrich A, Lamidey AM (1992) Tetrahedron Lett 33:487
16. (a) Nakagawa Y, Tsuno T, Nakajima K, Iwai M, Kawai H (1972) Bull Chem Soc Jpn 45:1162; (b) Nakajima K, Takai F, Tanaka T, Okawa K (1978) Bull Chem Soc Jpn 51:1577; (c) Korn A, Rudolph-Böhner S, Moroder L (1994) Tetrahedron 50:1717
17. Wakamiya T, Shimbo K, Shiba T, Nakajima K, Neya M, Okawa K (1982) Bull Chem Soc Jpn 55:3878
18. Kuyl-Yeheskiely E, Lodder M, van der Marel GA, van Boom JH (1992) Tetrahedron Lett 33:3013
19. Willems JGH, Hersmis MC, de Gelder R, Smits JMM, Hammink JB, Dommerholt FJ, Thijs L, Zwanenburg B (1997) J Chem Soc Perkin Trans 1963
20. (a) Stapley EO, Hendlin D, Jackson M, Miller AK (1971) J Antibiot 24:42; (b) Miller TW, Tristam EW, Wolf FJ (1971) J Antibiot 24:48
21. Molinski TF, Ireland CM (1988) J Org Chem 53:2103
22. Legters J, Thijs L, Zwanenburg B (1992) Recl Trav Chim Pays-Bas 111:75
23. (a) Davis FA, Reddy GV, Liu H (1995) J Am Chem Soc 117:3651; (b) Davis FA, Liang CH, Liu H (1997) J Org Chem 62:3796
24. Gentilucci L, Grijzen Y, Thijs L, Zwanenburg B (1995) Tetrahedron Lett 36:4665
25. Davis FA, McCoull W (1999) Tetrahedron Lett 40:249
26. Verstappen MMH, Ariaans GJA, Zwanenburg B (1996) J Am Chem Soc 118:8491
27. (a) Zwanenburg B, Philipse HJF, Verstappen MMH, Gieling RG (1997) Phosphorus Sulfur Silicon 120–121:453; (b) Verstappen MMH (1998) PhD thesis, University of Nijmegen
28. (a) Palacios F, Ochoa de Retana AM, Gil JI (2000) Tetrahedron Lett 41:5363; (b) Palacios F, Ochoa de Retana AM, Gil JI, Ezpeleta JM (2000) J Org Chem 65:3213
29. Nefkens GHL, Zwanenburg B (1983) Tetrahedron 39:2995
30. Legters J, Thijs L, Zwanenburg B (1992) Recl Trav Chim Pays-Bas 111:16
31. Legters J, Willems JGH, Thijs L, Zwanenburg B (1992) Recl Trav Chim Pays-Bas 111:59
32. Righi G, D'Achille R, Bonini C (1996) Tetrahedron Lett 37:6893
33. Church NJ, Young DW (1995) Tetrahedron Lett 36:151
34. Funaki I, Bell RPL, Thijs L, Zwanenburg B (1996) Tetrahedron 52:12253
35. Yamamoto Y, Chounan Y, Nishii S, Ibuka T, Kitahara H (1992) J Am Chem Soc 114:7652
36. Buijnsters PJJA, Feiters MC, Nolte RJM, Sommerdijk NAJM, Zwanenburg B (2001) J Chem Soc Chem Commun 269
37. Davis FA, Liang CH, Liu H (1997) J Org Chem 62:3796
38. Thijs L, Zwanenburg B, to be published
39. Legters J, van Dienst E, Thijs L, Zwanenburg B (1992) Recl Trav Chim Pays-Bas 111:69
40. García Ruano JL, Paredes CG (2000) Tetrahedron Lett 41:5357
41. Thijs L, Zwanenburg B, unpublished results
42. Cardillo G, Gentilucci L, Tolomelli A (1999) Tetrahedron 55:15151
43. Tomasini C, Vecchione A (1999) Org Lett 1:2153
44. Woydowski K, Liebscher J (2000) Synthesis 1444
45. Funaki I, Thijs L, Zwanenburg B (1996) Tetrahedron 52:9909
46. Funaki I, Dommerholt FJ, Zwanenburg B, unpublished results
47. Åhman J, Somfai P (1994) J Am Chem Soc 116:9781

48. Lindström UM, Somfai P (1997) J Am Chem Soc 119:8385
49. Tanner D, Kornø T, Guijarro D, Andersson PG (1998) Tetrahedron 54:14213
50. For a review, see Soai K, Niwa S (1992) Chem Rev 92:833
51. Lawrence CF, Nayak SK, Thijs L, Zwanenburg B (1999) Synlett 1571
52. ten Holte P, Wijgergangs JP, Thijs L, Zwanenburg B (1999) Org Lett 1:1095
53. Braun M (1996) Angew Chem Int Ed Engl 35:519
54. Andersson PG, Guijarro D, Tanner D (1997) J Org Chem 62:7364
55. Willems JG, Dommerholt FJ, Hammink JB, Vaarhorst AM, Thijs L, Zwanenburg B (1995) Tetrahedron Lett 36:603
56. Corey EJ, Bakshi RK, Shibata S (1987) J Am Chem Soc 109:5551
57. For some leading examples, see (a) Myers AG, Kung DW (2000) Org Lett 2:3019; (b) Myers AG, Kung DW (1999) J Am Chem Soc 121:10828; (c) Nicolaou KC, Koide K, Bunnage ME (1995) Chem Eur J 1:454
58. (a) Jurczak J, Golebiowski A (1989) Chem Rev 89:149; (b) Reetz MT (1991) Angew Chem Int Ed Engl 30:1531
59. Utsunomiya I, Fuji M, Sato T, Natsume M (1993) Chem Pharm Bull 41:854
60. Nayak SK, Thijs L, Zwanenburg B (1999) Tetrahedron Lett 40:981
61. Drewes SE, Khan AA, Rowland K (1993) Synth Commun 23:183
62. Haner R, Olano B, Seebach D (1987) Helv Chim Acta 70:1676
63. Nayak SK, Thijs L, Zwanenburg B (1998) Synlett 1187
64. Nugent TC, Hudlicky T (1998) J Org Chem 63:510
65. Lawrence CF, Thijs L, Zwanenburg, to be published
66. Davis FA, Reddy GV (1996) Tetrahedron Lett 37:4349
67. Sommerdijk NAJM, Buynsters PJAA, Pistorius AMA, Wang M, Feiters MC, Nolte RJM, Zwanenburg B (1994) J Chem Soc Chem Commun 1941, 2739
68. Sommerdijk NAJM, Buynsters PJJA, Akdemir H, Geurts DG, Pistorius AMA, Feiters MC, Nolte RJM, Zwanenburg B (1998) Chem Eur J 4:127
69. For reviews on azirine chemistry, see (a) Fowler FW (1971) Advances Heterocycl Chem 13:45; (b) Anderson DJ, Hassner A (1975) Synthesis 483; (c) Hassner A (1980) Heterocycles 14:1517
70. Verstappen MMH, Zwanenburg B, unpublished results
71. (a) Rayner CM (1997) Synlett 11; (b) Chuang TH, Sharpless KB (2000) Org Lett 2:3555 and refs. cited therein; (c) de Sousa SE, O'Brien P (1997) Tetrahedron Lett 38:4885; (d) Liu QY, Marchington AP, Rayner CM (1997) Tetrahedron 53:15729
72. (a) Crist DR, Leonard NJ (1969) Angew Chem Int Ed Engl 8:962; (b) Leonard NJ, Zwanenburg B (1967) J Am Chem Soc 89:4456

Synthesis of Medium-Sized Ring Lactams

Udo Nubbemeyer

Institut für Chemie – Organische Chemie, Freie Universität Berlin, Takustrasse 3, 14195 Berlin, Germany
E-mail: udonubb@chemie.fu-berlin.de

Medium-sized ring lactams (7- to 15-membered) find widespread use in organic chemistry as key intermediates in the synthesis of more complex structures or as core structures in natural product or pharmaceutically important compounds. Until now, the generation of such rings is still a challenge in organic synthesis. During the past decade an increasing interest has focused on the generation of cyclic peptides as well as hairpin and β-turn mimics and the medium-sized rings were thought to represent suitable fragments. Furthermore, a range of complex natural products is characterized by such lactam structure and their total synthesis is a broad field for organic chemists to test new strategies and to develop new methods.

The aim of the present chapter is to discus some modern strategies to synthesize seven- to fifteen-membered ring lactams, the scope is restricted to the generation to ring systems bearing a single lactam function. Although the range of reactions described concern only the generation of simple structures without complicated stereogenic properties, most of them seem to bear undiscovered potential with regard to stereoselective synthesis. However, the emphasis of the methods is focused on stereoselective processes. The review is subdivided in three major chapters of ring closure reactions and ring enlargements, additional information is given on cycloadditons and fragmentations.

Keywords: Medium-sized ring lactams, Macrocyclization, Ring expansion, Rearrangement, Fragmentation

1 Introduction 126

2 Ring-Closure Reactions 128

2.1 Ring-Closure Reactions by N-CO Bond Formation (Lactamization) 128
2.2 Ring-Closure Reactions by C-C Single Bond Formation 140
2.3 Ring-Closure Reactions by C=C Double Bond Formation 144

3 Cycloadditions 150

4 Ring Expansion Reactions by N Insertion 155

4.1 Beckmann Rearrangement 155
4.2 Schmidt Rearrangement 159
4.3 Oxaziridine Rearrangement 165

5 **Ring Expansion Reactions by C Insertion** 171

5.1 C3-Insertion 171
5.2 C4-Insertion 173

6 **Fragmentations** 187

7 **References** 193

Abbreviations

c	cyclo
Cy	cyclohexyl
DHP	dihydropyran
DPPA	diphenylphosphoryl azide
EDC	*N*-[3-(dimethylamino)propyl]-*N*-ethylcarbodiimide/HCl
HOAt	1-hydroxy-7-azabenzotriazole
HOBt	1-hydroxybenzotriazole
MPCA	monoperoxycamphoric acid
Ns	4-nitrophenylsulfonyl
PG	protective group
Pht	phthaloyl
PMP, PMB	4-methoxyphenyl, 4-methoxybenzyl
PyBop	1-Benzotriazolyloxy(tripyrrolidino)phosphonium hexafluorophosphate
PyBroP	bromo(tripyrrolidino)phosphonium hexafluorophosphate
TBS	TBDMS, *tert*-butyldimethylsilyl
TPS	TBDPS, *tert*-butyldiphenylsilyl

1 Introduction

The generation of medium-sized nitrogen heterocycles possessing defined constitutions and configurations is still a challenge in organic synthesis. On one hand, such rings are found as sub-units in natural and pharmaceutically important products (target molecules), on the other hand, the medium-sized and often constrained ring systems can serve as key intermediates in the synthesis of bicyclic amino compounds by selective transformations such as transannular ring-contractions, cycloadditions, etc.

Even when excluding cyclic peptide structures bearing more than one lactam unit in a medium-sized ring framework, a range of monolactams with interesting structural, biological and pharmaceutical properties are known. Focusing on natural products, the isolation of interesting compounds and the publication of the structural data have always induced synthetic efforts to achieve the total synthesis or the generation of some important analogs. During the past decade, a substantial number of new compounds has been described. While several total

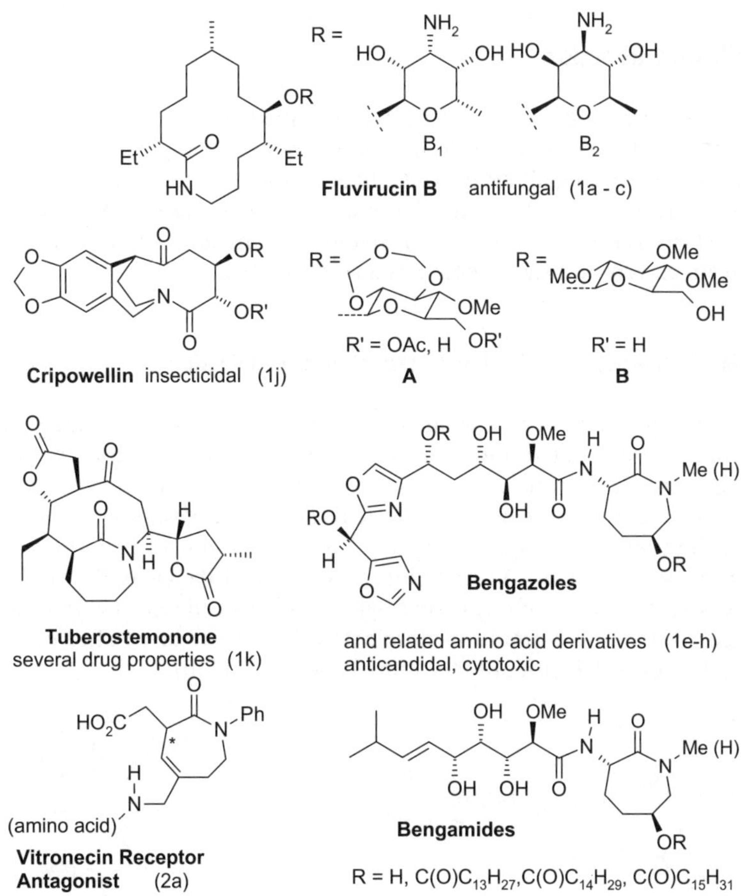

Fig. 1

syntheses have been successfully completed, a range of structures is still recognized as interesting targets. Some examples of natural products isolated in the 1990s are shown in Fig. 1 [1].

Furthermore, the de novo synthesis of peptidomimetics with potential properties concerning the construction of artificial turns and hairpins is still a tool of synthetic efforts. The generation of compounds characterized by specific receptor interactions in vitro and in vivo, respectively, is crucial to investigate biological and biochemical transformations as well as drug activities. Some representative examples are discussed in the following sections [2]. In a range of examples some general motivation on synthesizing medium-sized ring lactams is given.

2 Ring-Closure Reactions

Ring-closure reactions represent by far the most of the reactions described during the past decade to synthesize medium-sized ring lactams. Except for azepanone generation, all cyclizations needed more or less high dilution conditions to avoid any intermolecular reactions. The ease of the cyclization decreased from seven- to nine- or ten-membered ring formations and increased again from eleven- to fifteen-membered ring lactams.

2.1 Ring Closure Reactions by C-N Bond Formation (Lactamization)

The simplest way to generate lactams seemed to be the intramolecular reaction of ω-amino esters, but in most cases the direct conversion failed though the amide group is known to be more stable than an ester function. If at all, the lactam was found as a side product: the acid catalyzed cleavage of amino acetal **1** allowed the synthesis of optically active 2-amino-4,6-dihydroxypimelic acid **2**, the corresponding lactam **3** was found in less than 15% yield, Eq. (1) [3a]. In contrast, the cyclization of L-lysin methylester **4** has been reported to yield 71% of the corresponding 2-amino-ε-caprolactam **6** over two steps after protecting the α-NH_2 group of **5** with phthalic anhydride, Eq. (2) [3b].

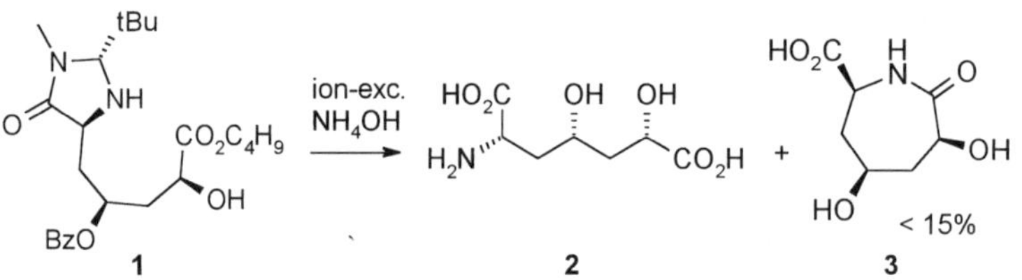

(1)

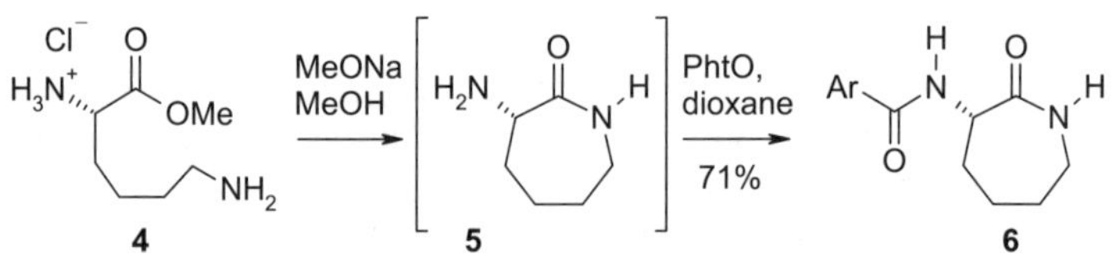

(2)

PhtO = Phthalic anhydride, Ar = 2-(hydroxycarbonyl)-phenyl

In analogy, Ugi et al. reported on a lactam formation by running a one-pot three components reaction: the condensation of L-lysine **7**, isobutyraldehyde and methyl isocyanide led to the corresponding α-amino-ε-caprolactam **9**, but the yield was not given. The authors presumed either a nucleophilic substitution of the ester **8** as the primary Ugi product by the amino function of the side chain or, alternatively, the nucleophilic attack of the NH_2-group on an intermediately formed *O*-acylamide and a subsequent rearrangement (Scheme 1) [4].

Under forced conditions, the reductive cyclization of a 6-oximeester at 150 °C gave a simple ε-caprolactam in about 60% yield [5a]. A preliminary chemoen-

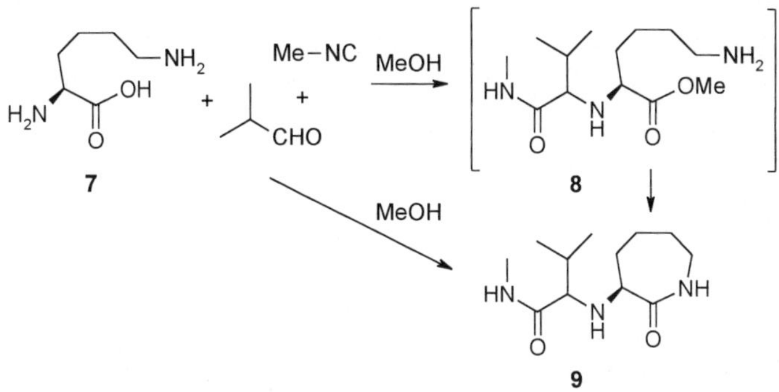

Scheme 1

zymatic approach to produce an azepinone from an aliphatic α,ω-dinitrile failed [5b]. On synthesizing fourteen-membered 12-azalactones in presence of a variety of enzymes, some eleven-membered ring contracted lactams were found as side products [5c].

The activation of either the carboxyl or of the amino function significantly increased the yields when running medium-sized lactam cyclizations. On synthesizing β-carboline alkaloids like manzamine C, the azacycloundecane ring fragment **12** was generated by a cyclization of the ω-amino acid **10** via an intermediately formed pentafluorophenylester **11** in 93% yield under high dilution conditions. The *Z*-configured 5,6-double bond in **11** supported the favorable pre-orientation of the carbon chains to achieve the smooth lactamization (Scheme 2) [6a].

The pentacyclic core **15** of more complicated manzamines was built up by a final ring closure of the thirteen-membered ring lactam in 50% yield. The tetracyclic system and the alkyne unit of the reactant **13** supposed a pre-orientation, an intermediately formed activated pentafluorophenyl ester **14** led to the desired

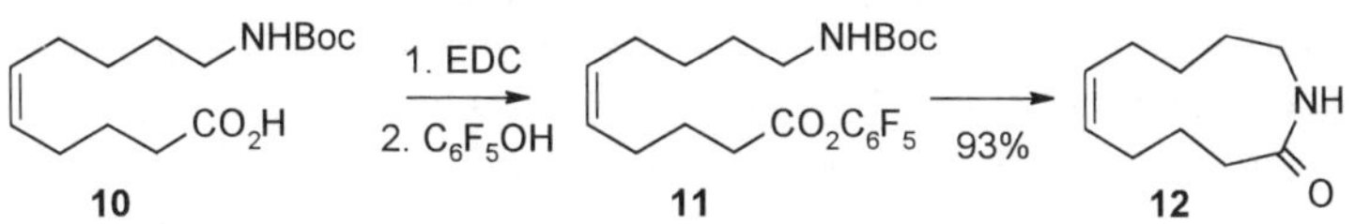

Scheme 2

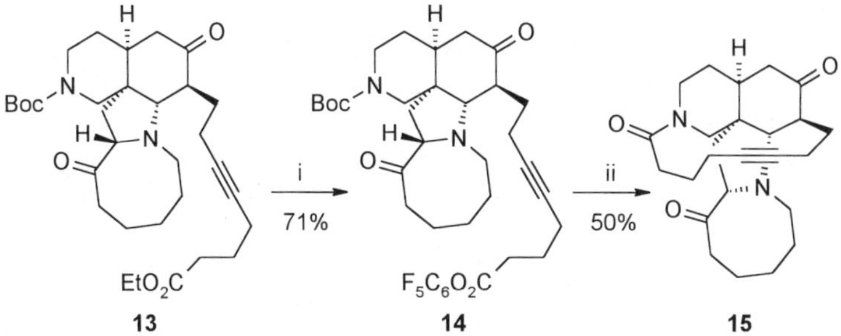

Scheme 3 i) 1. LiOH 2. DCC, C_6F_5OH ii) 1. TFA 2. $(i\text{-}Pr)_2NEt$

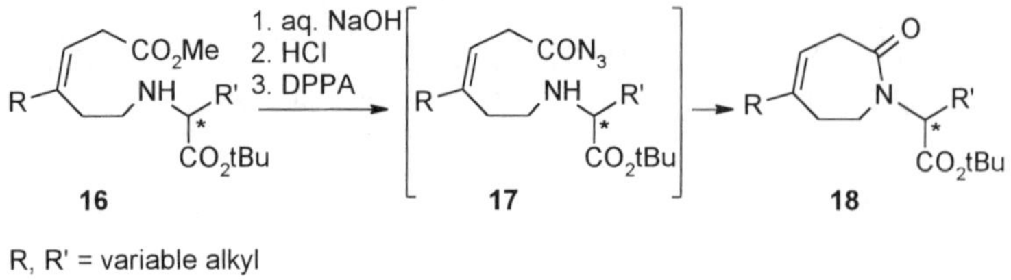

Scheme 4

lactam **15** after the removal of the Boc protective group under acidic conditions (Scheme 3) [6b, c]. A similar lactamization led to the formation of the seven-membered ring lactam in the course of a total synthesis of stenine [6d].

The synthesis of γ-turn mimetics basing on azepinones **18** succeeded by an in situ activation of the acids (after hydrolysis of **16**) with diphenylphosphoryl azide (DPPA) to afford **17**. A range of β,γ unsaturated lactams **18** was generated in about 70% yield, they were involved as key fragments in the synthesis of short peptide chains. Such peptides were tested concerning a HIV-1 protease inhibition activity [6d] or as angiotensin II receptor ligands with γ-turn mimetics (Scheme 4) [6e, f]. The cyclization of Boc-lysin **19** to **20** could be achieved in about 90% yield using an in situ PyBOP/$NaHCO_3$ activation of the carboxylic acid function, Eq. (3) [6g].

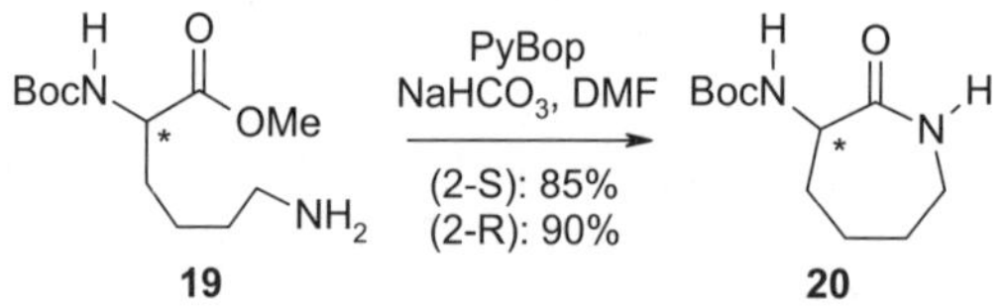 (3)

A similar pre-orientation involving unsaturated carbon chains was operative on generating twelve-membered enediyne **23** and arenediyne lactams **24** [7]. The *seco* methylesters **21** and **22** were cleaved with LiOH, the corresponding carboxylic acids underwent cyclizations after activation with 2-fluoro-pyridinium tosylate **25** [8]. Dimerization products were found as by-products (<10%). It should be pointed out, that the lactamization succeeded in a single step in about 75% yield by treating the *seco*-methylesters **21** and **22** with Me_3Al in refluxing methylene chloride. Obviously, the latter route was more convenient (Scheme 5).

The effect of micelles in the pre-organization of acyclic compounds in a thirteen-membered lactam formation using a modified Mukaiyama reagent was investigated by Rico [9].

Optically active, α-branched lactams **30** have been built by means of Meyers chiral auxiliaries [10]. The key step included the diastereoselective α-alkylations of the initially formed ω-*N*-sulfonamido oxazolines **26**. The *R* or *S* configuration in the product **27** was obtained reacting the appropriately configured intermediate aza enolates with alkyl halides, high diastereoselectivities have been reported. Several attempts to achieve a complete ring closure to the lactams **30** (via **29**) by an acidic cleavage of the oxazolines **27** failed. Varying mixtures of

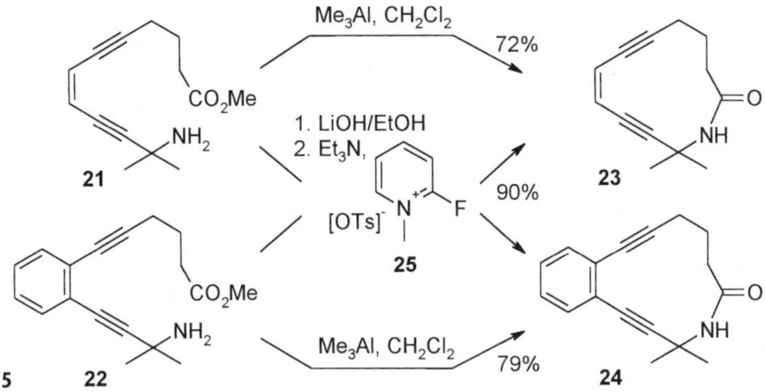

Scheme 5

lactams **30** and carboxylic acids **28** have been obtained which could be easily separated via column chromatography. Consequently, the crude mixture of **28** and **30** was treated with DCC to activate the carboxylic acid functions and a subsequent complete conversion to the corresponding lactams **30** occurred. The configuration of the stereogenic α-carbon had not been effected. The synthesis of seven and thirteen-membered ring lactams **30** succeeded in 30 to 70% yield (Scheme 6) (Table 1) [11a].

The in situ activation of 2-aminopimelic acid 1-amide with triethyl phosphite led to the corresponding seven-membered lactam under forced conditions in 52% yield [11b].

Activated carboxylic acids served as versatile precursors in lactam synthesis. Robl described some α-amino-ε-alkyl-ε-caprolactam syntheses, Eqs. (4–7)] [12]. Ring closure was induced after an EDC/HOBt activation of the acid function of **31** to form the 6-propyllactam **32** in 51% yield, Eq. (4). The cyclization of

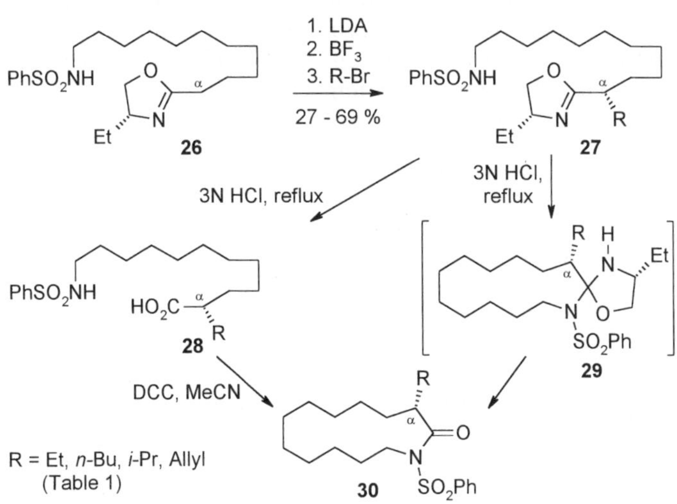

Scheme 6

Table 1. Ref. [11a]

Entry	R	Auxiliary[a]	Yield **30** (%)	D.r.	Config. α–C	Ring size Lactam **30**
1	Et	A	54	>95:5	*R*	13
2	Et	B	64	>95:5	*S*	13
3	*n*-Bu	B	69	>95:5	*S*	13
4	*i*-Pr	B	27	>95:5	*S*	13
5	Allyl	B	63	>95:5	*R*	13
6	Allyl	C	58	>95:5	*S*	7

[a] Structures of auxiliaries **A**–**C**:

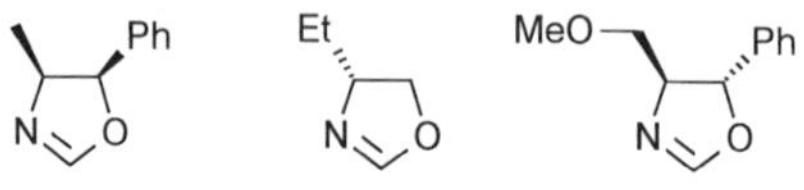

A **B** **C**

Auxiliaries Table 1

the corresponding 7-methyl compound **33** to **34** by a Lewis-acid catalysis has been reported to be successful, the yield was somewhat higher (57%). Eq. (5) [12a]. In accordance with the fact that the reactants **31** and **33** were built as a 1:1 mixture of diastereomers, the resulting lactams *trans*-**32/34** and *cis*-**32/34** were isolated in equal amounts. Additionally, tertiary amines as found in the 6,6-dimethyl-6-amino acid **35** underwent a smooth cyclization to **36** using HOBt/EDC conditions, Eq. (6) [12b]. In contrast, several spiro-alkylated azepinones **38** were synthesized by an acid mediated ring closure of the corresponding 6-aminoacid benzylester **37** using Me_3Al (spiro-cyclopentane adduct via a Beckmann rearrangement of the cyclohexanone oxime: see Sect. 4.1). The yields were reported to vary between 52 and 96% [12c]. The 2-amino function of the spiro cyclobutane lactam **38** (n = 2) was introduced later via enolate bromination and subsequent substitution with sodium azide. The lactams served as key compounds for vasopeptidase inhibitor investigations. A related cyclization led to the formation of a conformationally restricted arginine analog with some activity in enzyme inhibition [12d].

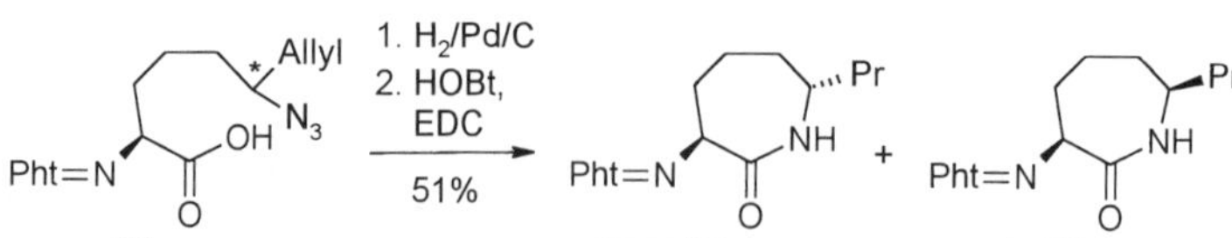

31 *trans*-**32** *cis*-**32** (4)

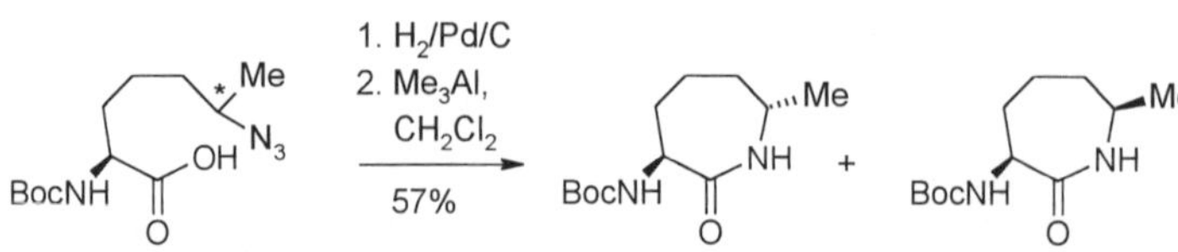

33 *trans*-**34** *cis*-**34** (5)

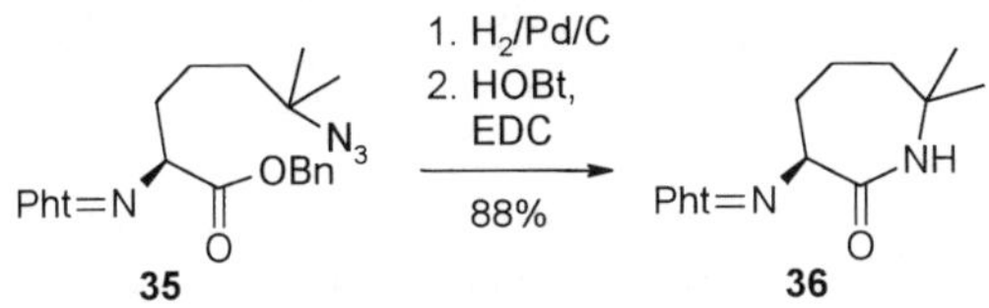

(6)

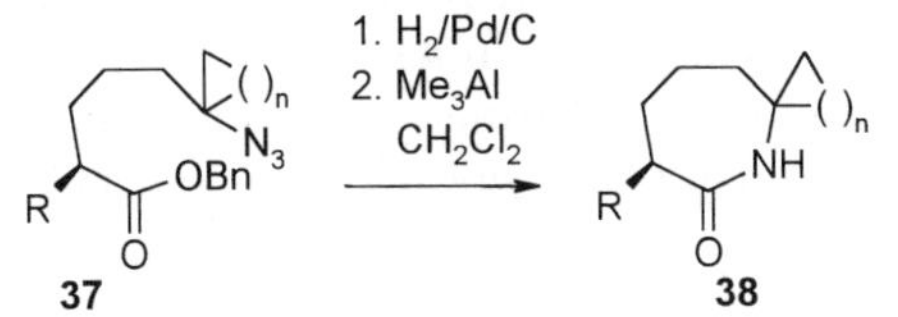

(7)

n = 1: 52% (R = NHBoc) n = 2: 96% (R = H)

On studying Pd-catalyzed vinylepoxide openings as key steps in macrolactam formations (see Sect. 2.2), Trost et al. published a synthesis of the fluvirucin B_1 aglycon. Following a convergent strategy, two halves of the molecule were coupled by the Pd-catalyzed allylation to give the Meldrum's acid derivative **39**. The ring closure was reported to be somewhat tricky. The best reaction conditions were found to involve activation in situ of the generated *seco* amino acid (hydrogenolytic cleavage of the benzyl ester; the double bond was not affected because of the sterical shielding) with PyBroP/DMAP to give the fourteen-membered ring **40** in 42% yield [13a]. A related cyclization started from the ketoacid **41**: after activation of the acid function with dipyridyl disulfide and Bu_3P, the azide was reduced with $SnCl_2$. The so formed amine induced an efficient macrocyclization to form the lactam **42** in 62% yield over both steps [13b]. Several final steps allowed the total synthesis of the fluvirucin B_1 aglycon from **40** and **42**, respectively (Scheme 7).

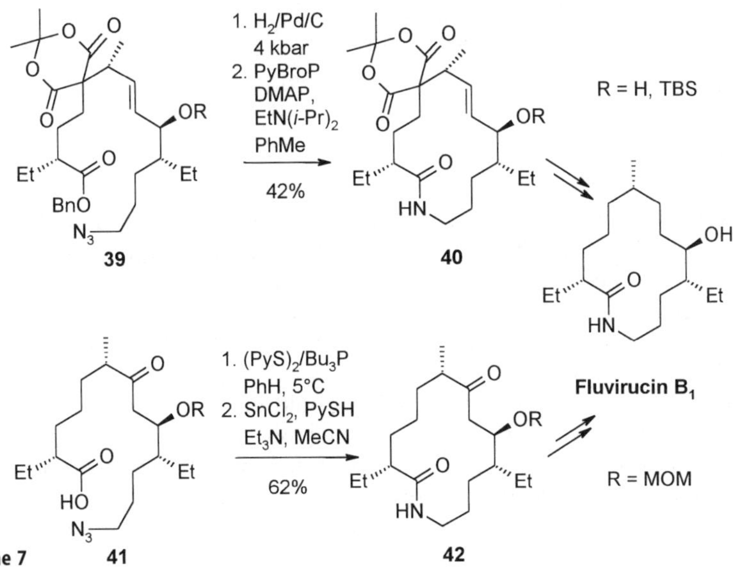

Scheme 7

Polymer bound HOBt as a catalyst was used in the synthesis of seven (23% yield), nine (13%) eleven (5%) and thirteen-membered ring lactams (34%) [13c].

Dutton reported on the synthesis of an ε-caprolactam analog of an anthelmintic cyclic peptide. The α-hydroxy-ε-caprolactam **44** was generated in an ex chiral pool synthesis staring from malic acid. The α-hydroxy carboxylic acid unit was protected as a dioxolanone in **43**. The protective group served simultaneously as the reactive function during cyclization: lactam **44** formation succeeded by ring opening of the dioxolanone **43** by the nucleophilic attack of the amino function, Eq. (8) [14].

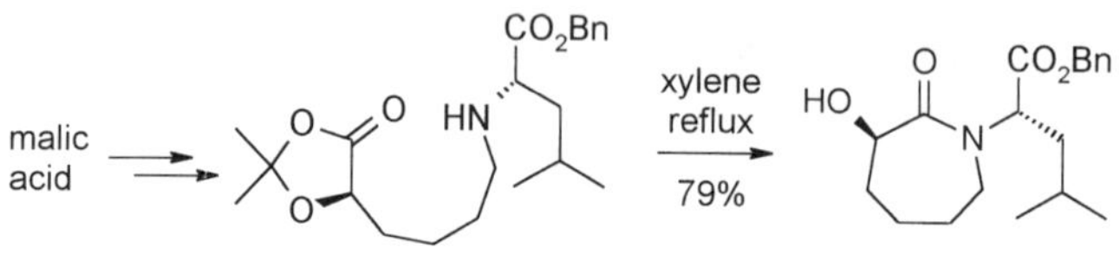

43 44 (8)

A novel method to generate multisubstituted eight-membered ring lactams has been developed as a ring contraction of nine-membered ring lactones. Initially, the medium-sized lactone ring **46** bearing an appropriately configured protected amino function and a *cis* double bond was built by a conventional Yamaguchi cyclization of **45** [15]. Then, the *N*-protective group was removed and the resulting aminolactone was heated to undergo a ring contraction to yield the azocinone **47** in quantitative yield. The so formed lactams bearing defined stereogenic centers were suitable for the elaboration to constrained dipeptides (Scheme 8) [16a].

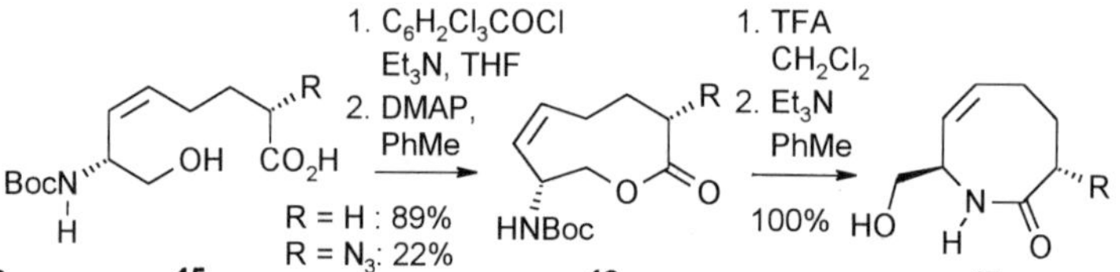

Scheme 8 45 46 47

A reverse route involved ring expansions to form azepinones **49** and **52**. Starting from ex chiral pool syntheses from D-galactono-1,4-lactone and related compounds, the terminal azides **48** and **51** were synthesized in several steps. After hydrogenation, the in situ generated amino functions induced the desired ring expansions replacing the five-membered lactones by the seven-membered ε-caprolactam units **49** and **52** in 65 to 98% yield, respectively [16b–d]. The *cis* arranged hydroxyl functions of **49** and **52** could be selectively protected as an acetonides **50** in 66% yield and **53** in 61% yield [16b, c] (Scheme 9). The azepinones served as conformationally rigid derivatives to determine the configuration of key intermediates in formal total syntheses of optically active balanols [16c].

A combination of N and CO activation was used by Vilarrasa to generate a range of medium-sized lactams **56**. ω-Azido carboxylic acids **54** were initially

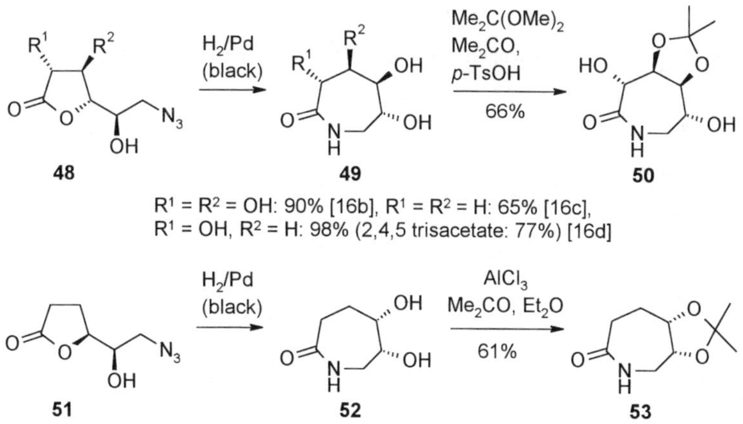

Scheme 9 [16c]: 85%, [16d]: 83% (4,5 bisacetate: 50%)

converted into the corresponding anhydride, thiopyridyl or trichlorophenylesters **55**. A subsequent Staudinger type phosphinimin generation induced the lactam formation in refluxing benzene. High dilution conditions were required to achieve satisfactory yields of the monomeric lactams **56**. In most runs the dimer was found as a side product, especially the ten to twelve membered rings suffered from this competing process. (Table 2) (Scheme 10) [17]. In contrast, the *in situ* activation of the ω-amino acid corresponding to **54** and the subse-

Table 2. Ref. [17]

Entry	X	n	R/R′	Ring size	Reagents	Solvent	Yield (%) 56	Yield (%) Dimer	Ref.
1	**A**	1	Me/H	10	Sn(II), PySH, Et_3N	MeCN	20	25	a
2	**A**	1	Me/H	10	PhSeH, Et_3N	MeCN	30	40	a
3	**A**	1	Me/H	10	$[Et_3NH]^+ [Sn(SePh_3)]^-$	MeCN	45	25	a
4	**A**	1	Me/H	10	DMAP, $[Et_3NH]^+ [Sn(SePh_3)]^-$	MeCN	45	25	a
5	**B**	1	Me/H	10	$[Et_3NH]^+ [Sn(SePh_3)]^-$	MeCN	45	20	a
6	**D**	1	Me/H	10	Bu_3P, DMAP	PhH	38	25	b
7	**C**	3	H/H	12	Bu_3P	PhH	41	44	b
8	**D**	3	H/H	12	Bu_3P	PhH	47	39	b
9	**D**	3	H/H	12	Bu_3P, DMAP	PhH	51	34	b
10	**E**	3	H/H	12	Bu_3P	PhH	51	34	b
11	**E**	3	H/H	12	Bu_3P, DMAP	PhH	57	35	b
12	**D**	4	*n*-Hex/H	13	Bu_3P, DMAP	PhH	61	30	b
13	**D**	5	H/Et	14	Bu_3P, DMAP	PhH	82	<2	b

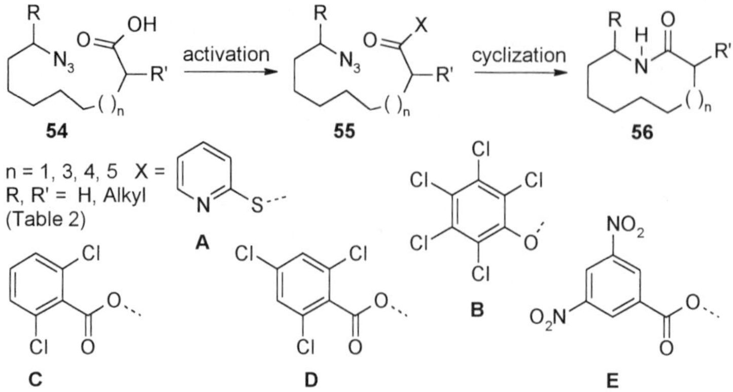

Scheme 10

quent cyclization to **56** gave only poor results (0–10% yield, up to 25% of the dimer).

The generation of ω-amino- or ω-amido-ketenes by means of photochemical or thermal 1,5-H-shift methods has been investigated to form a range of medium-sized heterocycles. A preliminary experiment was published by Hasegawa to convert a benzofuroxan into a 1*H*-azepine-2,7-dione, a highly reactive intermediate possessing two nitrile oxide functions was thought to be involved via a photochemically initiated ring-opening [18]. Diene ketenes **58** have been formed by a photochemical cleavage of 2-alkyl-2-acetoxycyclohexadienones **57**. An amino function attached to an ω-position of the side chain gave a subsequent lactamization to **59** in the presence of an appropriate base. All cyclizations were accompanied by the formation of the corresponding dimeric bislactam. The yields were observed to increase with increasing ring size ($n > 3$) (Scheme 11) [19].

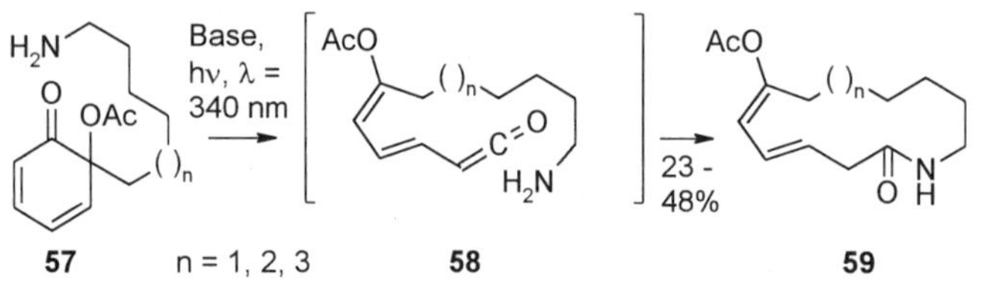

Scheme 11

A thermally 1,5 *H*-shift induced by heating 1-ethoxy alkynes **60** to about 150 °C led to ketenes **61**. Subsequent cyclizations to give lactams **62** were achieved by attaching a *N*-benzamido function at C-6 or C-7 of a linear alkyl chain. Azepanones **62** ($n = 1$) and Azecanones **62** ($n = 2$) were generated in 68% and 58% yield, respectively. The formation of larger rings ($n = 5, 9$) failed under the conditions reported (Scheme 12) [20].

A range of 6-substituted 2-amino-ε-caprolactams **66** has been synthesized by a sequence published by Robl et al. for investigating the generation of peptidomimetics. After Swern oxidation of the dipeptide **63**, the corresponding

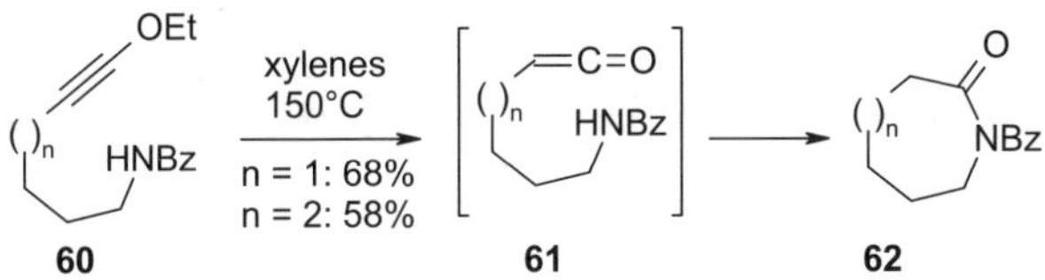

Scheme 12

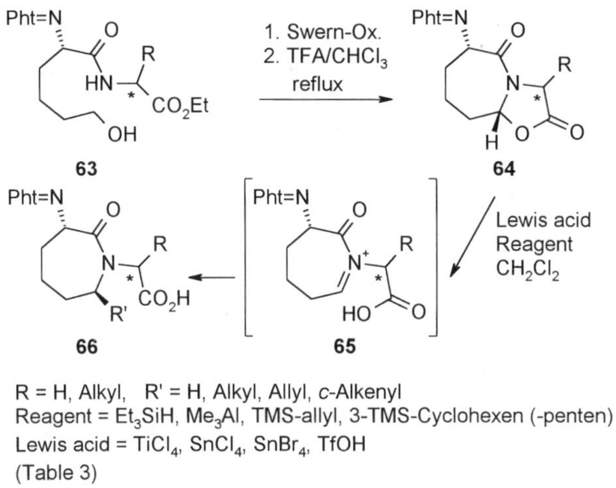

Scheme 13

aldehyde underwent acid-mediated cyclization to give a bicyclic amidoacetal **64** in 60 to 76% yield. In the presence of Lewis acids, an intermediate acyliminium salt **65** was formed, which was treated with a range of nucleophiles to yield the corresponding optically active 2-amino-ε-caprolactams **66** with moderate to high diastereoselectivities (Scheme 13) (Table 3). It should be taken in account that the stereogenic α-position of the oxazolinone-forming amino acid suffered from a partial epimerization depending on the acid involved (Table 3, entries 1, 2) [21]. The R-phenylalanin adduct **66** (α-R, Table 3, entry 6) was obtained in only 18% yield. The *S*-phenylalanin adduct **67** (R = β-Bn) suffered from an intramolecular Friedel Crafts type reaction: After generating the intermediate acyliminium ion in presence of TfOH, the phenyl substituent served as a nucleophile to form the tricyclic lactam **68** in 72% yield (Table 3, entry 13, Eq. (9). Conformationally fixed bicyclic compounds **72** were synthesized via a related pathway: Initially, the unsaturated pyrrolidine **71** (n =1) and piperidine **71** (n = 2) were formed after acid mediated cyclizations of the acyclic reactants **69** and **70**, respectively. The final bicyclization was induced by a 7-endo cation olefin

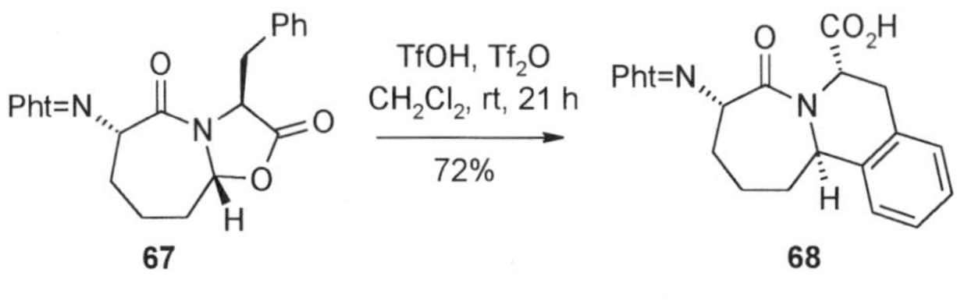

(9)

Table 3. Ref. [21]

Entry	Acid	Reagent	T [°C]	Time [h]	**64** R	Yield **66** [%]	R′	Ratio R β/α	Ratio R′ β/α	Ref.
1	TFA	Et_3SiH	70	20	β-Me	66	H	1:1	–	a
2	$BF_3 \cdot Et_2O$	Et_3SiH	25	64	β-Me	48	H	2:1	–	a
3	$TiCl_4$	Et_3SiH	25	18	β-Me	64	H	>98:2	–	a
4	$TiCl_4$	Et_3SiH	25	18	β-Bn	60	H	>98:2	–	a
5	$TiCl_4$	Et_3SiH	25	18	β-*i*-Pr	64	H	>98:2	–	a
6	$TiCl_4$	Et_3SiH	25	65	α-Bn	18	H	<2:98	–	a
7	$SnCl_4$	Me_3Al	25		H	73	Me	–	1.8:1	a
8	$SnBr_4$	TMS-Allyl	25	9	H	95	Allyl	–	>98:2	a
9	$SnCl_4$	3-TMS-1-cyclohexene	25	0.5	H	67	2-*cyclo*-hexenyl[a]	–	>98:2	a
10	$SnCl_4$	3-TMS-1-cyclopentene	25		H	65	2-*cyclo*-pentenyl[a]	–	>98:2	b
11	$TiCl_4$	TMS-Allyl	0	28	β-Me	63	Allyl	>98:2	>98:2	a
12	$TiCl_4$	TMS-Allyl	0		β-Et	30	Allyl	>98:2	>98:2	b
13	TfOH	–	25	21	β-Bn	72	Ph[b]	>98:2	>98:2	a

[a] Mixture of diastereomers concerning the unsaturated ring in R′, hydrogenation led to a single product (saturated R′), respectively.
[b] Fused with benzene ring system, Eq. (9): **67** → **68.**

reaction, the intermediately formed mixtures of halogenated or unsaturated azepinones were reduced by means of $(TMS)_3SiH$. The bicyclic derivatives **72** were isolated in 66% (n = 2) and 71% (n =1) yield, respectively (Scheme 14) [21c].

An auxiliary controlled enantioselective route to generate seven-membered ring lactams **75** used the α-alkylation of cyclic hydrazide derivatives **74**. Initially, 6-chloro hydrazides **73**, bearing the chiral information in the *N*-amino-pyrrolidine function underwent amidocyclization in the presence of a base. A subse-

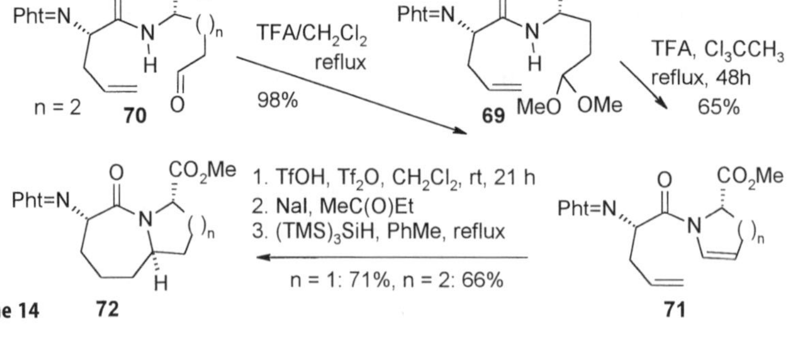

Scheme 14

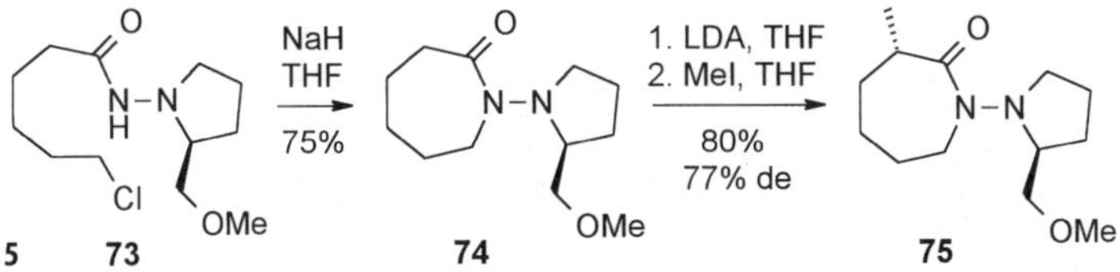

Scheme 15

quent α-deprotonation and enolate alkylation sequence gave the ε-caprolactam **75**, the N*H*-lactam was generated after a reductive cleavage of the hydrazine unit (Scheme 15) [22].

A somewhat unusual sequence to generate azepanones **80** involved the intramolecular addition of hydroxylamines to alkynes **76** to form cyclic nitrones **77**. A vinyl magnesium bromide addition at low temperatures and a reduction with $TiCl_3$ followed by *N*-Boc protection led to the azepane **78**. Double bond bromination and subsequent RuO_4 oxidation gave the lactam **79**. Several further steps allowed the generation of the lactam structure **80** proposed for *d*,*l*-acacialactam, but the spectral data of the synthetic material differed from that of the natural product (Scheme 16)] [23a, b].

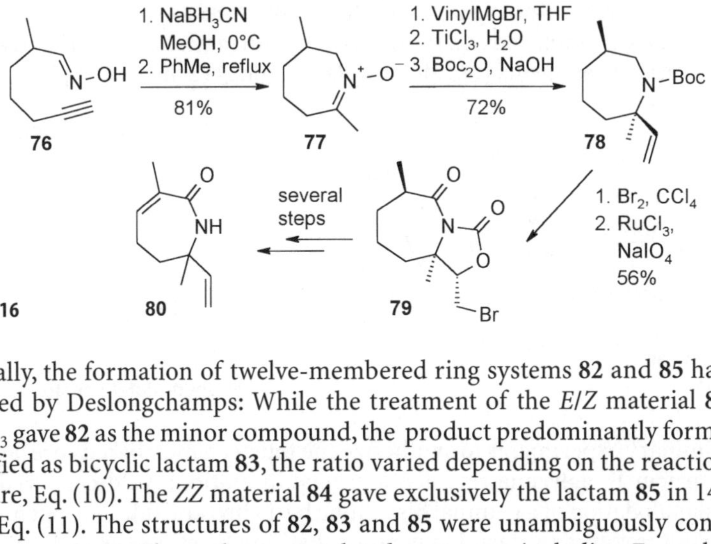

Scheme 16

Finally, the formation of twelve-membered ring systems **82** and **85** has been reported by Deslongchamps: While the treatment of the *E*/*Z* material **81** with Cs_2CO_3 gave **82** as the minor compound, the product predominantly formed was identified as bicyclic lactam **83**, the ratio varied depending on the reaction temperature, Eq. (10). The *ZZ* material **84** gave exclusively the lactam **85** in 14–19% yield, Eq. (11). The structures of **82, 83** and **85** were unambiguously confirmed via x-ray analyses. The authors postulated sequences including Favorskii type rearrangements and F_3C anion eliminations to explain the unexpected outcome of the macrocyclizations [23c].

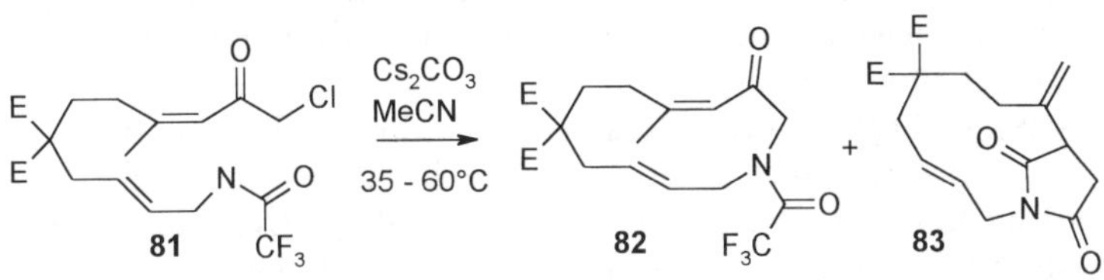

Yields **82/83** [%]: 35°C: 18/20, 45°C: 27/32, 60°C: 0/24 (10)

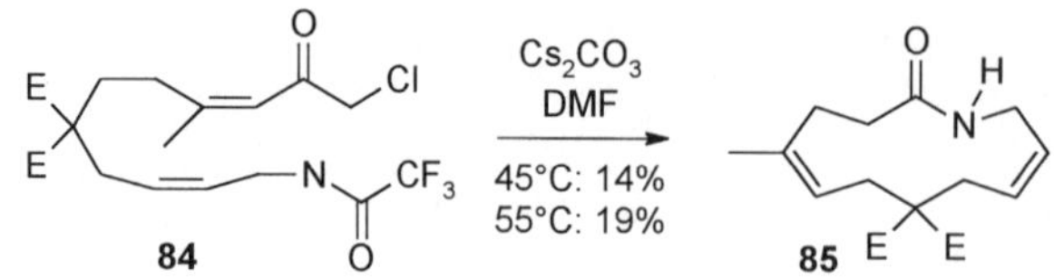

(11)

2.2 Ring Closure Reactions by C-C Single Bond Formation

Ring closures by C–C bond formation can be sub-divided in two strategic pathways: the first one generates a C–C single bond in the cyclization step, the second one leads to a new C=C double bond. Following the first strategy, the methodology is focused on radical cyclizations or metal-mediated allylations. During the last decade especially the second strategy has gained more and more importance because of extensive investigations on the field of ring-closing olefin metatheses. In contrast, the formerly employed Wittig olefinations and McMurry couplings became less important.

However, an almost classical attempt to generate an α,β-unsaturated ε-caprolactam **87** involved a Knoevenagel condensation of **86** as the ring forming step. The material was employed in a synthesis of conformationally restricted analogs of some anti cancer agents (tetrahydrofolate analog) (Scheme 17) [24].

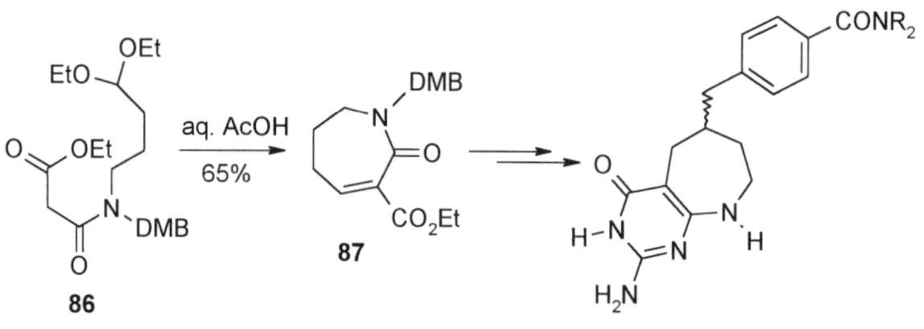

Scheme 17

Preliminary experiments to synthesize the fourteen membered core lactam of the fluvirucin B_1 aglycone revealed a crucial limitation: A Pd catalyzed intramolecular allylation of a geminal bissulfone by a vinyl epoxide ring opening succeeded; this involved the terminally non-substituted olefin **88** (R = H) to give the desired macrocycle **89** in 71% yield after PCC oxidation. In contrast, the alkylated analog **88** (R = Me) only underwent an isomerization to build the corresponding unsaturated ketone, the ring closure failed. The total synthesis of the natural product needed a different strategy as outlined previously (see Sect. 2.1, Scheme 7) (Scheme 18) [13a].

The synthesis of the CG525155 (a neutral endopeptidase inhibitor) required a Pd catalyzed Tsuji-Trost reaction as the key step following the strategy described by Johnson. Starting from the optically active allenyl amino acid methylester **91** (synthesized in several steps from **90**), the *seco*-derivative **92** as the crucial precursor was generated in several steps in high yield. The Pd (0)

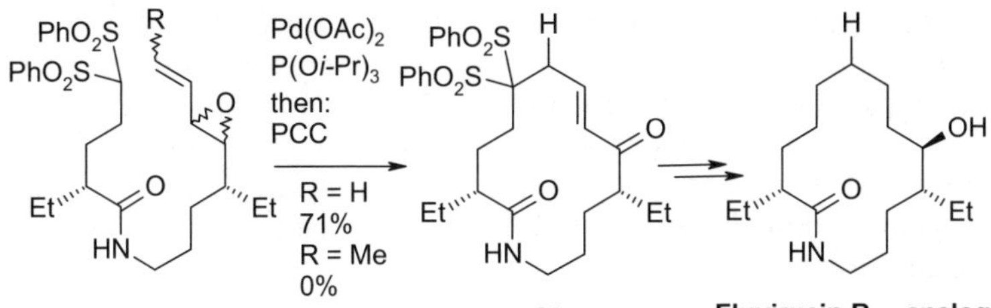

Scheme 18 88 89

mediated intramolecular allylation succeeded in 55 to 60% yield to give **93** under carefully optimized reaction conditions (dilution!) as a mixture of *E*- and *Z*-olefins (crude material, contained some decarboxylated derivative). Several final transformations allowed the synthesis of the desired biologically active material. (Scheme 19) [25a]. A more straightforward approach concerning the same target used a ring-closing cycloisomerization of a related allene **94** via a hypothetical palladium allyl complex **95**. On treating the allenylamido malonate **94** with a Pd(II) catalyst and a bidentate phosphine ligand, the ring closure succeeded in 86% yield. The corresponding lactam **96** was formed as a mixture of *E* and *Z* olefins, the subsequent hydrogenation formed a potential suitable key intermediate for the synthesis of the endopeptidase inhibitor (hypothetical, Scheme 19) [25b].

The synthesis of medium-sized lactams **100** has been attempted by means of amine-directed carbonylation of 5-*N*-benzylamino-1-pentene **97**. In presence of

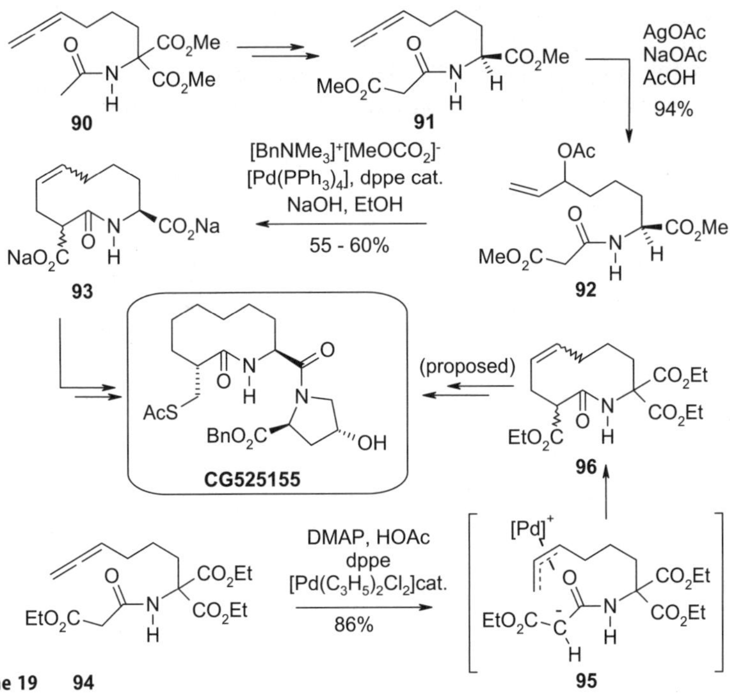

Scheme 19 94

stoichiometric amounts of $[RhCl(CO)_2]_2$ a chelate complex **98** was thought to be formed *in situ*. Under a CO atmosphere the six-membered ring **99** was formed predominantly (98:2). On employing a Rh(acac)-complex, the ε-caprolactam **100** still occurred as the minor product, but the ratio of **99**:**100** was found to be 78:22. In contrast, the corresponding hydrocarbonylations allowed the regioselective generation of azepanes with high yields using an H_2/CO atmosphere (Scheme 20) [26].

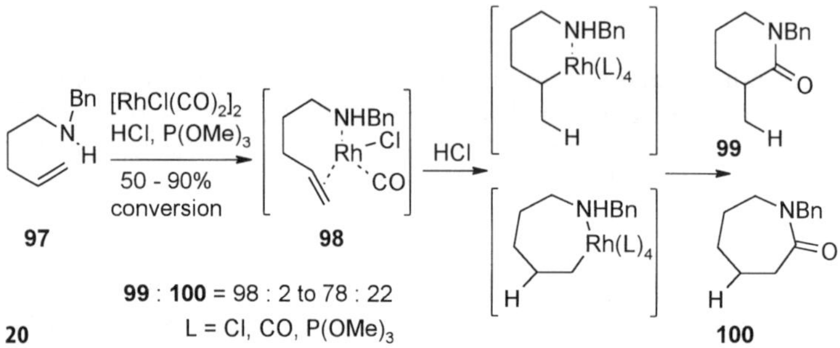

Scheme 20

Radical cyclizations are known to possess a high driving force but again, the formation of medium-sized rings is somewhat difficult. In most cases intermolecular competing processes could be observed. If radical additions to olefins are investigated, the formation of the smaller ring predominated (n-*exo*-trig exceeds n+1-*endo*-trig) with respect to the Baldwin rules. A convincing experiment has been published by Pattenden: Unsaturated *N*-chloroformyl amides **101** could be easily converted into the corresponding "Co-salophen" complexes **102**. A subsequent cyclization led to the six-membered ring **103**, the homologous azepinone **104** was only detected in trace amounts (Scheme 21) [27a].

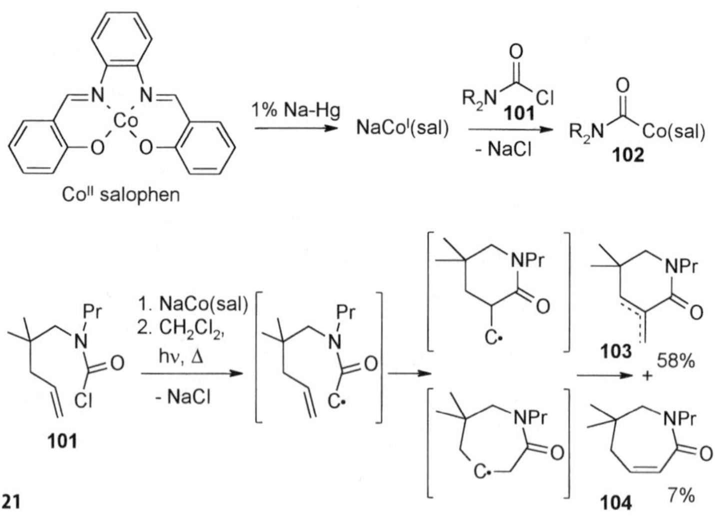

Scheme 21

A tandem reaction of sulfanyl radical addition-cyclization was used to generate lactams with up to six-membered rings. Preliminary investigations to synthesize larger lactams were reported to be mostly unsuccessful. A single experiment gave an azocinone **106** following an 8-*endo*-trig reaction of **105** but the yield was poor (19%). The formation of lager rings failed, only acyclic products such as **107** have been isolated, Eq. (12) [27b].

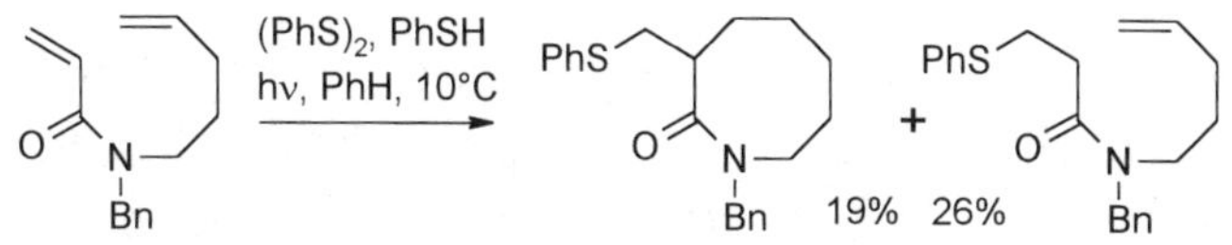

105 **106** **107** (12)

The synthesis of isoquinoline alkaloids through an azecinone **109** and a consecutive transannular cyclization was found to be an intriguing strategy for preparing the protoberberine basic nuclei. Initially, the ten-membered ring **109** should have been formed via a Pd(0) catalyzed reaction, but all attempts failed. In contrast, a radical cyclization generated the desired lactam: the almost rigid bromoacetylene **108** (R^1 = H) could be activated by means of the addition of AIBN and Bu_3SnH to induce a subsequent *endo*-selective attack of the aryl radical at the triple bond. The unsubstituted azecinone **109** could be isolated in 45% yield (R^2, R^3 = H) and 71% yield (R^2 or R^3 = OMe) as a mixtures of *E*- and *Z*-olefins. On reacting the corresponding silyl acetylene **108** (R^1 = TMS), the yield of **109** (R^1 = TMS) increased to 60 to 75% forming a single double bond isomer, the most electron-rich aromatic reactants (R^2, R^3 = OMe) gave the best results. Final transannular ring contractions allowed the regioselective formation of the tetracycles **110** and **111**, respectively (Scheme 22) [28].

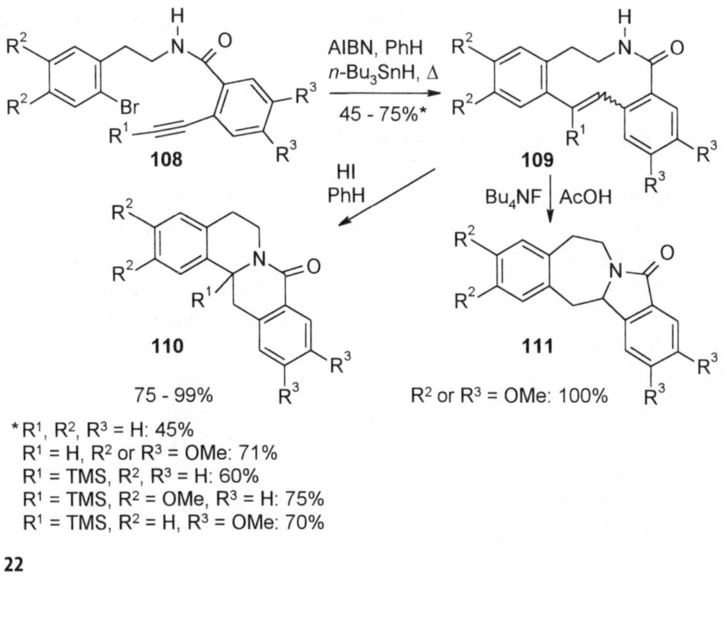

* R^1, R^2, R^3 = H: 45%
R^1 = H, R^2 or R^3 = OMe: 71%
R^1 = TMS, R^2, R^3 = H: 60%
R^1 = TMS, R^2 = OMe, R^3 = H: 75%
R^1 = TMS, R^2 = H, R^3 = OMe: 70%

Scheme 22

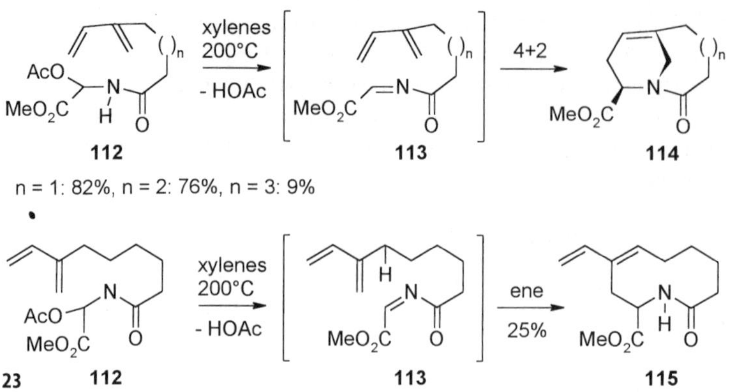

Scheme 23

A special ring closing reaction to generate medium-sized ring lactams has been described by Shea. Intramolecular imino Diels-Alder cycloadditions (type 2 Diels-Alder reaction) of *in situ* formed *N*-acylimines **113** (HOAc elimination from **112**) led to bicyclic [n.3.1] frameworks **114** with n varying between one and three. The so-formed bicycles **114** were characterized by a bridgehead double bond as well as a bridgehead lactam function. While the highly strained seven- and eight-membered ring lactams **114** (n = 1, 2) have been obtained in high yields, (82 and 76%, respectively), the azoninone **114** (n = 3) could be isolated in only 9%. The major competing process was found to be a new type of an intramolecular imino-ene reaction (type 3 ene reaction) to generate the less strained azecinone **115** bearing a *cis*-endocyclic double bond in about 25% yield. The NMR spectra of this compound were characterized by a significant mobility of the ring skeleton: At 150 °C a single set of peaks was observed, which changed to an unresolved broad set of signals at room temperature. At about –25 °C a double set of lines has been observed indicating two stable conformations of the ten-membered ring lactam **115** (Scheme 23)] [29].

3.3 Ring Closure Reactions by C=C Double Bond Formation

The formation of medium-sized ring lactams by means of ring-closing olefin metathesis is still one of the most intriguing strategies. Reaction conditions and catalysts tolerating a wide range of different functional groups and ring sizes have been developed during the last decade. The ease of generating suitable reactants, the reliability in running the cyclizations with high yield, and the increasing stereoselectivities for building the *E* or *Z* olefins characterized the metathesis as one of the most powerful methods to synthesize macrocycles [30].

The synthesis of *β*-turn mimics included a stereoselective generation of a rigid substance P analog bearing a ten-membered ring lactam. A classical approach to build the azecinone **117** via a nucleophilic attack of the amino group to one ester function of the malonate **116** in refluxing DMF succeeded in only 30% yield, the optical purity of the reactant could be maintained under the harsh cyclization conditions. After ring closure, the *trans* configured lactam **117** was

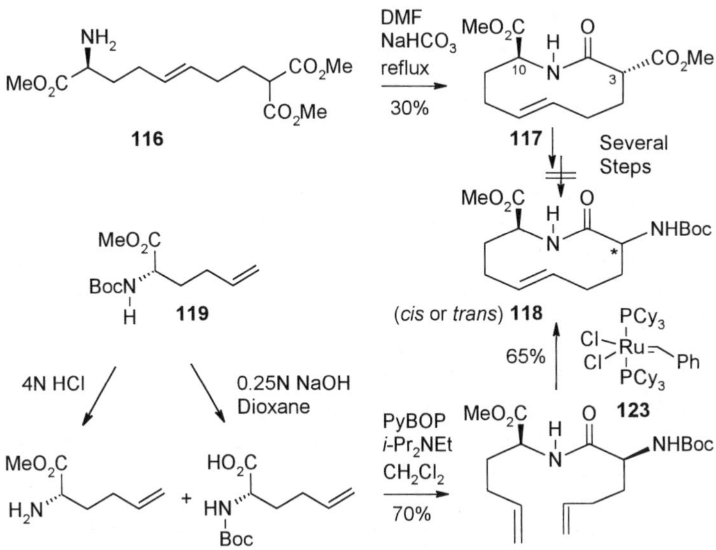

Scheme 24

formed, which should have been converted into the desired product *trans*-**118** in several steps. This latter sequence was abandoned because of severe problems on differentiating the ester functions at C-3 and C-10. Otherwise, the creation of an ideal type I β-turn mimetic needed the generation of the diastereomeric *cis* lactam **118**, but the related synthesis via the condensation route failed. Then a convergent sequence was developed using twofold the reactant amino acid **119:** After peptide coupling of the sub-units **120** and **121** to afford **122**, a ring closing metathesis with Grubb's catalyst **123** gave the desired *cis* azecinone *cis*-**118** in 65% yield along with about 10% of the dimer under carefully optimized high dilution conditions. The double bond was found to be *E* configured. The latter sequence pointed out that the metathesis path represented the exceptionally more straightforward route to generate the medium-sized ring (Scheme 24) [31a].

The cyclization of some enantiopure amino acid-derived precursors **124** and **126** to give the corresponding optically active 6-methoxycarbonyl ε-caprolactams **125** and **127** has been reported by Rutjes. The use of Grubb's carbene **123** led to the unsaturated azepinones in moderate to high yields. While the ring closing metathesis involving two electron rich olefins (**126**) smoothly gave the lactam **127** in 65% yield [Eq. (14)], the analogous reaction employing an electron deficient double bond as found in **124** suffered from the predominant formation of the dimer, only 15% of **125** was isolated [Eq. (13)]. Preliminary experiments to synthesize the homologous azocinone failed, Eqs. (13, 14) [31b].

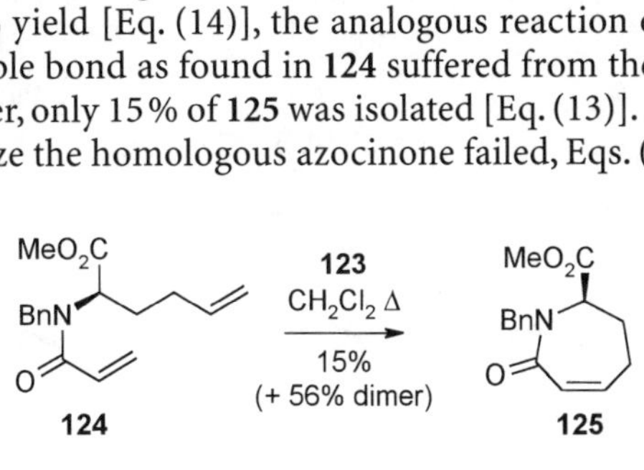

(13)

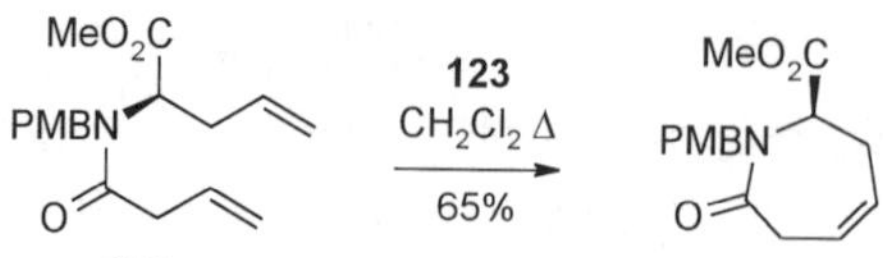

126 127 (14)

The fluvirucins B_1 and B_2 were isolated in 1990 as a new class of effective antifungal agents with some activity against the influenza A virus [1]. The heterocyclic core is a fourteen-membered lactam bearing four stereogenic centers in defined positions. The Hoveyda group developed a convergent enantioselective synthesis employing an olefin metathesis with Schrock's carbene **128** as the ring-closing reaction. Initial experiments showed that the defined substitution pattern of the *seco*-derivative **129** was necessary to achieve high yields in the cyclization step, thus implying an efficient pre-organization of the open chain reactant. Without any side chain, only 2% of the desired macrocycle could be isolated. The reaction of two simple vinyl units (absence of the methyl group) gave the lactam in about 40% yield, but the process suffered from the formation of up to 20% of the dimer. Under optimized high dilution conditions at elevated temperatures the cyclization of **129** to **130** succeeded in about 90% yield using Schrock's catalyst **128** (20 mol%), the use of the Grubb's carbene **123** exclusively led to the formation of the dimer. A single double bond isomer could be detected. The facile formation of the aglycon **130** (R = OTBS) suffered from the fact that the final introduction of the carbohydrate fragment $\mathbf{B_1}$ or $\mathbf{B_2}$ failed. This problem was solved by introducing the glycoside prior to the metathesis without any decrease in yield during the cyclization step. The generation of the last stereogenic center could be achieved by catalytic hydrogenation with high diastereoselectivity. Summing up these results, each ring-closing metathesis needs a careful optimization of all reaction parameters to guarantee a maximum yield of a medium-sized ring (Scheme 25) [32].

Another natural product synthesis using a metathesis as the key step has been reported to generate the ABCD ring system **133** of manzamine A. A multistep

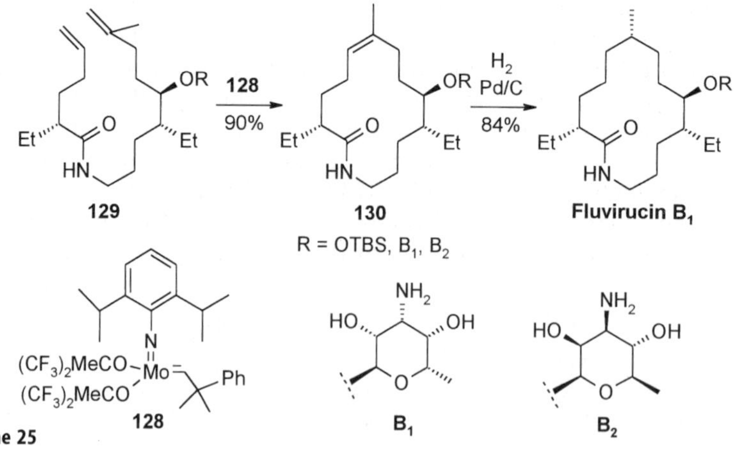

Scheme 25

sequence led to the tetracyclic framework **131**, the metathesis involving Grubb's catalyst **132** allowed the formation of the thirteen-membered ring system **133** bearing the desired *Z* olefin. Obviously, the defined structure of the reactant induced an efficient pre-organization of the unsaturated side chains supporting the ring closure (Scheme 26) [33a, b]. Additionally, the eight-membered E ring has been generated via an olefin metathesis. Grubb's catalyst **132** was used to achieve the desired ring closure of **134** to **135** as one of the final steps in the total synthesis. Somewhat earlier, Schrocks carbene **128** was employed to form the same ring of manzamine A in about 64% yield (**136** to **137**), but the molecule **137** did not yet bear the thirteen-membered D-ring framework (Scheme 26) [33b, c].

A systematic investigation of the ring-closing metathesis of **138** in the synthesis of a range of fourteen-membered ring lactams **139** and lactones has been reported by Weiler. The geometry of the resulting double bond was determined, the position of the olefin was broadly varied. The ratios obtained were compared to that derived from molecular mechanics calculations, Eq. (15), Table 4 [34].

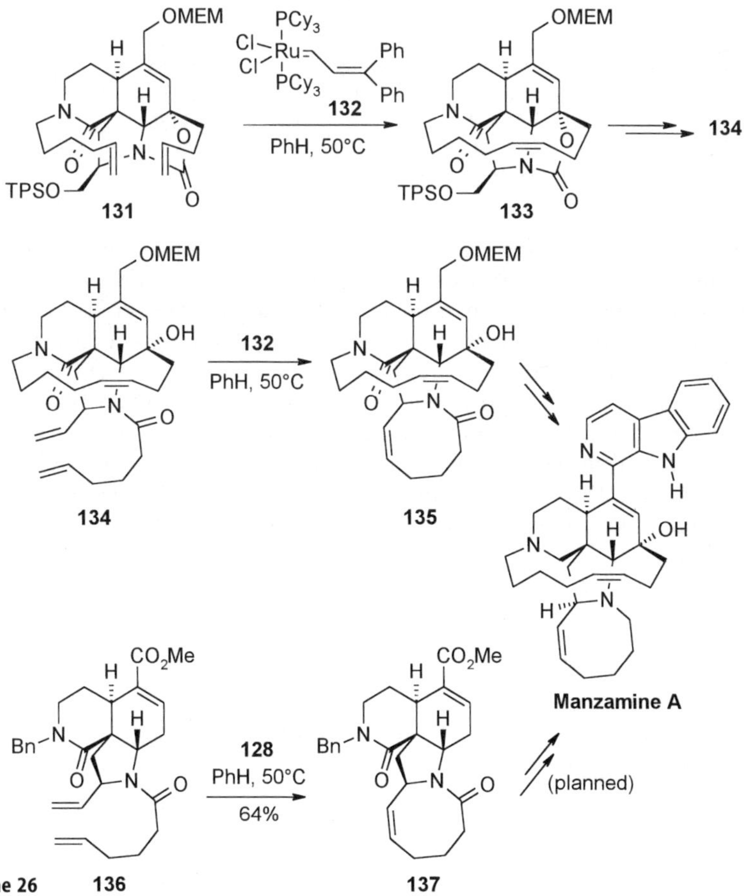

Scheme 26

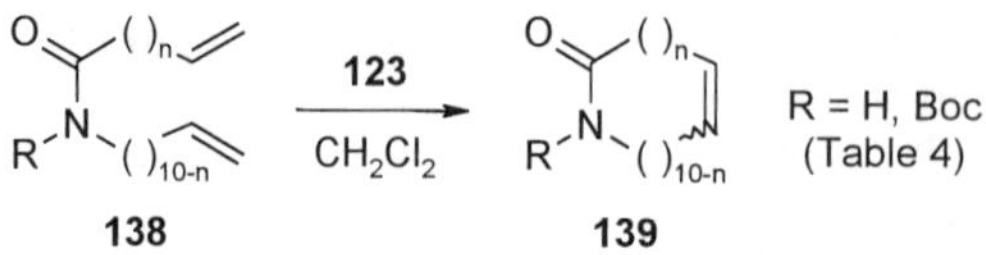

(15)

Table 4. Ref. [34]

Entry	R	n	Time (h)	Yield **139** (%)	Recovered **138** (%)	*E/Z* ratio (detected)	*E/Z* ratio (calculated)
1	H	1	24	0	82	–	99:1
2	H	2	24	11	75	80:20	99:1
3	H	3	25	39	32	11:89	17:83
4	H	4	4	74	9	58:42	57:43
5	H	5	1	87	12	>99:1	97:3
6	H	6	7	86	0	54:46	48:52
7	H	7	22	47	39	16:84	16:84
8	H	8	20	32	60	72:28	95:5
9	H	9	26.5	7	83	>99:1	63:67
10	Boc	1	24	0	93	–	99:1
11	Boc	2	24	31	41	82:18	99:1
12	Boc	3	10	71	8	20:80	17:83
13	Boc	7	8	62	11	13:87	16:84
14	Boc	8	17	57	5	64:36	95:5
15	Boc	9	24	20	67	>99:1	63:67

Solid phase ring-closing metathesis has experienced growing interest in combinatorial chemistry. The easy work-up (simple filtration and washing steps), the complete conversion of the reactants (large excesses of reagents were allowed) and the option to automate the process led to intensive investigations to synthesize a great number of Freidinger lactams and related azepinones **141** from **140,** Eq. (16) [35]. Most of the preliminary experiments focused on the development of suitable linkers and on testing a range of resins. The catalyst of choice was found to be a Grubb's carbene **123.** Compared to the well known solution reactions the yields of azepinone **143** decreased (97% versus 5 to 54%), the reaction times increased using **140** (→ **142**) on solid phase techniques, Eq. (17) [35a].

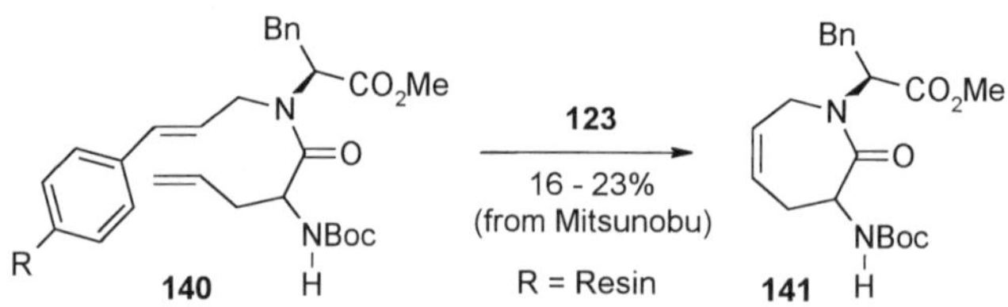

(16)

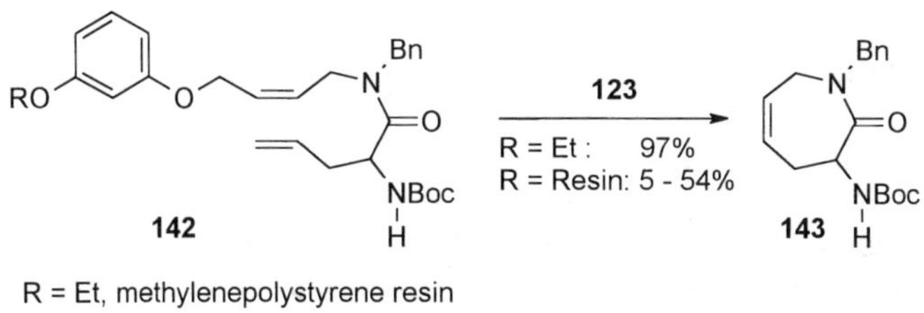

(17)

Until now, the most efficient approach to synthesize Freidinger lactams **147** started from a resin-bound cinnamylamine **144.** A Fukuyama-Mitsunobu reaction to **145** followed by sulfonamide cleavage and a consecutive appropriate acylation built up the diene **146,** which underwent ring-closing metathesis involving Grubb's catalyst **123** to generate the desired lactams **147** (Scheme 27, Table 5) [35d].

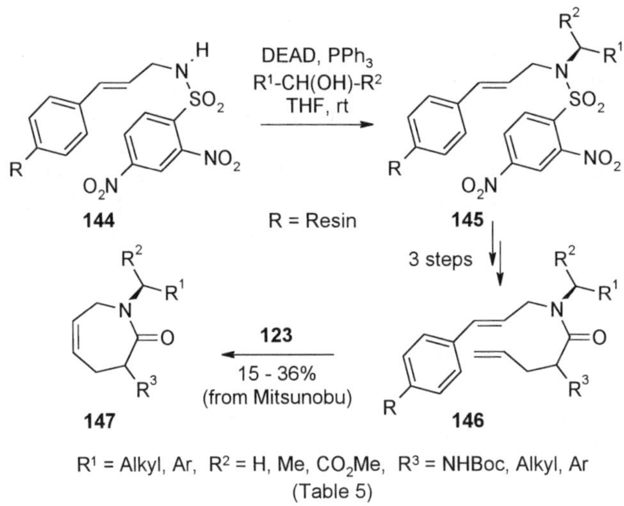

Scheme 27

Table 5. Ref. [35d]

Entry	R^1	R^2	R^3	Yield **147** (%)
1	Ph	H	NHBoc	36
2	3-CF_3-Ph	H	NHBoc	31
3	3-Cl-Ph	H	NHBoc	31
4	PMP	H	NHBoc	30
5	*n*-Pr	H	NHBoc	34
6	Me	Me	NHBoc	29
7	Ph	CO_2Me	NHBoc	23
8	Ph	H	(C_6H_{11})-CH_2CH_2	34
9	Ph	H	*i*-Pr	35
10	Ph	H	Ph	14

3
Cycloadditions

The use of cycloadditions for synthesizing medium-sized ring lactams is more or less restricted to the generation of seven-membered rings. The simplest method to generate azepinones seems to be the [6 + 1] reaction of a 1,6-dicarboxylic acid chloride **148/152** and a phosphinimine **149**, the in situ formed chloro enamine **150** underwent a Chapman rearrangement to give a cyclic imide **151/153** (Scheme 28)] [36].

The generation of dinucleophiles and dielectrophiles to induce cycloadditions seemed to be an attractive strategy to synthesize medium-sized rings. The 2-tosylmethyl-propenyl-1-amine **154** was used as an appropriate precursor of the bis-donor. After in situ generation of the corresponding dilithium derivative, treatment with a range of bis-acceptors led to the azepines and azocines, respectively, by a double S_N2 process. In contrast, the synthesis of an azepinone **156** needed a two-step sequence: After mono-deprotonation and a stereoselective Michael addition to methyl crotonate the corresponding *threo* adduct **155** was formed. A second lithiation induced the final cyclization to give the desired *trans*-ε-caprolactam **156** in 48% yield (Scheme 29)] [37].

A radical cyclization of a 2-chloroacyl enamine **157** was used to synthesize 2-substituted pyroglutamates **160**. Usually, the radical **158** undergoes an initial 5-*endo* cyclization (path a) and the resulting intermediate **159** attacked electrophiles like methyl acrylate to give the pyroglutamate **160**. Unexpectedly, the reaction with methyl methacrylate took another course and a seven-membered

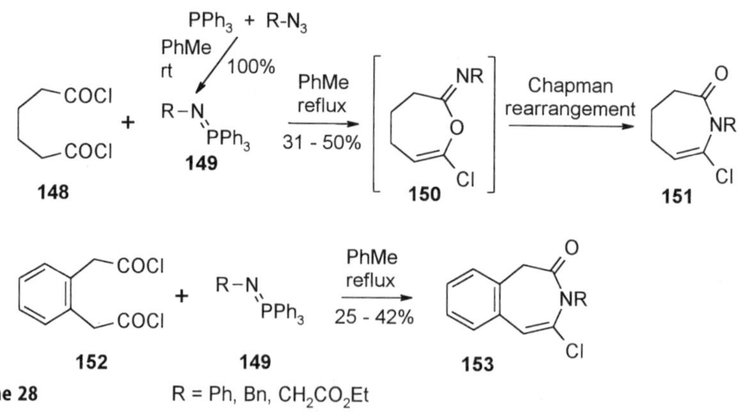

Scheme 28

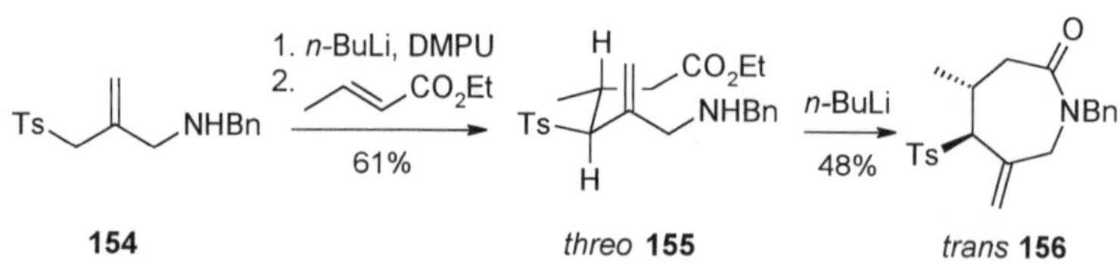

Scheme 29

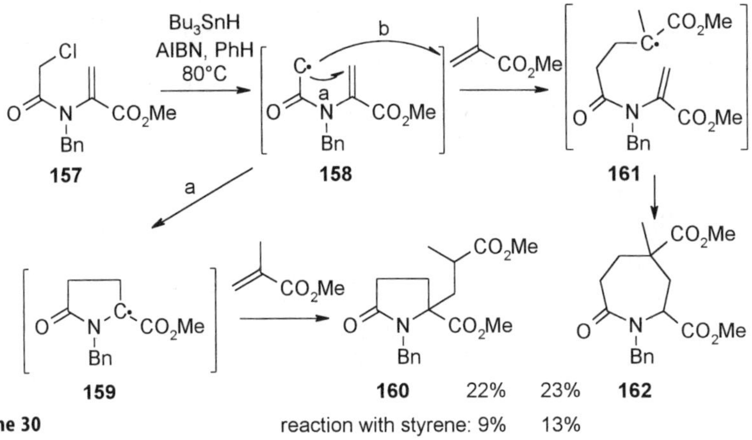

Scheme 30

ring **162** was formed in a competing [5 + 2] process. The reaction path (path b) could be attributed to an initial intermolecular addition of the unsaturated ester to give **161** and a final 7-*endo*-trig cyclization to deliver the azepinone **162** as a single diastereomer in 23% yield and the pyroglutamate **160** in 22% yield. The related addition of styrene gave diastereoselectively the phenyl analog of **162** in 13% yield (Scheme 30) [38].

[4 + 3] Cycloadditions were efficiently carried out by annulations of alkynyl and alkenyl Fischer carbene complexes to α,β-unsaturated imines (azadienes). The reaction mechanism involving the akynyl carbenes **164** was thought to be a two step [4 + 3] cycloaddition rather than a [2 + 1] cycloaddition and a consecutive [3,3] rearrangement. The electron deficient azadiene **163** started with a nucleophilic attack of the nitrogen on the carbene C to give **165** followed by a ring closure to yield the η^2-chromium complex **166,** which underwent an unprecedented 1,2-migration to generate the η^1-metal complex **167** as a stable intermediate (structure determined via x-ray analysis). Finally, removal of the chromium and trapping with protons or iodine gave the azepinones **168** in high yields (Table 6, Scheme 31) [39a].

Table 6. Ref. [39a]

Entry	R^1	R^2	R^3	T [°C] (time (h))	Yield **167** (%)	R^4	Yield **168** (%)
1	Ph	*n*-Pr	Ph	20 (3)	90	H	93
2	Ph	*n*-Pr	Ph	20 (3)	90	I	93
3	Ph	Bn	Ph	20 (3)	80	H	87
4	Ph	Bn	Ph	20 (3)	80	D	90
5	Ph	*i*-Pr	Ph	20 (3)	85	H	90
6	Me	*n*-Pr	TMS	20 (24)	60	H	85
7	Ph	*n*-Pr	TMS	20 (3)	–	H	64
8	Ph	*n*-Pr	*t*-Bu	50 (36)	–	H	57

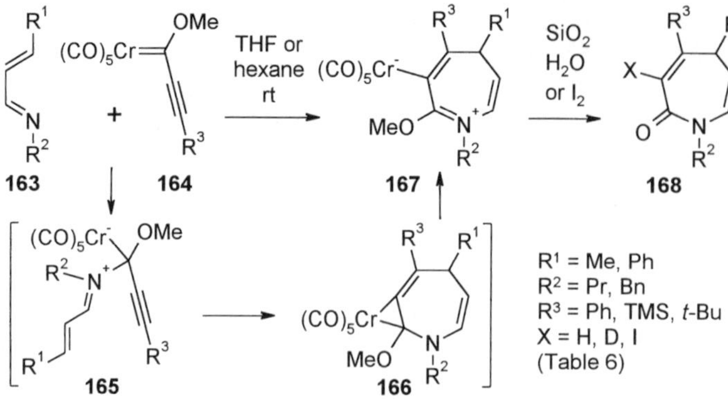

Scheme 31

Analogously, alkenyl Fischer carbene complexes **169** reacted with 1-azadienes *Z*-**170** to give unsaturated seven-membered lactim ethers **173**. In contrast to the cycloadditions with alkynyl carbenes, two stereogenic centers were generated regio- and stereoselectively. The reaction mechanism was explained to follow a two step [4 + 3] annulation: Initially, the imine nitrogen of **170** attacked the carbene **169** in the regiochemistry determining step to generate a hypothetical zwitterion **171**. After *Z*/*E* epimerization of the imine double bond, the olefins underwent cyclization to afford **172** (elimination of $Cr(CO)_5$) arranging the most bulky substituent in a quasi equatorial position of a crown-type transition state to generate exclusively the *trans* configured seven-membered lactim ether **173**. While the unsaturated imines *Z*-**170** (X = H) reacted at low temperatures, the corresponding oximes *E*-**170** (X = OH) needed refluxing THF to start the

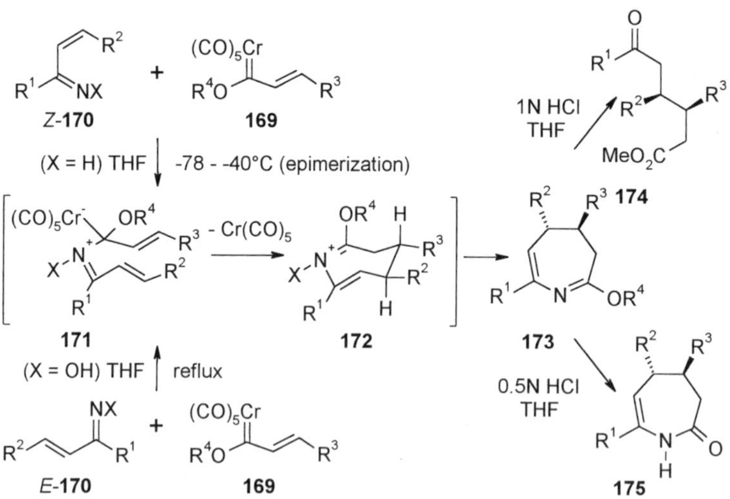

Scheme 32 R^1 = H, Alkyl, Aryl, R^2 = Alkyl, Aryl, NH*t*-Bu, R^3 = Ph, 2-(furyl), R^4 = Me, (+)-, (-)-menthyl (Table 7)

Table 7. Ref. [39b]

Entry	X	R^1	R^2	R^3	R^4	T [°C] time (h)	Yield **173**	Yield **174**	Yield **175**
1	H	*c*-Pr	NH-*t*-Bu	Ph	Me	40 (3)	90		
2	H	*c*-Pr	NH-*t*-Bu	2-furyl	Me	40 (3)	80	75[a]	
3	H	Et	NH-*t*-Bu	Ph	Me	40 (3)	91		
4	H	4-Tol	NH-*t*-Bu	2-furyl	Me	40 (3)	62		
5	H	Ph	NH-*t*-Bu	Ph	Me	40 (3)	70		
6	H	4-Cl-Ph	NH-*t*-Bu	Ph	Me	40 (3)	52		
7	OH	H	Me	Ph	Me	60 (20)	83	64	32
8	OH	H	Me	2-furyl	Me	60 (20)	85		
9	OH	H	Ph	Ph	Me	60 (44)	57		
10	OH	H	Ph	2-furyl	Me	60 (44)	57		
11	OH	H	(*E*)-1-propenyl	2-furyl	Me	60 (65)	52		
12	OH	H	Me	Ph	(–)-menthyl		87 (70:30)	63	34
13	OH	H	Me	2-furyl	(–)-menthyl		90 (72:28)	59	32
14	OH	H	Me	Ph	(+)-menthyl		80 (70:30)		

[a] Yield of the α,β unsaturated ketone after β elimination of *tert*-butylamine.

cyclization. Furthermore, at least two equivalents of the carbene **169** were required to complete the removal of the metal from **172** to form the lactim ether **173**. The cleavage in the presence of HCl gave the medium-sized ring lactams **175** and varying amounts of ring opened esters **174** (Scheme 32, Table 7) [39b].

The high simple diastereoselectivities observed running the [4+3] cycloadditions raised the question concerning the induction of chirality. Preliminary experiments involving chiral menthyloxy Fischer carbenes **169** (R^4 = (–)-menthyl) resulted in the formation of the diastereomeric lactim ethers **173–1** and **173–2** in a 7:3 ratio, which could be separated by means of a crystallization. A final acidic hydrolysis gave the enantiomerically pure ε-caprolactams **175*** and ent-**175*** and the acyclic esters, respectively. No signs of racemization have been detected, Eqs. (18, 19) [39b].

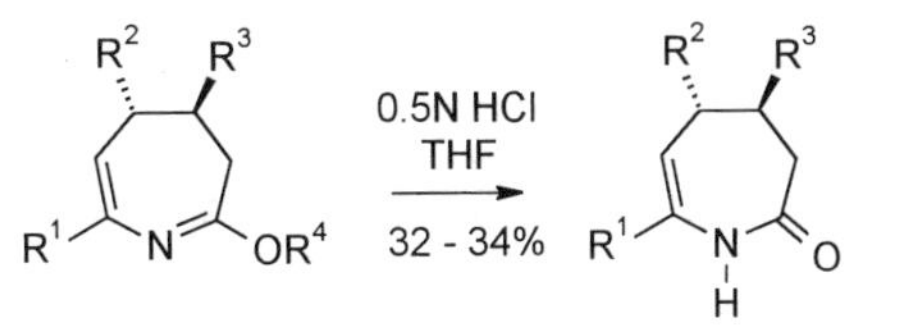

173-1* **175***

R^1 = H, R^2 = Me, R^3 = Ph, 2-(furyl), R^4 = (-)-menthyl

(18)

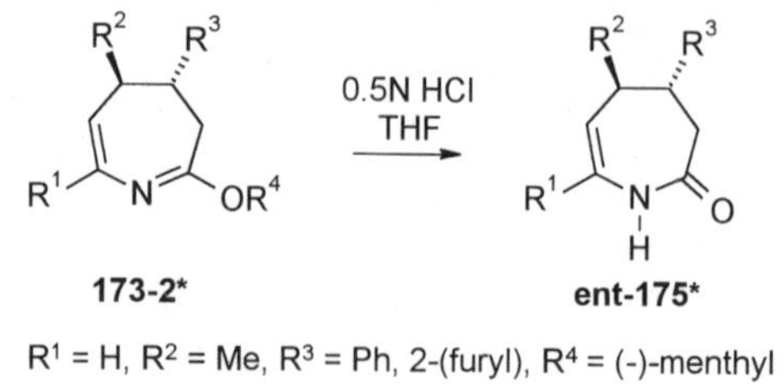

(19)

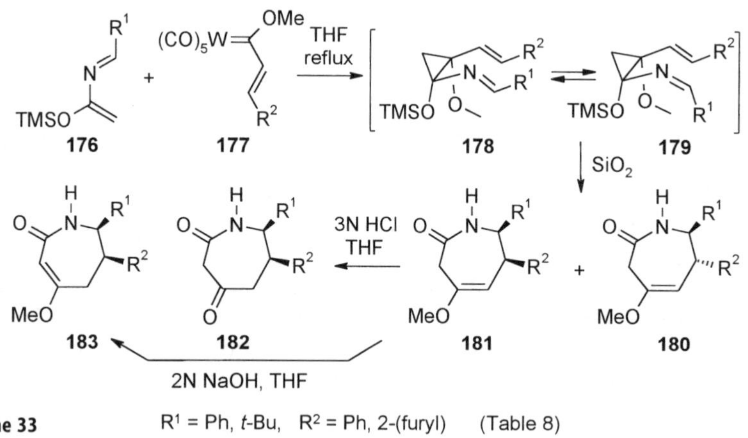

Scheme 33

Table 8. Ref. [39c]

Entry	R^1	R^2	Yield **181** (%)	Yield **180** (%)	Yield **183** (%)	Yield **182** (%)
1	Ph	Ph	42	38	-	-
2	*t*-Bu	Ph	94	-	95	94
3	*t*-Bu	2-furyl	88	-	93	96

4 Ring Expansion Reactions by N Insertion

The nitrogen insertion reactions use ketones as precursors to generate the medium-sized ring lactams. Due to the fact, that the ring system involved expands by only a single additional atom, the methods are restricted to more or less readily accessible cyclic ketones. Some difficulties in synthesizing medium-sized lactams are shifted to the question how to build up the appropriate medium-sized ketones. Consequentially, most medium-sized ring lactams synthesized via *N*-insertion were the seven-membered azepinones allowing one to start from easily accessible cyclohexanones.

En gros, the N insertion reactions can be subdivided into a Beckmann type and a Schmidt-type rearrangement part. Furthermore, some photochemical rearrangements of chiral oxaziridines are known to generate a range of optically active lactams.

4.1 Beckmann Rearrangement

Until now, classical Beckmann rearrangements found widespread use to generate non-*N*-alkylated ε-caprolactams starting from cyclohexanones. Generally, the C–C bond anti with respect to the oxime OH function migrated, but harsh conditions allow the oxime to epimerize in the course of the reaction. With the intention to investigate new inflammatory process inhibitors, a range of imino homopiperidinium salts **187** has been synthesized. The yields of **186** varied between 40 and 80% and the reactions were reported to be not regioselective because of an unselective formation of oxime **185**. Rearranging 2-substituted cyclohexanones **184**, the formation of 6-substituted ε-caprolactams **186** predominated (Scheme 34) [40].

The mostly used harsh reaction conditions for running the Beckmann rearrangement have been found to be incompatible with a range of functional groups. Thus, several efforts to develop milder variants led to the investigation of catalytic methods. Using sub-stoichiometric amounts of $SbCl_5$ and $AgSbF_6$ in refluxing acetonitrile, the rearrangement of the trimethylsilyl oxime **189** (R′ = TMS) gave the corresponding thirteen-membered ring lactam **190** (n = 7, entry 8, Table 9) in 77% yield, a catalytic cycle was postulated by the authors [41 a]. The reaction of cyclohexanone oximes **189** in refluxing nitromethane in the presence of a perrhenate and TfOH to give the ring expanded azepinones **190**

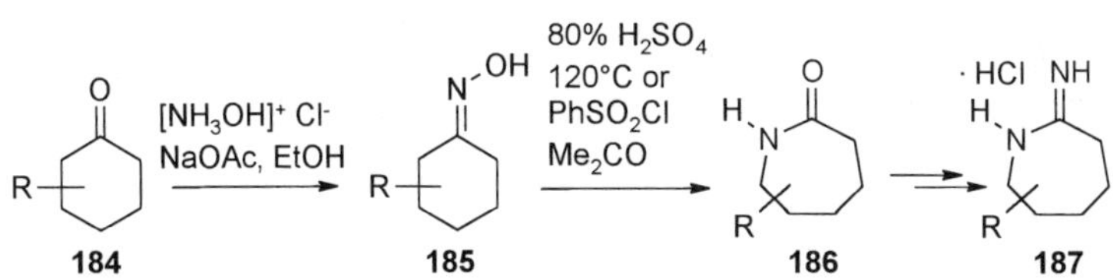

Scheme 34 R = Me, Et, *n*-Pr, 2-Ethylpropyl, *n*-Bu, Allyl, Ph

Table 9. Ref. [41]

Entry	n (R)	**189** R′	Method	Solvent	Time [min]	Yield **190** (%)	Ref.
1	1 (H)	H	0.2 eq [Bu_4NReO_4], 0.2 eq TFA	$MeNO_2$	60	85	b
2	1 (Ph)	H	0.2 eq [Bu_4NReO_4], 0.2 eq TFA	$MeNO_2$	60	91	b
3	1 (H)	SO_3H	SiO_2, Microwave	–	10	86	c
4	2 (H)	SO_3H	SiO_2, Microwave	–	20	72	c
5	3 (H)	SO_3H	SiO_2, Microwave	–	15	65	c
6	6 (H)	SO_3H	SiO_2, Microwave	–	15	72	c
7	7 (H)	SO_3H	SiO_2, Microwave	–	20	82	c
8	7 (H)	TMS	0.2 eq $SbCl_5$-$AgSbF_6$	MeCN	180	77	a
9	8 (H)	H	P_2O_5, MesOH (Eaton's reagent)	–	–	91	d

increased the yield from originally 42% to 85 to 91% (entries 1, 2, Table 9) [41 b]. Olah's one pot procedure [42] of oxime formation and consecutive rearrangement could be carried out as a microwave mediated solid phase process: ketone **188** and hydroxylamine-*O*-sulfonic acid were absorbed on silica gel. After microwave irradiation for an appropriate time, the corresponding lactams **190** were extracted with acetone in high yields (entries 3–7, Table 9) [41 c]. The Baldwin group started the total synthesis of the cytotoxic sponge alkaloid motuporamine B with a Beckmann rearrangement of **189** (n = 8), the use of Eaton's reagent led to the formation of the desired fourteen- membered lactam **190** in 91% yield (Table 9, entry 9) (Scheme 35) [41 d].

Most attention has been directed to the regiochemical outcome of the Beckmann rearrangements involving asymmetric ketones. As a general rule, the group that migrates is opposite to the leaving group on the oxime nitrogen. The ratio of the lactams can be predicted from the ratio of the starting *syn*- and *anti*-oximes. When α,β-unsaturated ketones are subjected to the rearrangement conditions, the migration of the alkyl group often predominated independently of the ratio of the oximes **191** and **193**. Such an observation presupposed that the isomerization of the oximes should have been faster than the rearrangement to give **192**. On suppressing any isomerization by conversion of the OH function into the corresponding sulfonate **194** and **195** (tosylate, mesylate), the isomer **195** with the *anti* arrangement of sulfonate and alkenyl group was cleaved to give the corresponding ketone, no **196** was formed in significant amounts. The *syn* epimer **194** underwent the rearrangement to give the lactam **192** [43 a, b].

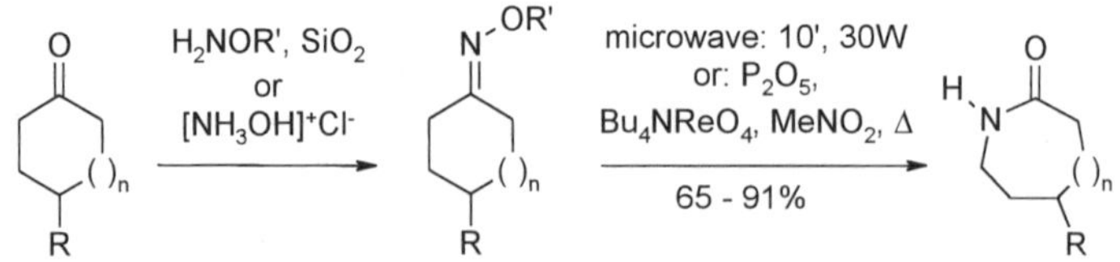

188 **189** **190**

Scheme 35 R = H, Ph R' = H, SO_3H, TMS n = 1, 2, 3, 6, 7, 8 (Table 9)

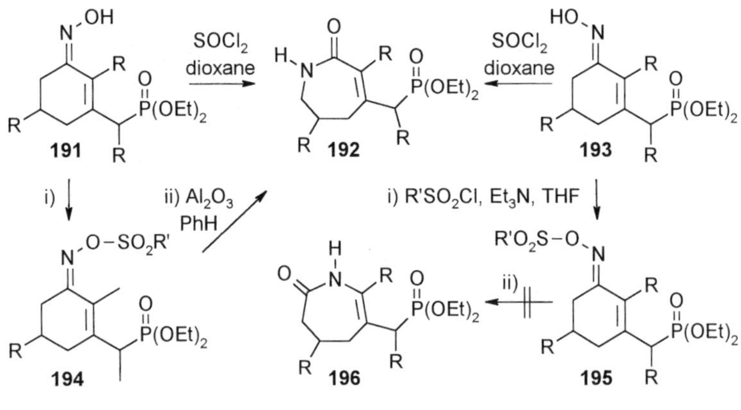

Scheme 36 R = H, Me, R' = Me, Tol yield **192**: 35 - 60%

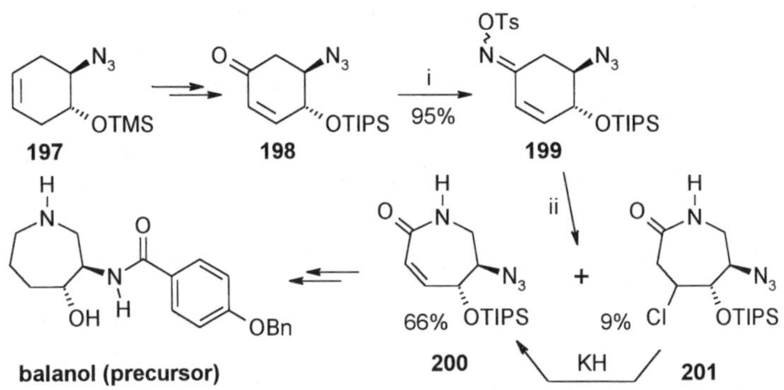

i) $[NH_3OH]^+Cl^-$, Py, CH_2Cl_2, then TsCl, Py, CH_2Cl_2 ii) $TiCl_3(O$*i*-Pr), CH_2Cl_2, rt

Scheme 37

Regioselective Beckmann rearrangements were used as key steps in the synthesis of phosphonoalkyl azepinones (Scheme 36) [43b] and in a formal total synthesis of the protein kinase C inhibitor balanol (Scheme 37): the optically active azide **197** derived from cyclohexadiene mono-oxide was converted into ketone **198** in several steps. After preparation of the oxime tosylates **199** (2.3:1 mixture), a Lewis acid mediated regioselective Beckmann rearrangement gave the lactams **200** and **201** in 66% and 9% yield, respectively. Lactam **201** underwent a β-elimination to give additional **200**, which served as a key intermediate in a balanol precursor synthesis (Scheme 37) [43c].

The regiochemical course reacting saturated ketones depended on the substitution pattern of the α-positions. In most cases, the intermediate oxime had an *anti* *N*-OH function with respect to the chain branched α-position. Consequently, the more substituted alkyl group preferentially migrates. This advantage was utilized for synthesizing the spiro α-amino-ε-caprolactam (**202** $\rightarrow$ **203**, Scheme 38) [12c], the Mexican bean beetle azamacrolide allomone (**205** $\rightarrow$ **206**, Scheme 39) [44a], in a key step of the chiral synthesis of benzomorphanes

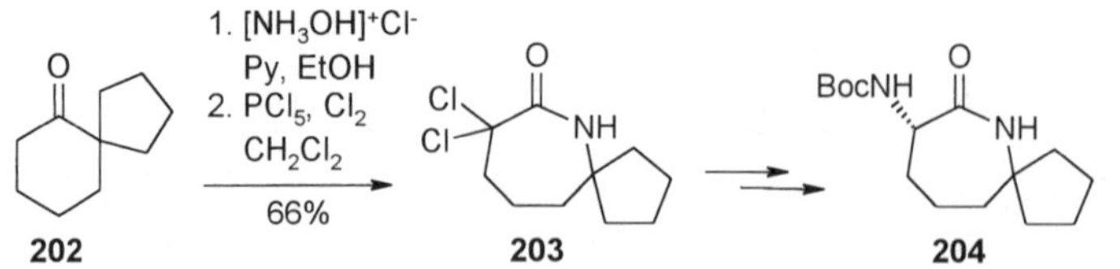

Scheme 38

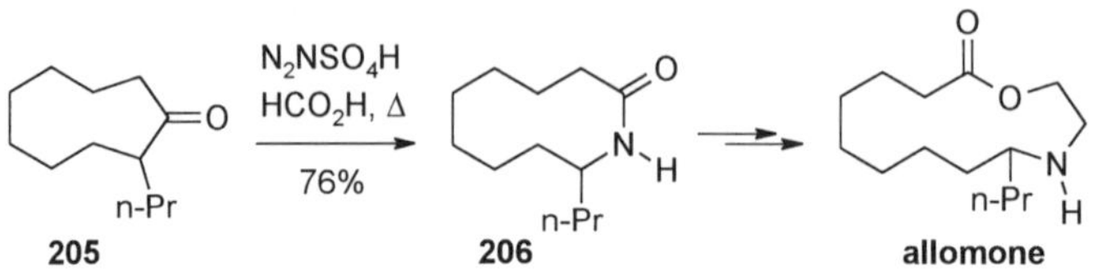

Scheme 39

(potential analgesic properties) (**207** → **209**, Scheme 40) [44b] and as initial ring enlargement (**210** → **212**) generating the optically active ε-caprolactam **212**, which had been converted into a bicyclic turn dipeptide employing a final ring closing olefin metathesis (Scheme 41) [44c].

The situation became less predictable when reacting an α,α′-disubstituted cyclooctanone **213**: The competition of phenyl and methyl substituents with respect to the space filling properties led to a 1:2 mixture of the corresponding oximes **214** and **215**. The Beckmann rearrangement of the oxime **215** gave the desired β-turn mimetic precursor **216**, but further transformations of the olefin of **216** with retention of the nine-membered ring structure failed because the highly efficient transannular participation of the nitrogen could not be avoided. In contrast, the ring expansion process succeeded again with significantly higher selectivity on reacting the cyclooctanone **217** (absence of the methyl group) via oxime **218**. The resulting azoninone **219** was characterized by a com-

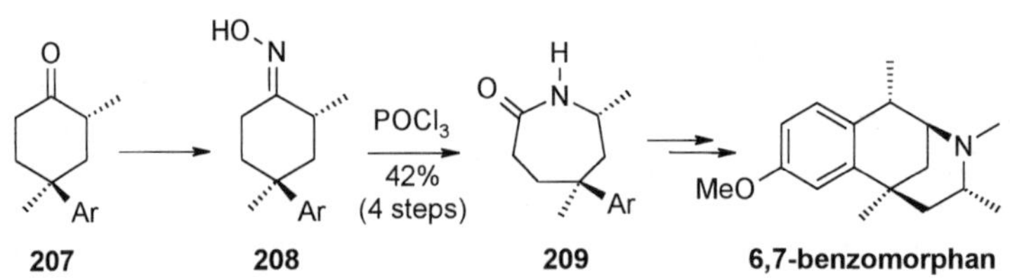

Scheme 40

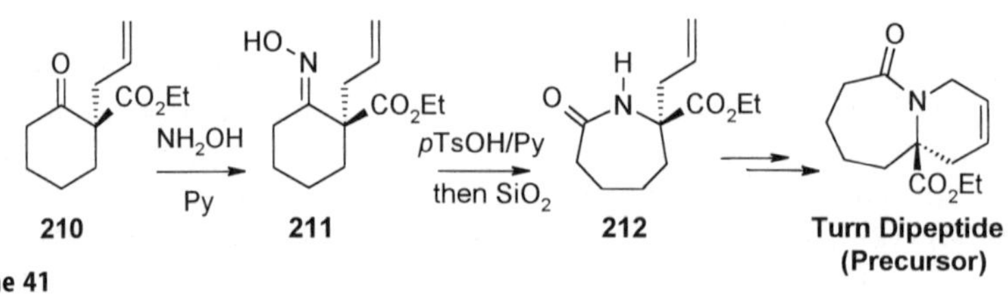

Scheme 41

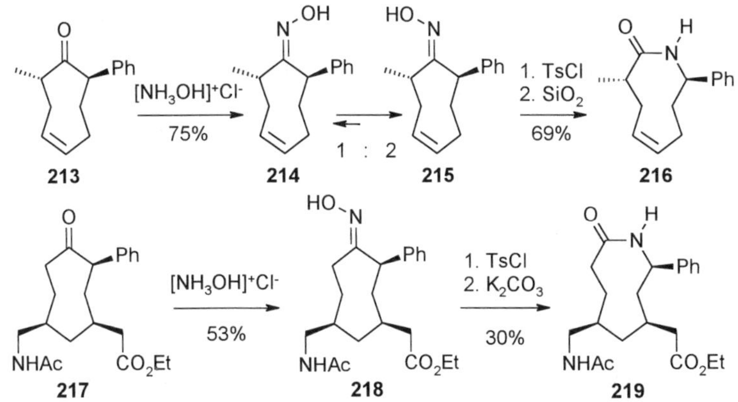

Scheme 42

plicated ^{1}H NMR spectrum at room temperature indicating the coexistence of several stable conformations (Scheme 42) [45a].

Some efforts to change the regiochemistry of the Beckmann rearrangement by inducing migration of the less alkylated position gave only disappointing results. In the process of generating chiral bases, the desired precursor lactam **221** was obtained as the minor compound (**221**:**222** = 1:2) from **220** in spite of extensive variations of the reaction conditions, Eq. (20) [45b].

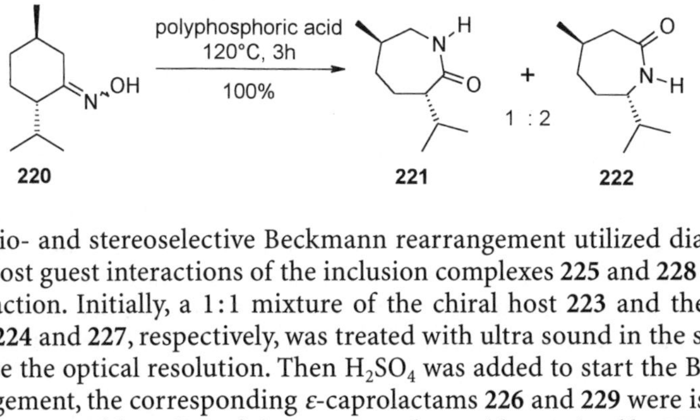

(20)

A regio- and stereoselective Beckmann rearrangement utilized diastereoselective host guest interactions of the inclusion complexes **225** and **228** in a solid state reaction. Initially, a 1:1 mixture of the chiral host **223** and the racemic oximes **224** and **227**, respectively, was treated with ultra sound in the solid state to induce the optical resolution. Then H_2SO_4 was added to start the Beckmann rearrangement, the corresponding ε-caprolactams **226** and **229** were isolated in 68% and 64% yields and ee of about 80% and 69% (determined by HPLC analysis on chiracel OC) (Scheme 43) [46].

4.2 Schmidt Rearrangement

In addition to the Beckmann reaction, the Schmidt rearrangement is used to generate *N*-alkylated lactams, too. Alkyl azides **231** react with the cyclic ketones (and aldehydes) in the presence of proton or Lewis acids. On running the intermolecular reactions, in most cases symmetric ketals **230** have been converted

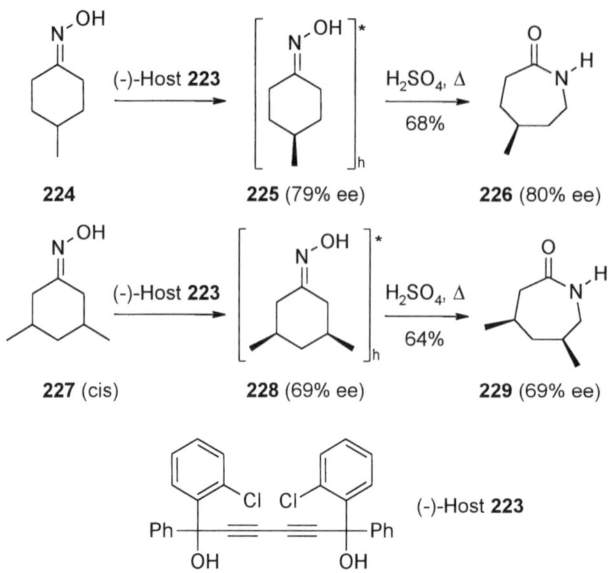

Scheme 43

into the corresponding lactams **232** [47a]. Alternatively, the azide could be replaced by an alkylated hydroxylamine *O*-sulfonate **231** (X = OSO_3H). A one pot procedure was developed to convert cyclic ketones into the corresponding *N*-methyl lactams **232**. The reaction path was described as a sequence of ketalization, oxonium ion formation (**A**) and an addition of the activated hydroxylamine to generate aminal (**B**) followed by the rearrangement to give the imidate ion (**C**). A final dealkylation with sodium iodide yielded the desired lactam **232** [47b, c]. Silylenol ethers were used as starting materials to synthesize non alkylated lactams **232** (Beckmann products, R^3 = H) by means of a Schmidt type process. Initially, a regioselective acid catalyzed addition of trimethylsilyl azide **232** (R^3 = TMS) to the electron rich double bond (enol of **A**) gave an azidohydrin **B**. Then, a photolysis induced the rearrangement to give the corresponding lactams in high yields (75 to 89%) (Scheme 44) [47d].

Non-symmetrical ketones suffer from the fact that the regiochemistry cannot be predicted. Since both adjacent C–C bonds migrate, the yield is decreased,

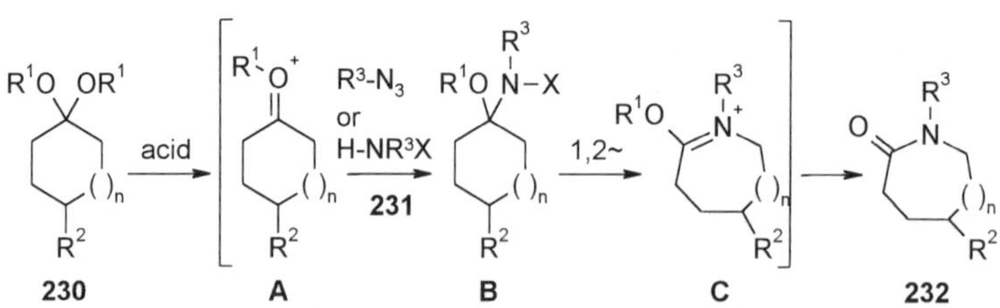

R^1 = H, Me, Et, TMS, TIPS R^2 = H, *n*-alkyl, Bn R^3 = Me, OMe, TMS, 2e⁻
X = ONs, OMe, N_2^+ n = 1, 2, 3, 5, 6

Scheme 44

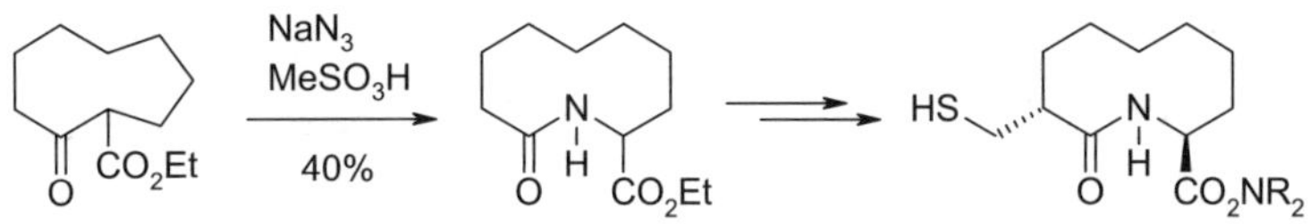

Scheme 45

when only one regiomer is required. Although the reaction was involved as a key step in a neutral endopeptidase inhibitor synthesis (**233** → **234**, Scheme 45), the yield was reported to be only 40% [48a, b]. The Aubé group investigated the Schmidt reaction of a range of ketones such as **235** in the presence of $TiCl_4$. The yields achieved varied between 20% and 95% of **236** and **237** depending on the substitution pattern of the reactant, and the regioselectivities reported were poor, Eq. (21) [48c]. Cyclohexanone **238** was converted into the corresponding azepinone **239**, but the yield and the regioselectivity have not been published, (Eq. 22) [48d].

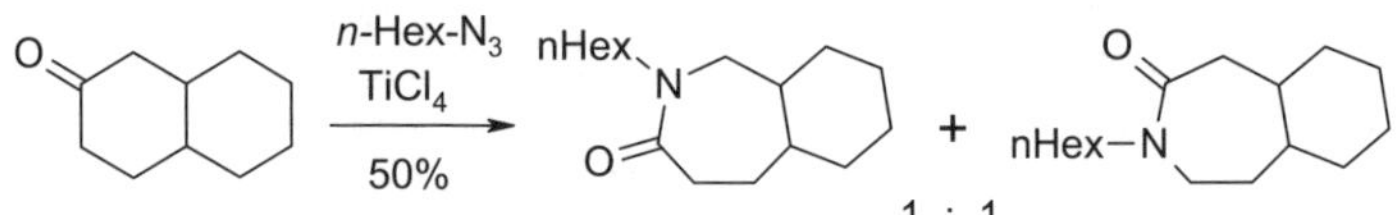

(21)

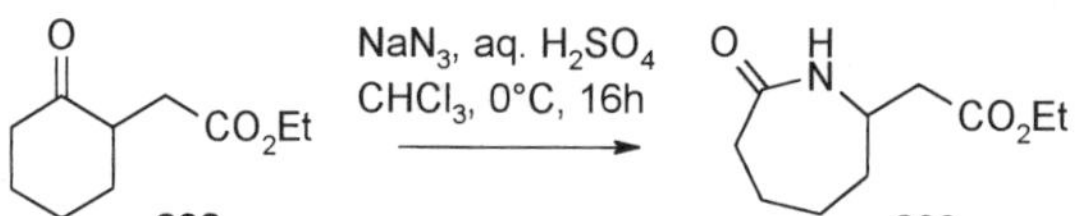

(22)

Considering the fact, that *N*-alkylated lactams were synthesized via the Schmidt rearrangement, 2- or 3-hydroxyalkyl azides **240** were found to be superior reaction partners for ketones **241**. Now, the primary reaction products were not the lactams **243** but the corresponding iminium ethers **242**, which could be transformed into the azepinones **243** by addition of nucleophiles in moderate to high yields. In most attempts the nucleophile attacked the position α to the oxygen of **242**, a S_N-type reaction led to the generation of the monocyclic lactams **243**. Only a minority of nucleophiles added to the activated carboxylic function to maintain the bicyclic system [49a]. The efficiency of the process strongly depended on the distance between azide and OH function: A smooth reaction to give the lactams **243** was observed, if the intermediate bicyclic system **242** bears a combination of a seven- and a six- or a five-membered ring, respectively, larger bicyclic iminium ether arrays severely decreased the yields of the lactams **243** [49b]. Furthermore, the formation of 2- and 3-hydroxyalkyl lactams suffered from the presence of some side products such as **244**, the ratio was found to be controlled by the pH of the medium used for the hydrolysis of the imidate salt and the ring size. Strong basic conditions predominantly led to the desired medium-sized rings **243**, milder bases favored the formation of the lactone **244**. Increasing the ring size of ketone **241**, the formation of lactone **244** became more

Table 10. [49]

Entry	m	n	R^1	R^2	Yield [%] ($NaHCO_3$)		Yield [%] (KOH)		Ref.
					243	**244**	**243**	**244**	
1	1	0	H	H	97	0	98	0	a/b/d
2	1	1	H	H	98	0	95	0	a/b/d
3	1	2	H	H	34	–	–	–	b
4	1	3	H	H	33	–	–	–	b
5	1	1	H	Me	83	–	–	–	b
6	1	1	H	*t*-Bu	88	–	–	–	b
7	1	1	H	Ph	82	–	–	–	b
8	1	1	H	X[a]	79	–	–	–	b
9	2	0	H	H	27	43	68	0	c/d
10	2	1	H	H	21	70	99	0	c/d
11	1	1	Br	H	0	67	43[b]	0	c
12	3	0	H	H			29	0	d
13	3	1	H	H			3	28	d
14	4	0	H	H	26	64	88	0	d
15	4	1	H	H	46	49	23	40	d
16	5	0	H	H	30	64	51	19	d
17	5	1	H	H	29	67	25	54	d
18	7	0	H	H	8	79	50	30	d
19	7	1	H	H	5	93	36	45	d

[a] 1,2-ethylidenedioxy.
[b] +9% α,β-unsaturated Lactam (after elimination of HBr).

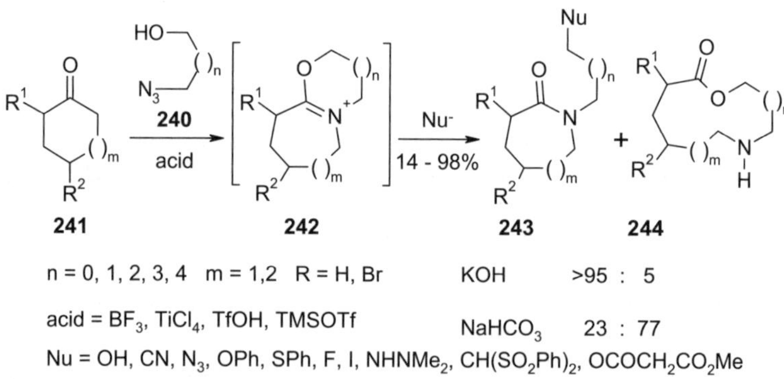

Scheme 46

and more important because of the tendency to generate less constrained larger rings. Detailed information is given in Table 10 (Scheme 46) [49c, d].

In contrast to the intermolecular Schmidt rearrangement, the intramolecular variant did not suffer from the regiochemical difficulties [50]. Only a single regioisomeric lactam **246** was obtained from ketone **245**, respectively, the one bearing the fused ring junction. No bridged lactams were found, the regioselective migration of the on-tether substituent was rationalized by an antiperiplanar

Table 11. [50]

Entry	Reactant	n	m	R^1	R^2	R^3	Acid	Product	Yield [%], (ee [%])
1	**245**	1	0	H	H	H	TFA	**246**	0
2	**245**	1	1	H	H	H	TFA	**246**	85
	245 (dimethylketal)	1	1	H	H	H	TFA	**246**	85
3	**245**	1	1	CO_2Et	H	H	TFA	**246**	93
4	**245**	1	1	H	Me	H	TFA	**246**	74
5	**245**	1	2	H	H	H	TFA	**246**	0
6	**245**	1	2	H	H	H	BF_3	**246**	29
7	**245**	1	2	H	H	H	$TiCl_4$	**246**	91
8	**245**	1	3	H	H	H	$TiCl_4$	**246**	0
9	**245**	2	1	H	H	H	TFA	**246**	80
10	**245**	2	2	H	H	H	$TiCl_4$	**246**	0
11	**245**	3	1	H	H	H	TFA	**246**	96
12	**245**	7	1	H	H	H	TFA	**246**	89
13	**245**	1	1	Me	H	H	TFA	**246**	87 (89)
14	**245**	1	1	SPh	H	H	TfOH	**246**	42
	245 (ethylideneketal)	1	1	SPh	H	H	TfOH	**246**	94
15	**245**	1	1	H (α)	H	*t*-Bu	TFA	**246**	96
16	**245**	1	1	H (β)	H	*t*-Bu	$TiCl_4$	**246**	92
17	**253**	1	1	H	H	H	TfOH	**254**	0
18	**247**	1	1				TfOH	**248**	82
19	**249**	1	1				TfOH	**248**	95
20	**253**	1	1	H	H	H	TFA	**254**	0
21	**253**	1	1	H	H	H	TMSOTf	**254**	0
22	**250**	1	1				TfOH	**251**	45
23	**252**	1	1				TFA	**251**	89
24	**255**	1	1				TFA	**256**	85

arrangement of the involved orbitals during nitrogen loss passing through a chair-like transition state. Investigations of the role of the acid showed that mostly the TFA or the $TiCl_4$ mediated reactions gave the best results, up to 95% yield was achieved. The length of the tether between ketone and azide was found to be crucial for the reaction, the distance of four C atoms ($n = 1$) represented the optimum, Eq. (23), (Table 11, entries 2, 3, 4, 7, 13). First experiments to test the stereochemical outcome of the intramolecular Schmidt reaction revealed retention of the configuration on rearrangement of an optically active ketone (Table 11, entry 13). Furthermore, enolizable ketones (Table 11, entries 15, 16) were converted into the corresponding azepinones. No crossover products between the two possible reaction products were observed, thus proving a completely diastereoselective rearrangement. It should be pointed out, that the conversion of an enolizable cyclohexanedione failed, the related alkylated analog **247** gave a smooth reaction to generate the lactam **248** (Scheme 47) [Table 11, entries 18, 19]. Investigating the scope of the intramolecular Schmidt rearrangement, some unsaturated ketones were subjected to the optimized reaction conditions. While the less reactive aryl ketone **250** were able to be reacted under forcing conditions to **251** (Scheme 27), (Table 11, entry 22), the rearrangement

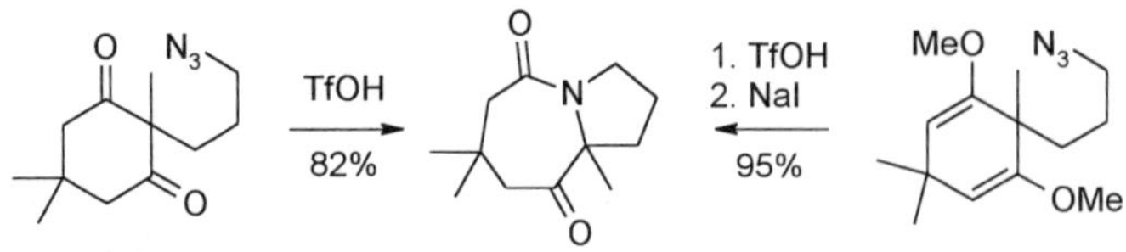

Scheme 47 **247** **248** **249**

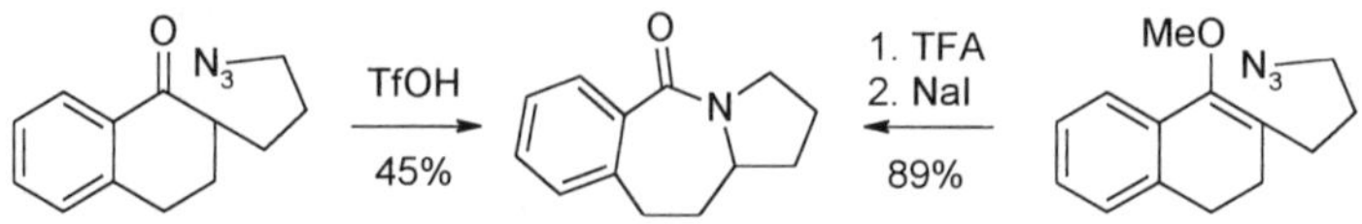

Scheme 48 **250** **251** **252**

of cyclohexenones **253** to **254** failed, Eq. (24), (Table 11, entries 17, 20, 21) [50a]. A stereoselective Schmidt rearrangement of **255** was used to generate a tetracyclic Erythrina alkaloid derivative **256,** Eq. (25), (Table 11, entry 24) [50b].

Several experiments were repeated starting from the corresponding dimethyl or ethylidene ketals as well as the corresponding enol ethers (Table 11, entries 2, 14, 19, 23), some examples were outlined in Eq. (23) (**245** → **246**) and the Schemes 47, 48, (**249** → **248** and **252** → **251**). In some cases, the yields were reported to increase significantly compared to the previously used ketones (Table 11, entries 14, 19, 23) [50c].

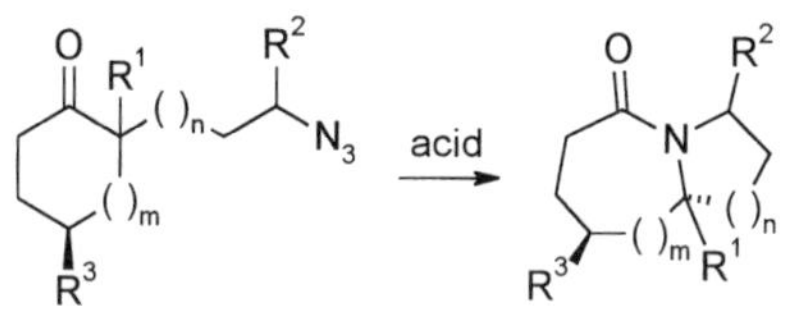

(23)

245 **246**

R^1 = H, Me, SPh, CO_2Et, R^2 = H, Me, R^3 = H, *t*-Bu
n = 1, 2, 3, 7 m = 0, 1, 2, 3 (Table 11)

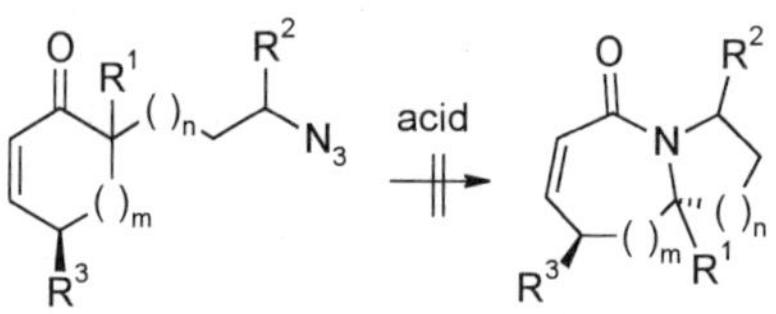

(24)

253 **254**

R^1, R^2, R^3 = H; n, m = 1 (Table 11)

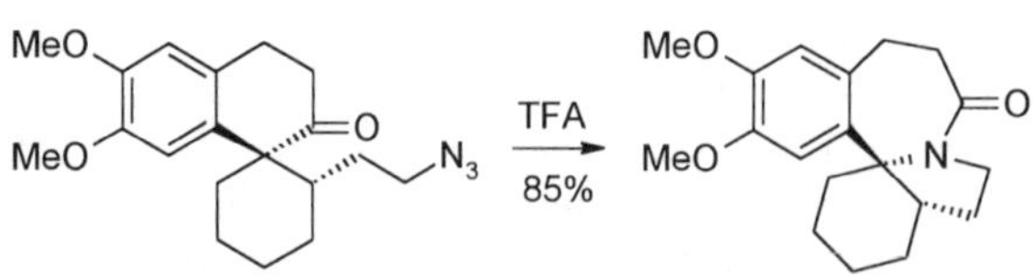

(25)

255 **256**

4.3 Oxaziridine Rearrangement

With the intention to investigate an asymmetrical nitrogen insertion, a new process to convert prochiral ketones into the corresponding ring-expanded optically active medium-sized ring lactams was developed [51]. Firstly, the ketones **257** were treated with chiral *R*- or *S*-phenylethylamine **258** to give the corresponding imines **259**. A consecutive oxidation either with *m*-CPBA or with optically active (+)-MPBA gave the stereochemically enriched axially dissymmetric spiro oxaziridines **260** (**A-D**) in 70 to 97% yields. Then, a stereoelectronically controlled photochemical rearrangement was carried out as the key step to generate the corresponding ring-expanded lactams **261** and **262** (Scheme 49).

For the synthesis of the spiro-oxaziridines, the reaction mechanism had be explained either as a concerted one step oxidation process or a fast two step sequence via a consecutive *O*–N and a *O*–C bond formation without molecular relaxations during the course of the ring formation. The combination of two senses of stereochemical control had to be assessed to explain the axially disymmetrical outcome of the oxidations (Fig. 2). On one hand the equatorial attack of a bulky oxidizing reagent was preferred (intra-annular stereocontrol: path a or c, Fig. 2), on the other hand, a minimized allylic 1,3-strain should place the benzylic hydrogen in the plane of the C=N double bond. Then, the oxidizing agent should attack *anti* with respect to the bulky phenyl substituent (extra-annular stereocontrol: path **a** or **b**, Fig. 2). Consequentially, path **a** should predominate, paths **b** and **c** were partially disfavored and path **d** represented the least favored attack of the oxygen. **A/D** and **B/C** bear the same sense of axial chirality. A double diastereoselection should have been operative using a combination of chiral amine **231** and chiral oxidizing agent but surprisingly, the effects observed were apparently non-significant on varying the configuration of the amine (entries 2 and 3, Table 11). Obviously, the chiral interactions of the incoming reagent and the cyclohexane predominated, the influence of the optically active phenylethyl side chain was found to be more or less neglictable. However, the predominant formation of one spiro oxaziridine bearing a defined stereogenic nitrogen center could be achieved reacting a range of ketones using a single readily available oxidizing agent. The heating in toluene effected an epimerization of the nitrogen center to give a 1:1 mixture of the spiro oxaziridines **A/C** and **B/D**. If necessary, the diastereomers were separated by means of column chromatography (Fig. 2).

The photolysis (UV 254 nm) of the chiral oxaziridines **260** resulted in the formation of chiral medium-sized ring lactams **261** and **262**. Because of the fact that the carbon substituent *anti* to the lone pair on the oxaziridine nitrogen preferentially migrated to a formally electrondeficient nitrogen atom, the axially chiral information of the system **260** can be transferred nearly completely to the formation of the optically active lactams **261** and **262**. The level of stereoselectivity was found to be high for either process, provided, that the purity of the reactant oxaziridine **260** was adequate. The limiting selectivity appeared to be ≥76% as found when investigating the rearrangements of several isolated and purified oxaziridines **260**: Pure **A** and **D** (R^1 = Me) gave lactam **261** in 86 and 92% de, **A**

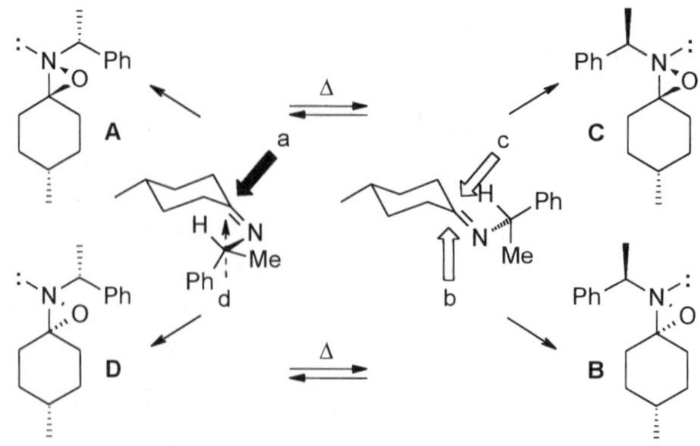

Fig. 2

(R^1 = Ph) gave **261** in 80% de. On reacting **B** or **C** (R^1 = Me), lactam **262** was formed in 76 and 90% de, **B** (R^1 = Ph) gave **262** in 80% de, respectively. Detailed information is summarized in Table 12 (entries 1–19) [51a, c].

Following the same sequence of imine formation, oxidation and photolytic rearrangement as described above, the reaction of unsymmetrical ketones resulted in regioselective processes. The choice of the absolute configuration of the phenylethylamine **258** involved allowed to predict the predominantly formed ring-expanded lactam **261** or **262** (Table 12, entries 20–22). In analogy, the (3*R*)-cyclohexanone (Table 12, entries 21, 22) was rearranged using D or L phenylalanine methyl ester as the chiral auxiliary. The yields and diastereoselectivities obtained were comparable with those of the phenylethylamine series [51d].

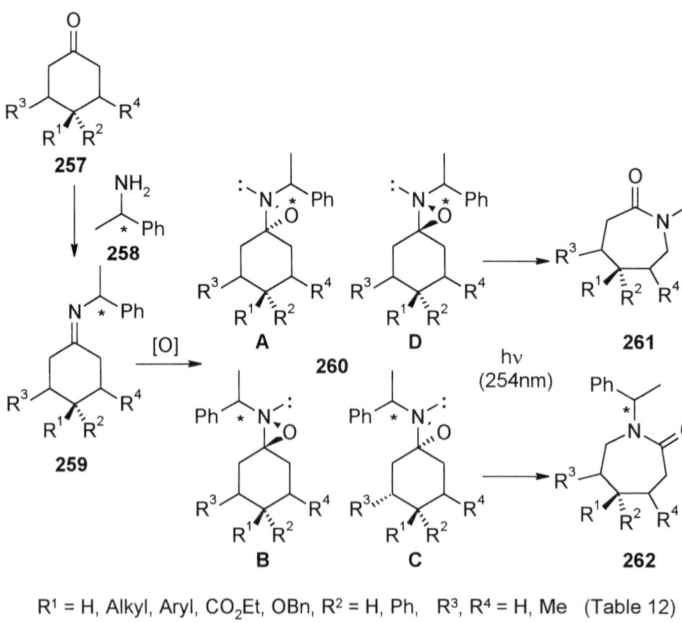

Scheme 49

Table 12. Ref. [51]

Entry	R^1	R^2	R^3	R^4	**258**	Method[a]	**260**: ratio (A+D):(B+C)	**261/262** Yield (%)	**261/262** Ratio	Ref.
1	Me	H	H	H	*R/S*	A	67:33	77	74:26	a
2	Me	H	H	H	*R*	B	84:16	61	84:16	a
3	Me	H	H	H	*S*	B	86:14	–	–	a
4	Et	H	H	H	*R/S*	A	66:34	76	79:21	a
5	Et	H	H	H	*R*	B	84:16	71	84:16	a
6	*t*-Bu	H	H	H	*R/S*	A	77:23	69	80:20	a
7	*t*-Bu	H	H	H	*R*	B	86:14	70	88:12	a
8	Ph	H	H	H	*R/S*	A	75:25	55	80:20	a
9	Ph	H	H	H	*R*	B	85:15	87	88:12	a
10	OBn	H	H	H	*R/S*	A	64:36	74	59:41	a
11	OBn	H	H	H	*R*	B	68:32	45	59:41	a
12	EtO_2C	H	H	H	*R/S*	A	60:40	67	54:46	a
13	EtO_2C	H	H	H	*R*	B	65:35	80	63:37	a
14	Me	Ph	H	H	*R/S*	A	59:41	79	62:38	a
15	Me	Ph	H	H	*R*	B	74:26	89	70:30	a
16	H	H	α-Me	α-Me	*R/S*	A	85:15	78	84:16	a
17	H	H	α-Me	α-Me	*R*	B	94:6	45	84:16	a
18	H	H	α-Me	α-Me	*R*	A	>80 **C**	54	0:100	c
19	H	H	α-Me	α-Me	*S*	A	>80 **A**	53	100:0	c
20	H	H	β-Me	α-Me	*R*	A	1	28	–	c
			α-Me	β-Me			1	32	–	c
21	H	H	β-Me	H	*R*	A	>90 **D** (1:<10)	80	89:11	b
22	H	H	β-Me	H	*S*	A	90 **B** (4:6)	78	15:85	b

[a] Methods: A: oxidation with *m*-CPBA, B: oxidation with (+)-MPCA.

Finally, the chiral auxiliary was removed by a Birch reduction or a catalytic hydrogenation. After ring opening several optically active 6-aminohexanoic acids served as linkers in cyclic peptides as β-turn mimetics (Table 12, Scheme 49) [51c].

In contrast, ketones and the corresponding imines **263** bearing α-substituents behaved somewhat differently. On the one hand, the C(R^1,R^2) group mostly served as the key stereocontrol element for the oxaziridine **264** synthesis determining the configuration of the nitrogen: The bulky phenylethyl substituent in **A** was arranged *anti* with respect to the branched α-position causing the predominant formation of lactam **266** after the rearrangement step. Lactams **267** resulting from the conversion of the corresponding oxaziridines **B** were isolated without exception as minor compounds (Table 13, entries 3–5). On the other hand, the stereogenic centers of the imines **263** might undergo epimerization

Table 13. Ref. [51b]

Entry	Imine **263**			Method[a]	Oxaziridines **264**			**265** Yield (%)	Lactams		
	R^1	R^2	ratio		Yield (%)	Ratio[b]			Yield (%)	Ratio	
						A	B			**266**	**267**
1	Me	H	1	A	81	38	–	–	62	1	–
	H	Me	1			46	–		57	1	–
2	Me	H	1	A	81	37	–	–	62	1	–
	H	Me	0			48	–		57	1	–
3	OMe	H	1	B	95		38	–	60	–	1
	H	OMe	1			51			50	64	32
4	Ph	H	1	B	56	16	25	24	–	–	–
	H	Ph	1			53			64	92	8
5	Ph	H	1	B	54	14	25	23	–	–	–
	H	Ph	0			60			64	92	8

[a] Methods: A: oxidation with *m*-CPBA, B: oxidation with (+)-MPCA.

[b] Only the major diastereomers are given. Additionally, the formation of several minor oxaziridines occurred in all attempts.

during the course of the reaction. Such equilibration could be effected intentionally by heating the imines **263** or the oxaziridines **264**. The α-hydroxylated spiro oxaziridines of type **265** were found as side products. Obviously, any intermediate generation of an enamine type structure caused some oxidation of the electron rich double bond. The ease of the enolization (acidity of the α position) directly influenced the product ratio: i.e. the β-iminoester **263** ($R^1 = CO_2Et$, Scheme 50) had completely been converted into the oxaziridine of type **265**. Some results were summarized in Table 13 (Scheme 50) [51b].

The exchange of the chiral phenylethyl amine against an optically active amino acid fragment **269** allowed the synthesis of conformationally restrained dipeptidyl lactams **271** and **272** including the so called Freidinger lactams as

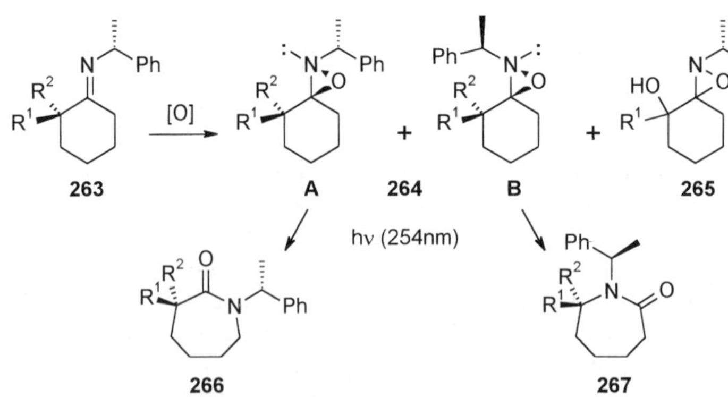

Scheme 50 R^1, R^2 = H, Me, OMe, Ph, (Table 13)

building blocks for peptidomimetics. Protected α-amino cyclohexanones **268** and the related cycloheptanone analogs were converted into the imines using appropriate amino acid esters **269.** An oxidation with *m*–CPBA gave the oxaziridines **270** in moderate to high diastereoselectivities, a complete conversion into **A** (R^1 = Bn, Table 14, entry 1) was observed using mild reaction conditions and low temperatures. The selectivity was explained by a matched combination of the stereogenic centers during the course of the oxidation step (in analogy to the reasons outlined above, Fig. 2). In contrast, the more or less complete racemization of the original stereogenic center of the ketone to generate **A** and **C** (or **B** and **D**, respectively) could be achieved, if harsh reaction conditions were used (entries 4–7, Table 14). The same result was obtained by heating the imines or the oxaziridines in toluene. Then, the diastereomers formed had to be separated. The epimerization was found to be somewhat faster when reacting the D configured esters because of the mismatched combination of the stereogenic properties (entries 3 and 5, Table 14). The photolytic rearrangement led to the corresponding ring-expanded lactams **271** and **272** in up to 72% yield. Furthermore, an azocinone (one additional ring atom in **271**) was synthesized with the aid of *S*-phenylalanine in about 48% yield. The usability of the 2-aminoazepinones was illustrated by the subsequent synthesis of potential angiotensin converting enzyme inhibitors (ACE-inhibitors) (Table 14, Scheme 51) [51d].

A regio-and stereoselective rearrangement of *N*-phenylspirooxaziridines **274** was reported by Suda [51e]. A Mn(III) tetraphenylporphyrin complex [Mn(tpp)Cl] served as activating reagent. Six- to eight-membered ketones **273** (n = 1, 2, 3) were converted into the corresponding seven- to nine-membered lactams **275** and

Table 14. Ref. [51d]

Entry	Ester **269**		Oxaziridines **270**			Lactams	
	R^1	R^2	**270**	Ratio	Yield (%)	No.	Yield (%)
1	Bn	H	A	>95:5	72	**271**	72
2	Bn	H	A	1	72	**271**	72
			C	1		**272**	63
3	H	Bn	B	93	61	**271**	68
			D	7		–	–
4	*i*-Bu	H	A	77	62	**271**	61
			C	23		–	–
5	H	*i*-Bu	A	50	60	**271**	59
			B	50			
6	*i*-Pr	H	A	48	66	**271**	59
			B	13			
			C	39		**272**	55
7	CH_2CO_2Me	H	A	57	70	**271**	44
			A+B/C	43		**271**	35
						272	17

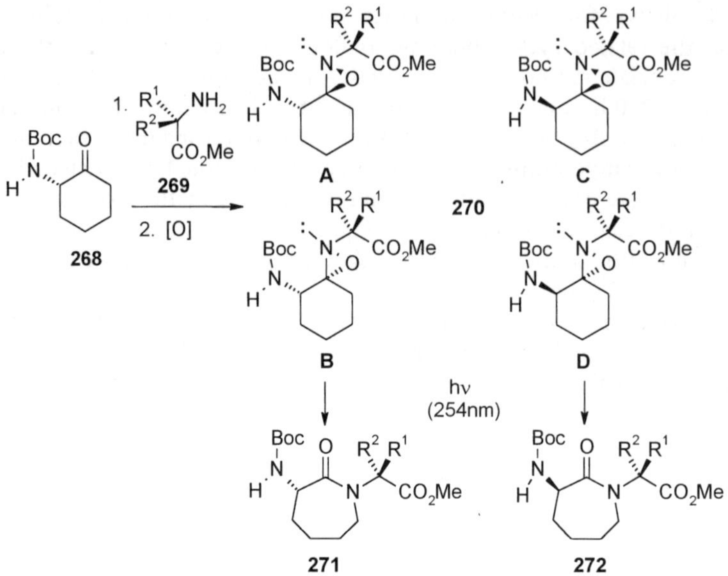

Scheme 51 R^1 = H, *i*-Pr, *i*-Bu, Bn, CH_2CO_2Me , R^2 = H, *i*-Bu, Bn (Table 14)

276, respectively. Furthermore, unsymmetrically substituted spirooxaziridines **274 A** and **B** (R^1 or $R^2 \neq 1$ H) gave the azepinones **275** and **276** with high diastereoselectivities starting from defined reactants **274**: Again, the C-C bond anti with respect to the nitrogen lone pair tended to migrate (**A** → **275**, **B** → **276**) (Scheme 52).

Finally, an acid catalyzed rearrangement of an *o*-(2-arylphenyl)-hydroxylamine followed by a ring enlargement to yield some unsaturated 7-arylazepinones should be mentioned (not shown). Until now, scope and limitations of that process are not known [52].

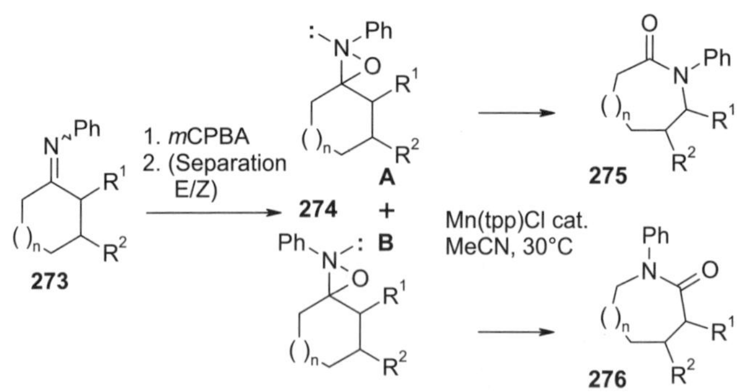

Scheme 52

R^1, R^2 = H: n = 1: 97%, n = 2: 72%, n = 3: 90% (**275** = **276**)
274 A, R^1 = Me, R^2 = H, n = 1: **275/276** = 98/2 (98%)
274 A, R^1 = H, R^2 = Me, n = 1: **275/276** = 94:6 (96%)
274 B, R^1 = Me, R^2 = H, n = 1: **275/276** = 2:98 (94%)

5 Ring Expansion Reactions by C Insertion

In contrast to the nitrogen insertion reactions allowing an expansion to lactams by only a single additional ring atom, the C insertions are characterized by a higher flexibility. Sigmatropic rearrangements use N, O and S heterocycles as precursors to generate the medium-sized ring lactams. Due to the fact, that the ring systems involved are expanded by three or four additional C-atoms, starting from more or less easily accessible three to eight membered heterocyclic reactants hold the prospect of an efficient synthesis of medium-sized ring lactams with up to eleven ring atoms. Furthermore, the well known stereochemical characteristics of the sigmatropic processes, due to the highly ordered transition states, promise the transfer of chiral information to the ring-expanded lactams. Until now, by far the majority of ring enlargements reported used a Claisen rearrangement as the key step, the related (hetero-) Cope reactions were only found in a few attempts to generate medium-sized lactams.

En gros, the C insertion reactions can be classified as C3 type and as C4 type rearrangements. Furthermore, a two step C4-type process involving chromium carbene addition and a CO insertion has been reported.

5.1 C3-Insertion

C3 expansions used cyclic ketene aminals **279** as key intermediates to generate seven to ten-membered rings [53]. Racemic and optically active 1,n-amino acids (n = 2, 3, 4) were used as starting materials by Holmes. After conversion into the corresponding aldehydes, a subsequent vinylmagnesium chloride addition gave the amino allyl alcohols **277**. The ketene aminal function was generated by an initial transacetalization process with α-phenylseleno acetaldehyde diethylacetal to give the selenomethyl aminoacetal **278**. Then, a three step sequence led to the desired ring-expanded lactams **280**: Firstly, the seleno function was oxidized

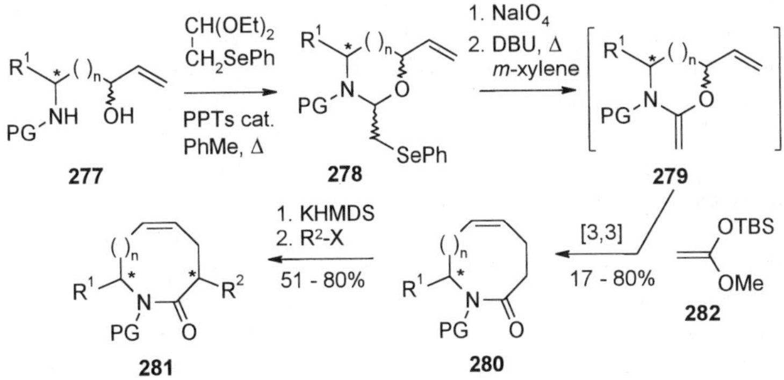

R^1 = H, Me, *i*-Bu, TBSOCH$_2$, *t*-BuO$_2$C, R^2 = N$_3$, SePh, PG = Ts, Cbz, X = Cl, 2,4,6-Tri-(*i*-Pr)-phenylsulfonyl, n = 0, 1, 2, 3 (Table 15)

Scheme 53

to the corresponding selenoxide. Secondly, an elimination in the presence of an appropriate base led to the ketene aminal **279**, which underwent an immediate thermal Claisen rearrangement (*m*-xylene, reflux) to generate the lactam **280** (Scheme 53).

Because of the fact that the stereogenic center adjacent to the nitrogen function was not directly involved in the reorganization of the carbon skeleton during the course of the rearrangement, the stereochemical information was completely retained on synthesizing the lactam **280**. Furthermore, the new double bond was found to be *cis* indicating the participation of a chair-like transition state. Passing through a boat-like transition state would have formed the corresponding lactam bearing an *E* olefin, but presumably, the high activation barrier for such a process prevented the generation of these products (Fig. 3).

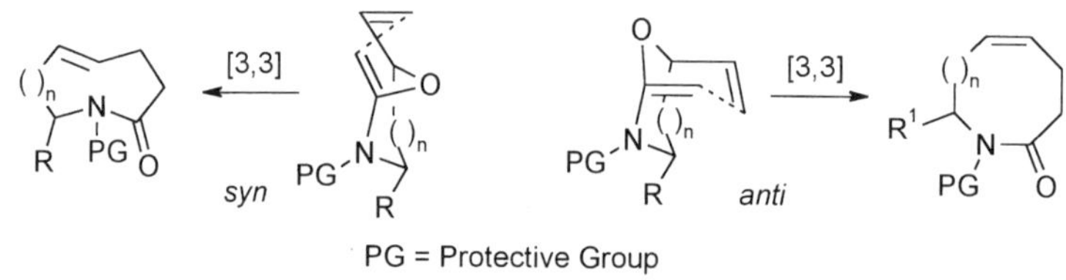

Fig. 3

Some comments concerning this sequence were worthy of mention: the nitrogen should be protected as a carbamate or an sulfonamide to guarantee an efficient transacetalization (→ **278**) in high yields. Amides (RHN-CO) gave no ketene acetals. The optical purity of the lactam **280** is not influenced during the rearrangement. The first series suffered from moderate yields on running the cyclizations. A detailed analysis of the elimination – rearrangement step made it clear that the in situ formed electron rich ketene aminal double bond of **279** partly underwent a re-addition of the PhSeOH to give a hypothetical hemiacetal **283**. This hemiacetal eliminated either α-phenylselenobenzaldehyde regenerating the starting amino alcohol or an alternative H_2O elimination forming a selenoketene aminal **284** (*Z*), which underwent a consecutive rearrangement to give the α-phenylselenyl lactam **281** (α,ω-cis, Scheme 54). With the intention of suppressing the competing electrophilic addition, an excess of a silyl ketene acetal **282** derived from methyl acetate was added as a scavenger to trap the PhSeOH. Using such a modification, the yield of the rearrangement products **280** increased significantly. Additionally, the α-phenylseleno acetic acid methylester derived from **282** was isolated in significant amounts improving the efficiency of the scavenger variant. Several medium sized lactams **280** were used in α-functionalizations to generate the corresponding amino and phenylseleno lactams **281**, respectively, which served as versatile key fragments for peptide mimetic syntheses [53f]. So far, the exclusive use of terminally unsubstituted olefinic units **279** (except the phenylselenyl side products **284** generating the azocinones) has been reported, the synthesis of more complex lactams using more elaborated olefin systems should be an intriguing extension of the strategy (Table 15, Schemes 53, 54) [53].

Table 15. Ref. [53]

Entry	n	R^1	PG	**278** Yield (%)	**280** Yield (%)	R^2	Lactam **281** Yield (%)	Lactam **281** Ratio (α:β)	Ref.
1	0	β-*i*-Bu	Cbz	94	38	N_3	65	1:-	a,f
						PhSe	76	1:-	a,f
2	0	β-*i*-Bu	Ts	91	39				a
3	0	(±)-*i*-Bu	Cbz	82	44				b
4	1	(±)-Me	Cbz	90	51–57	N_3	80	4:1	b,f
						PhSe	62	1:1	
5	2	(±)-*t*-$BuCO_2$	Cbz	72	75	N_3	55	1.4:1	b,f
6	0	β-$TBSOCH_2$	Cbz	68	21				b
7	0	β-$TBSOCH_2$	Cbz	58	31[a]				e
8	3	H	Cbz	58	78[a]				e
9	1	β-Me	Cbz	-	58				d
10	1	β-Me	Cbz	-	80[a]				d,e
11	1	β-Me	Cbz	-	17[b]	PhSe	51	9:1	d
12	1	β-Me (1,3 *syn*)	Cbz	90	48–51				e
13	1	β-Me (1,3 *anti*)	Cbz	92	43–72				e

[a] Optimized rearrangement conditions in presence of excess silylketene acetal.
[b] Rearrangement in the presence of dihydropyrane, re-addition of PhSeOH.

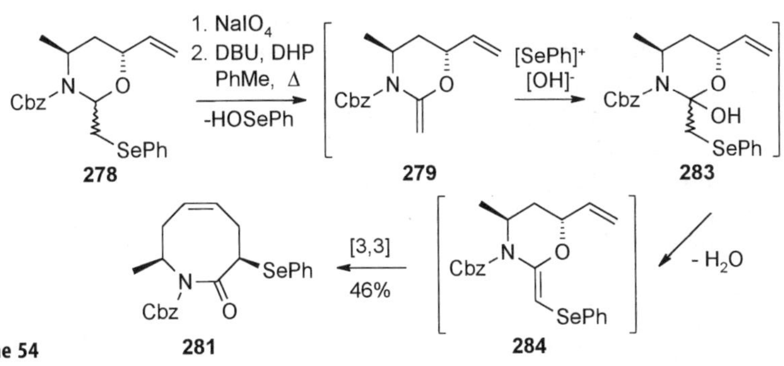

Scheme 54

5.2
C4-Insertion

C4-insertions by means of a sigmatropic rearrangement process have been described using either a thermal Cope reaction, anionic Claisen amide enolate and zwitterionic aza-Claisen rearrangements.

In contrast to the synthesis of carbocyclic rings, the Cope rearrangement has been used sparsely for generating azepinones. Recently, the enantioselectivity of the conversion of 2-aza-divinylcyclopropane **286** has been investigated. The synthesis started from the optically active cyclopropanecarboxylic acid (90% ee), which had been converted into the isocyanate **286** by initial azidation to **285** and a consecutive Curtius rearrangement. Furthermore, the conditions of the iso-

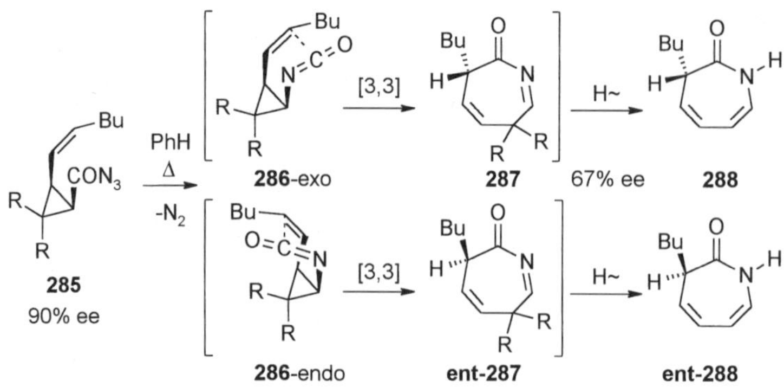

Scheme 55 R = H: 60% yield, R = Me: no reaction

cyanate formation effected the aza Cope rearrangement to give the azepinone **288** after a final migration of the imine double bond in **287**. Surprisingly, the ee detected was only 67% indicating a partial epimerization in the course of the reaction. Since the Curtius side product and the azepinone **288** were found to be stable under the reaction conditions and a biradicalic reaction path could be widely excluded, two competing transition states **286**-*exo* and **286**-*endo* of the 3,3 sigmatropic process were suggested to be responsible for the loss of chiral information. With the intention of suppressing the passing through of an *endo* transition state conformation **286**-*endo* the dimethyl analog (R = Me) was synthesized but the rearrangement of this material failed. Still, the reaction mechanism is unproven (Scheme 55) [54].

Amide enolate Claisen rearrangements served as powerful key reactions to synthesize optically active azepinones **291**. The stereochemical information of an easily formed C–N bond was completely transferred to a C–C bond by means of the highly ordered cyclic *endo* transition state **290**. Furthermore, the defined enolate geometry of the in situ formed ketene aminal double bond caused a high internal asymmetric induction leading to a predominant relative configuration of the newly generated stereogenic centers.

In the rearrangement of divinylaziridines **289**, the participation of a boat-like transition state **290** explained the stereochemical outcome of the reactions to give the azepinones **291** in 73 to 85% yield. The divinylaziridines **289** were synthesized via ex-chiral pool sequences starting from optically active α-amino acids, Table 16, Eq. (26) [55].

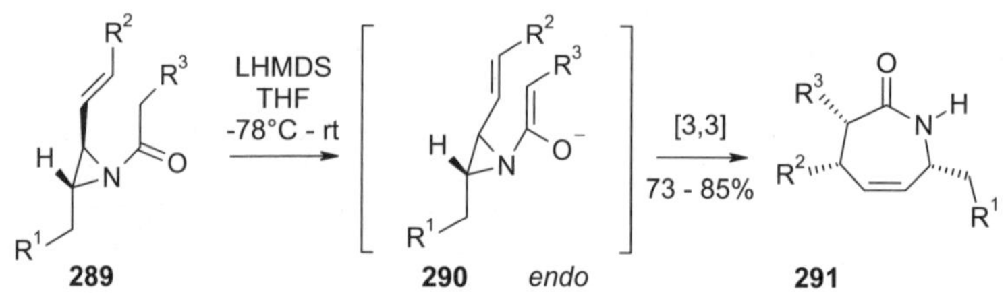

R^1 = H, OBn, Bn, R^2 = H, Me, OBn, R^3 = H, Me, OBn, NHBoc (Table 16) (26)

Table 16. Ref. [55]

Entry	1	2	3	4	5	6	7
R^1	OBn	Bn	Bn	Bn	Bn	OBn	H
R^2	H	H	H	H	H	β-Me [a]	α-CH_2OBn [b]
R^3	H	H	Me	OBn	NHBoc	H	H
yield **291** (%)	83	83	85	81	76	73	73

[a] Rearrangement of *Z*-olefin.
[b] Rearrangement of *E*-olefin.

The synthesis of azecinones **294** started from 2-vinyl piperidines **292**. An amide enolate aza-Claisen rearrangement led to the corresponding ten-membered ring lactams. On reacting terminally unsubstituted olefins, a complete 1,4-chirality transfer was observed. The stereochemical outcome of the process was rationalized by the participation of a chair-like transition state **293** minimizing repulsive interactions. Furthermore, the amide enolate in **293** should have the *Z* configuration. In contrast to the related Ireland ester enolate rearrangements, the aza variant required higher temperatures (PhMe reflux) to give the products. One optically active azecinone **294** served as key intermediate in an asymmetric total synthesis of fluvirucin A_1, Eq. (27) [56].

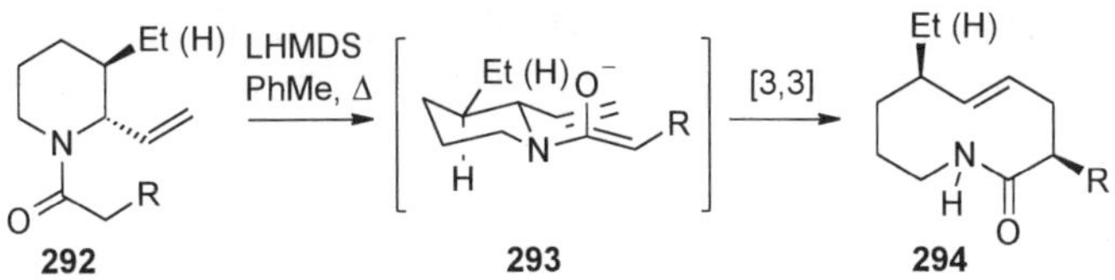

(27)

R = H (H): 40%, R = Me (H): 75%, R = OMe (H): 84%, R = Me (Et): 74%

The aza-ketene Claisen rearrangement could be described as an intermolecular variant of the 3,3 sigmatropic process. An initial addition of an electron deficient ketene to the nitrogen of an allylamine **295** generated a hypothetical zwitterion **296**, which immediately underwent a 3,3 sigmatropic rearrangement to form the γ,δ-unsaturated amide function in **297**. Due to the charge neutralization as the major driving force of the process, the reaction required comparatively low temperatures of 0 to 62 °C.

Starting from *N*-benzyl-2-vinyl pyrrolidine (n = 0) and piperidine (n = 1) **295**, respectively, the ketene Claisen rearrangement using an in situ generated dichloroketene led to the corresponding azoninone and azecinone **297** (R^1 = Bn) in 64 and 96% yield. Replacing the N protective group by the more electron rich PMB substituent (R^1 = PMB), the yields of **297** were observed to decrease to 52 and 54%, respectively. Though the double bond included in the medium sized ring was found to be exclusively *E* configured, both rearrangements suffered from a complete loss of chiral information because of the use of terminally symmetrically substituted olefins in **295** (=CH_2) and ketenes (=CCl_2) as reactants. The NMR spectra of the azecinone **297** (n = 1) were characterized by the coexistence of two conformers. In contrast, the nine-membered ring **297** (n = 0) proved to be a single species. Both medium-sized lactams were used in transannular ring contractions to yield the corresponding quinolizidinones **299/302** (n = 1) and indolizidinones **299** (n = 0), respectively. The *E* double bonds suffered from an external attack of an electrophile (I^+, $PhSe^+$, Me_3Si^+) and the resultant onium ion underwent a regio- and stereoselective addition of the N center of the lactam to give an acylammonium salt **298**. Then, the benzyl group was removed (→ **299**) by a von-Braun type degradation to form the corresponding benzyl halide. Surprisingly, the relative configuration of bridgehead hydrogen and the adjacent substituent were found to be *trans* on synthesizing quinolizidinones **302**. After dechlorination with Zn/Ag/HOAc to give the lactams **300**, the transannular ring contraction of the azoninone (n = 0) took the expected path generating indolizidinone **299** (R^2 = H). In contrast, the analog reactions involving the azecinone **300** (n = 1) gave bicyclic system **302** by passing the hypothetical acylammonium ion **301** as a quasi *syn* adduct of E and N at the double bond. Finally, the quinolizidinone **302** (E = I) was employed as a key intermediate in a total synthesis of D,L-epilupinine (Scheme 56, Table 17) [56].

The mild reaction conditions and the obviously high potential driving force of the ketene Claisen rearrangement recommended the use of the process for more complex systems. The first series of this type of reaction suffered from severe limitations. On the one hand, only electron-deficient ketenes added to the allylamines, and useful yields of the lactams had exclusively been achieved by employing dichloroketene [57, 58a]. On the other hand, the rearrangement was restricted to either monosubstituted olefins in the amino fragment or the

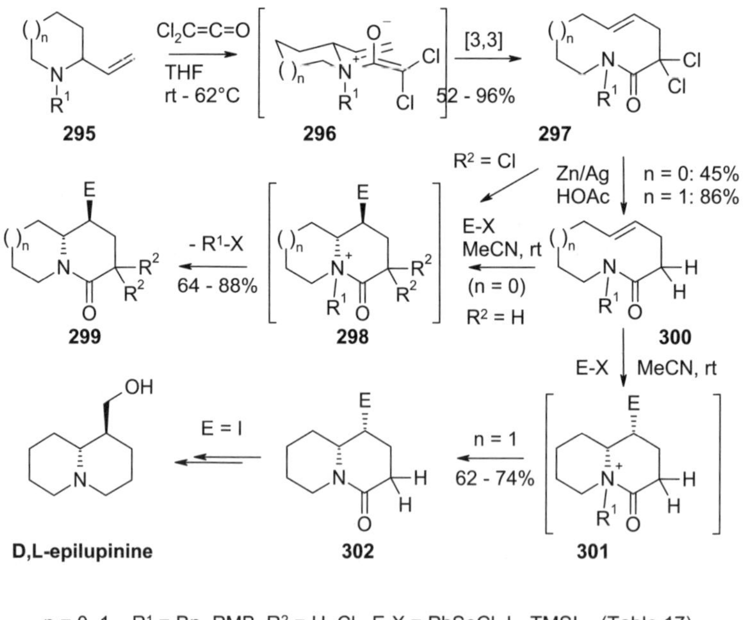

n = 0, 1 R^1 = Bn, PMB, R^2 = H, Cl, E-X = PhSeCl, I_2, TMSI (Table 17)

Scheme 56

Table 17. Ref. [57]

Entry	R^1	n	Yield (%) **297** (**300**)	Yield (%) **299** [**299**], (**302**)[a]		
				E = I	E = PhSe	E = TMS
1	Bn	0	64 (45)	88 [87]	79 [64]	72
2	PMB	0	52	–	–	–
3	Bn	1	96 (86)	85 (62)	84 (74)	–
4	PMB	1	54	–	–	–

[a] Yield: **299**: R^2 = Cl, [**299**]: R^2 = H, (**302**): R^2 = H.

driving force had to be increased by a loss of ring strain during the process. Furthermore, two competing processes need to be mentioned (Fig. 4):

1. The tertiary amines **303** and the acid chlorides **304** (X = Cl) initially formed acylammonium salts **305**, which underwent a von Braun type degradation by an attack of the nucleophilic chloride ion at the allyl system to give allyl chlorides **306/307** and carboxylic acid amide functions.
2. Acyl chlorides **304** led to the corresponding ketenes **308** while the allylamines were deactivated as ammonium salts **309** (Schotten-Baumann conditions).

Three changes concerning the processing led to a pioneering exceeding of the limitations for converting 2-vinylpyrrolidines into the corresponding azoninones:

1. The addition of stoichiometric amounts of a Lewis acid, especially trimethyl aluminum to the reaction mixture: A range of α-substituted carboxylic acid halides **304** (X = Cl, F) as precursors of the ketenes could be used overcoming the restriction concerning the ketene component but until now, the rearrangement failed using α,α difunctionalized carboxylic acid halides. The Lewis acid might have increased the acidity of the α-protons by interacting with the carbonyl group facilitating the formation of the intermediate zwitterions **310** and/or the Lewis acid had stabilized the zwitterionic intermediate **310**, thereby suppressing the elimination of ketene **308**. Furthermore, allylamines **303** bearing 1,2-disubstituted double bonds could be successfully rearranged overcoming a restriction concerning the carbon framework [58b, c].
2. Replacement of the acyl chlorides **304** (X = Cl) by the corresponding acyl fluorides **304** (X = F) as the substitutes of the ketenes: The von Braun type degradation observed as a major competing reaction was efficiently suppressed. The fluoride counter ion was known to be less nucleophilic but more basic. Furthermore, the potential formation of a stable Al-F bond should have eliminated the fluoride as a latent nucleophile. The acyl fluorides **304** were found to be less reactive as compared to the corresponding acid chlorides causing some difficulties in the rearrangement with *n*-alkyl carboxylic acid derivatives. Such transformations needed longer reaction times, and the yield of the corresponding rearrangement products is moderate [58d].

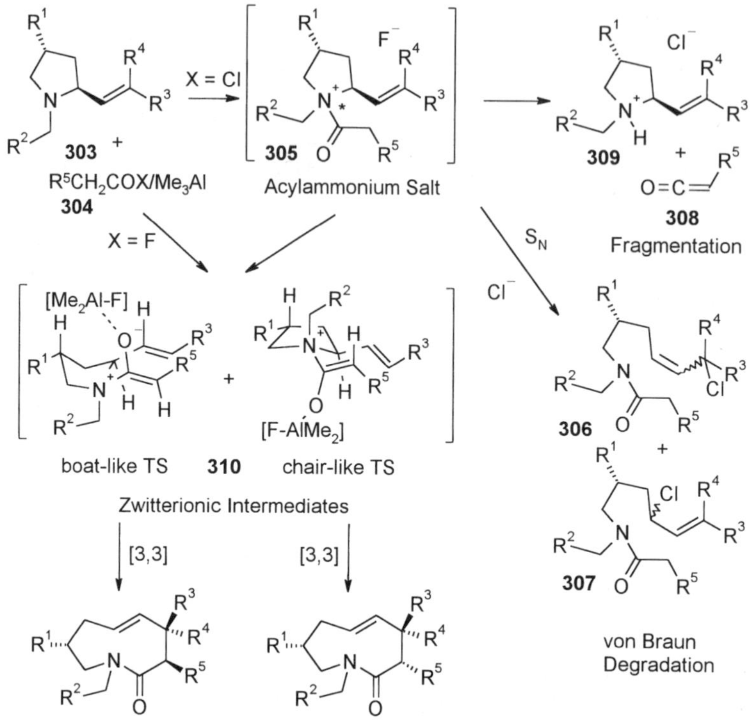

Fig. 4 311 312 (R¹ - R⁵ see Table 18)

3. The use of a second base to trap all proton acids generated during the course of the rearrangement: In most cases, a two-phase system of solid potassium carbonate as a suspension in dichloromethane or chloroform gave the best results [58a–e].

Employing the optimized reaction conditions, the stereochemical advantages of the Claisen rearrangements were combined with an efficient synthesis of the azoninones **311** and **312** bearing defined *E* configured double bonds in the medium-sized rings. As is known for all Claisen rearrangements, a complete 1,3-chirality transfer was observed on treating *E*-allylamines **303** (R^1, R^4 = H) with acetyl chloride **304** (R^5 = H) [58a]. Both enantiomers of the core framework were constructed starting from the same L-(-)-proline derivative choosing either an *E* (R^4 = H) or a *Z* (R^3 = H) allylamine **303**. Furthermore, a high internal asymmetric induction could be observed involving α-substituted acyl halides **304** ($R^5 \neq 1$ H) in the synthesis of the lactams. In most cases the diastereomeric excess was >5:1 in favor of the 3,4-*trans* lactam **312** (Entries 4–14, Table 18). The phenylacetyl halide rearrangement (R^5 = Ph, entry 7, Table 18) only gave a nearly equal mixture of *cis* and *trans* azoninones **311** and **312** (R^5 = Ph). The stereochemical outcome of the rearrangement of **303** (R^1 = H) was explained by the participation of a chair-like transition state **310** ($c\alpha$) with minimized repulsive interactions and a defined *Z* enolate geometry (as known for all amide enolates)

[58b, d]. However the participation of the chair-like transition state **310** (cβ) could not be excluded: both **310** (cα) and (cβ) resulted the same diastereomer *pS*-**312**! Surprisingly, the rearrangement of the 4-*tert*-butyldimethylsilyloxy-2-vinylpyrrolidines **303** (R^1 = OTBS, R^3, R^4 = H) took another course. The stereochemical outcome had to be rationalized by the participation of a boat-like transition state **310** (bβ) to give the 3,8-*trans* lactams **311** (R^1 = OTBS, entries 15–19, Table 18). The corresponding *cis* product **312** (R^1 = OTBS) resulting from

Table 18. Ref. [58]

Entry	R^1	R^2	R^3	R^4	R^5	X	Yield (%)	Ratio **312:311**	Ref.
1	H	H	CO_2Et	H	H	Cl	70	–	a
		Ph				Cl	60	–	b
2	OTBS	H	CO_2Et	H	H	Cl	53	–	a
3	OTBS	H	CO_2Et[a]	H	H	Cl	47	–	a
4	H	H	CO_2Et	H	Me	Cl	77	>95:<5	b
		Ph[b]				F	73	>6:<1	d
5	H	Ph[b]	CO_2Et	H	CH_2CH_2Cl	F	51	>6:<1	d
6	H	H	CO_2Et	H	$CH{=}CH_2$	Cl	80	>95:<5	b
		Ph[b]				F	72	>3:>1	d
7	H	H	CO_2Et	H	Ph	Cl	32[c]	45:55	b
		Ph[b]				F	79	1:2	d
8	H	H	CO_2Et	H	Cl	Cl	72	>95:<5	b
		Ph				Cl	22	90:10	b
		Ph				F	81	90:10	d
9	H	H	CO_2Et	H	OBn	Cl	68	80:20	b
		Ph				Cl	30	80:20	b
10	H	H	CO_2Et	H	NPht	Cl	35[c]	>94:<6	b
11	H	Ph	CH_2OBn	H	Cl	Cl	11	87:13	b
12	H	Ph[b]	H	Me	CH_2CH_2Cl	F	51	>6:1	d
13	H	Ph[b]	H	Me	Ph	F	87	1:>4	
14	H	Ph	H	Me	Cl	Cl	9	1:1	d
						F	91	3:1	d
15	OTBS	H	H	H	Ph	Cl	17[c]	1:>10	c
		Ph				Cl	26[c]	1:>10	c
		Ph				F	95	1:>10	e
16	OTBS	H[a]	H	H	Cl	Cl	22	1:>10	c
17	OTBS	H	H	H	Cl	Cl	29[c]	1:>10	c
		Ph				Cl	20[c]	1:5	c
		Ph				F	92	1:>10	e
18	OTBS	Ph[b]	H	H	OBn	F	73	1:>10	e
19	OTBS	Ph	H	H	NPht	Cl	17	1:>10	c

[a] Reaction with 2*R*-vinylpyrrolidine.
[b] Rearrangement with acid chloride failed.
[c] Up to 50% of the reactant recovered.

the expected chair-like intermediate **310** (cβ) had only once been isolated as a minor compound (entry 17, Table 18). The completeness of the 1,4 chirality transfer should be pointed out. Obviously, the configuration of the intermediately generated stereogenic ammonium center in **310** had to be considered: on rearrangement of the 2-vinylpyrrolidines **303** (R^1 = H), the *N*-acylation should have been directed by the adjacent side chain to the opposite face of the five membered ring to give **310** (α) (1,2 *anti* induction, as found analyzing appropriate acylammonium salts). Consequently, the rearrangement proceeded via a chair-like transition state **310** (cα) as known for the acyclic 3,3 sigmatropic reaction leading predominantly to lactams **312**. In contrast, the *N*-acylation of the 2,4-*trans* disubstituted pyrrolidines **303** (R^1 = OTBS) is directed by the bulky silylether to generate a *syn* arrangement of the vinyl and acyl group in an intermediate ammonium salt **310** (β) (1,3-*anti*-induction, 1,2-*syn*). Then, an appropriate conformation to undergo a Claisen rearrangement presumably was the boat-like form **310** (bβ) with minimized 1,3 repulsive interactions resulting in the lactams **311**. However, the 2,4-*cis* disubstituted pyrrolidine **313** (R^1 = OTBS, R^3, R^4 = H) gave the expected lactam diastereomer **312** via a chair-like transition state conformation **314**, Eq. (28), Table 18 entry 16, Scheme 57 [58c–e].

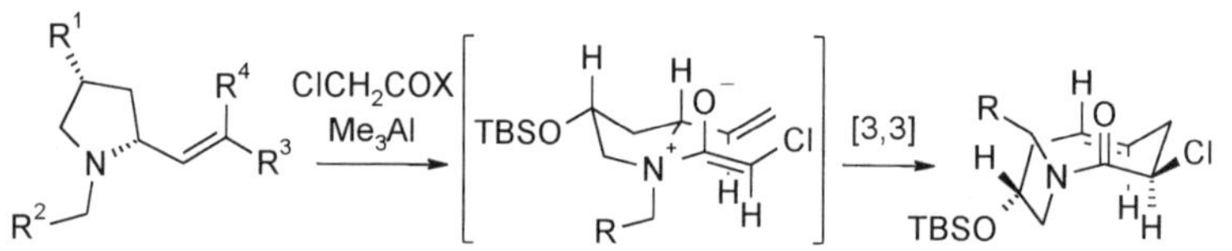

313 **314** **312**

R^1 = OTBS, R^2, R^3, R^4 = H, R^5 = Cl, X = Cl (28)

The lactam and the olefinic unit characterized the heterocyclic cores **311** and **312** as constrained ring systems, the conformations of which were found to be strongly dependent on the substitution pattern and the relative configuration of the stereogenic centers. The planar chiral properties of the medium-sized rings with internal *trans*-double bonds have to be taken in account for analyzing the nine-membered rings. The rearrangements of the 2*S* vinylpyrrolidines **303** passing through a boat-like transition state **310** (b) effected initially the formation of the medium sized ring with *pS*-arrangement of the *E*-double bond (*pS*-**311**). This planar diastereomer *pS*-**311** was obviously unstable: NMR and NOE analyses indicated the coexistence of one preferred *pS*-**311** (1) and at least one additional minor conformation *pS*-**311** (2) (vide infra: Scheme 58) as a highly flexible equilibrium of some arrangements of the lactam function. Finally, the epimerization (flipping of the *E* double bond) to give the *pR* arrangement *pR*-**311** of the olefin with respect to the ring generated the most stable and rigid conformation. Preliminary force field calculations of the azoninones **311** and molecular mechanics calculations of the related *E*/*Z*-1,5-nonadiene confirmed these observations [59]. In contrast, the lactams **312** (R^4 = H) generated via chair-like zwitterions **310** (c) were found to be generated directly in a stable *pS* arrangement of the *E* double bond *pS*-**312** (Schemes 57, 58). Nevertheless, a high activation barrier had to be passed to achieve the change of the planar chiral

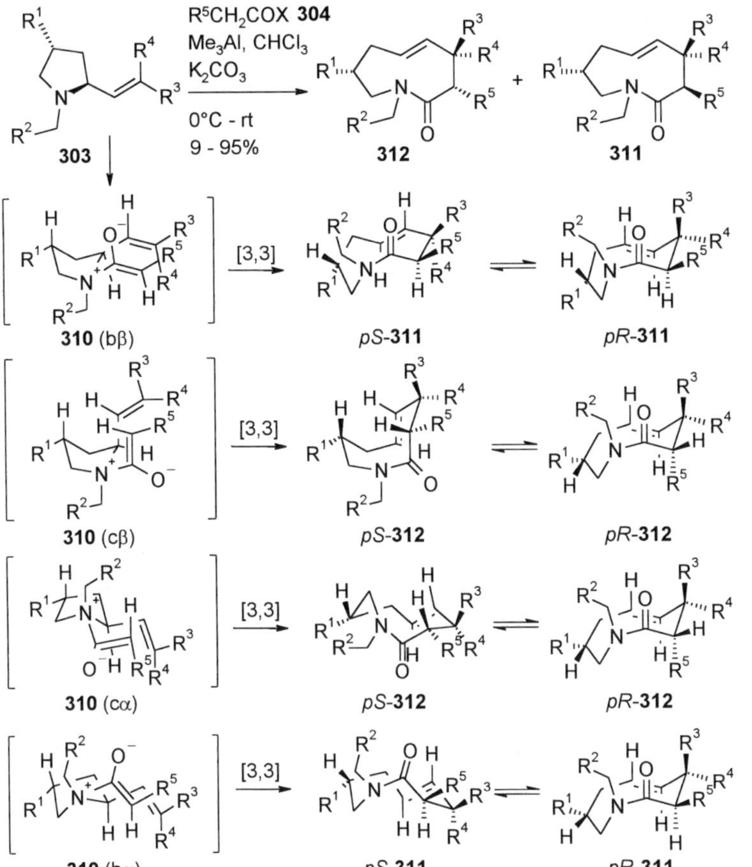

R^1 = H, OTBS, R^2 = H, Ph, R^3 = H, CO_2Et, CH_2OBn, R^4 = H, Me,
R^5 = H, Alkyl, Vinyl, Ph, Cl, OBn, NPht, X = Cl, F (Table 18)

Scheme 57

information (*pS*-**311** → *pR*-**311**). This fact allowed the isolation and the characterization of the conformers of the nine-membered rings (Schemes 57, 58) [58c–e].

The planar diastereomeric azoninones *pS*-**311** and *pR*-**311** (R^3, R^4 = H) were subjected to cycloadditions to synthesize the azonanones **315** to **318** (Table 19, Scheme 58). Low temperature reactions (≤25 °C) allowed an almost complete conversion of the planar chiral information of the reactants **311** into new chiral centers of the products: Cyclopropane, aziridine and oxirane annulated azonanones **317/318** were synthesized as well as dihydroxylated (protected) nine-membered ring lactams **315/316**. If the reaction required higher temperatures, the diastereoselectivity decreased severely because of the competing flipping of the double bond (*pS* → *pR*) with respect to the ring (Table 19, entry 4/5, 2nd reaction) (Scheme 58) [60a].

Table 19. R^1 = OTBS, R^2 = Ph, R^3, R^4 = H [60]

Entry	Lactam 311	R^5	Yield azonanones [%] 317/318				315/316	
			E = CH_2	E = C $(CO_2Et)_2$	E = N-*p*Ts	E = O	E = 2 × OH[a]	E = 2 × OH[b]
1	*p*S	Ph	–	–	58/–	100/-	95/-	–
2	*p*R	Ph	–	–	–	13/82	–	–
3	*p*R	OBn	–	–	–	–/91	–	–
4	*p*S	Cl	92/-	12/18	56/–	92/–	71/–	70/–
5	*p*R	Cl	9/91		–/52	–/86	17/55	10/88

[a] Diol protected as acetonide.

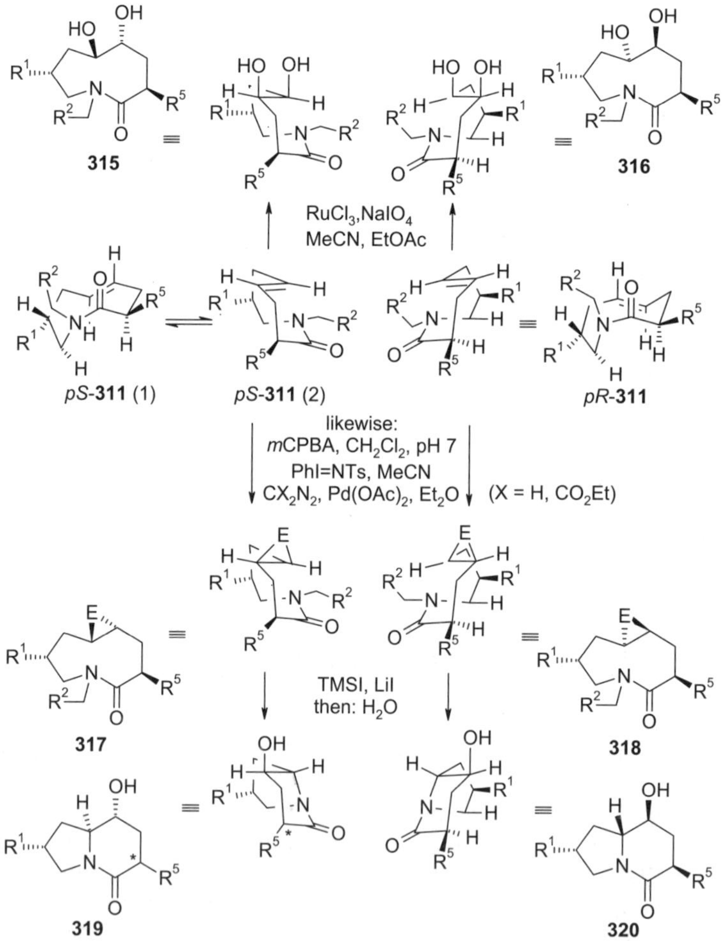

Scheme 58 R^1 = OTBS, R^2 = Ph, R^3, R^4 = H, R^5 = Ph, Cl, OBn, E = CH_2, $C(CO_2Et)_2$, N-*p*Ts, O, Yields: 30 - 100% (Tables 19, 20)

The oxirane functions of the epoxy azonanones **317** and **318** underwent regioselective openings and consecutive transannular ring contraction sequences to generate the 8-hydroxy indolizidinones **319** and **320** with a complete stereoselectivity and a high regioselectivity (Table 20, entries 13–16). Always the exclusive formations of δ-valerolactams in **319/320** was found running the reactions at rt, though the reactant azonanones **317/318** were characterized by high conformational mobility of the amide function with diastereomeric properties. As supposed, the conversions of the 5*S*,6*S*-azonanones **318** yielded exclusively bicycles **320** because of their short *N*-C6 distances as determined via X-ray analyses and some force field optimized calculations (Table 20, entries 14/15). In contrast, the 5*R*,6*R*-azonanones **317** used different reaction paths to generate the bicycles **319** (Table 20, entries 13, 16). The direct conversion of the conformation determined via X-ray and NOE analyses bore a slightly shorter N-C5 distance implying the formation of (some) γ-butyrolactam product. Obviously, the lactam function of the azonanones **317** was characterized by some flexibility to generate at least one additional significantly more reactive conformer resulting from *pS*-**311** (2) with a shorter *N*-C6 distance to induce an efficient δ-valerolactam **319** formation. The so formed defined hydroxylated indolizidinones **319** and **320** should serve as useful key intermediates in the synthesis of leguminose type alkaloids and pumiliotoxins (Scheme 58) [60].

First investigations had shown that the planar diastereomeric azoninones *pS*-**311**, *pR*-**311** and *pS*-**312** underwent regio- and diastereoselective ring contractions to indolizidinones **323**–**326** (Scheme 59). The rigid conformations of the unsaturated lactams involved caused a defined anti attack of an external electrophile at the unshielded face of the double bond (*pS* $\rightarrow$ re-attack, *pR* $\rightarrow$ *si*-attack) and the intramolecular trapping of the resultant cation **321** and **322** by the lactam nitrogen (*pS* $\rightarrow$ *si*-attack, *pR* $\rightarrow$ *re*-attack). The so formed intermediate *N*-benzyl acylammonium ion underwent an immediate von Braun degradation to give the indolizidinones **323**–**326**. In contrast to the Curtin-Hammett-Principle, lactams allowed defined additions to the double bond with respect to the predominant conformation. The planar chiral information of the nine-membered ring **311** and **312** (*pS* or *pR*) could be transferred into defined stereogenic centers of **323**–**326** by means of the ring contraction. While the reactions of the rigid type azoninones *pS*-**312** (R^4 = H) and *pR*-**311** yielded exclusively bicycles **325** and **326** with δ-valerolactam function, respectively, the *pS*-**311** rings used different reaction paths to generate two further products **323** and **324**. Obviously, the lactam function of the kinetically formed azoninones *pS*-**311** showed some flexibility to generate at least two reactive conformations *pS*-**311** (1, as found in the NMR spectra) and *pS*-**311** (2, hypothetical) with diastereomeric properties under the reaction conditions. The variation of the transannular reaction conditions allowed one to pick out predominantly one of these conformations yielding either the series **323** or series **324**, respectively (Scheme 59) [58c, e]. One indolizidinone **325** (R^1, R^4, R^5 = H, R^3 = CO_2Et) was employed as a key intermediate in a total synthesis of a dendroprimine (indolizidine alkaloid) [60b].

An alternative pathway using a zwitterionic aza-Claisen rearrangement to generate azoninones has been described by Hegedus [61]. 2-Vinylpyrrolidines

Table 20. (R^2 = Ph, R^4 = H) [58, 60]

Entry	Lactam	R^1	R^3	R^5	Method[a]	Yield Indolizidinones [%] **323/324/325/326**			**319/320**	Ref.
						E = I	E = Br	E = PhSe	(E = OH)	
1	*pS*-**312**[b]	H	CO_2Et	H	B	–	–	–/22/–/–	–	58b
2	*pS*-**312**	H	CO_2Et	H	B	–	–	–/70/–/–	–	58b
3	*pS*-**312**	H	CO_2Et	Cl	B	–	–	–/70/–/–	–	58b
4	*pS*-**311**	OTBS	H	Ph	A	47/-/16/-	40/–/–/–	44/–/–/–	–	58e
5	*pS*-**311**	OTBS	H	Ph	B	–/–/99/–	11/–/44/–	–/–/95/–	–	58e
6	*pR*-**311**	OTBS	H	Ph	A or B	–/–/–/49	–/–/–/72	–/–/–/69	–	58e
7	*pS*-**311**	OTBS	H	OBn	A	12/–/24/–	16/–/–/–	38/–/8/–	–	58e
8	*pS*-**311**	OTBS	H	OBn	B	–/–/70/–	–	–/–/74/–	–	58e
9	*pR*-**311**	OTBS	H	OBn	A or B	–/–/–/40	–	–/–/–/20	–	58e
10	*pS*-**311**	OTBS	H	Cl	A	–	25/–/–/–	19/–/1.5/–	–	58e
11	*pS*-**311**	OTBS	H	Cl	B	–/–/15/–	–/–/82/–	–/–/81/–	–	58e
12	*pR*-**311**	OTBS	H	Cl	A or B	–/–/–/64	–	–/–/–/64	–	58c, e
13	**117** (E = O)	OTBS	H	Ph	B	–	–	–	60/–	60
14	**118** (E = O)	OTBS	H	Ph	B	–	–	–	–/32	60
15	**118** (E = O)	OTBS	H	OBn	B	–	–	–	–/57	60
16	**117** (E = O)	OTBS	H	Cl	B	–	–	–	47/–	60

[a] Method A: Addition of the lactam to the reagent at –20 °C; Method B: Addition of the reagent to the lactam at rt.
[b] R^2 = H (two step ring contraction).

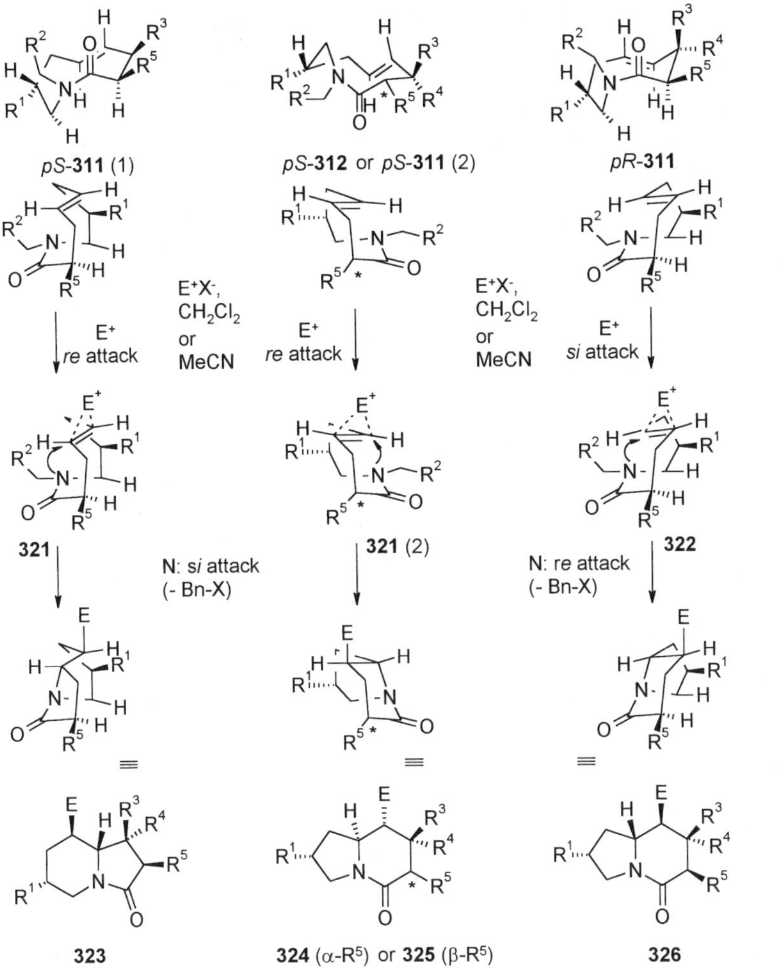

R^1 = H, OTBS, R^2 = H, Ph, R^3 = H, CO_2Et, R^4 = H, R^5 = H, Alkyl, Ph, Cl, OBn,
E = Br, I, SePh, X = Cl, Br, I   Yields: R^2 = H: 30 - 55 %, R^2 = Ph: 16 - 95%
(Table 20)

Scheme 59

327 and chromium carbene complexes **328** underwent photochemical reactions in presence of a Lewis acid to give the corresponding nine-membered ring lactams **329** bearing *E* double bonds in up to 71% yield. Although reactants and products suffered from some instability towards Lewis acids, the presence of the zinc chloride or dimethylaluminium chloride was mandatory to start the rearrangement. In contrast to the classical ketene Claisen process, electron rich ketene equivalents such as alkoxy or amino ketenes could be used, since the donor substituents stabilized the chromium carbene complex **328**. Furthermore, α,α disubstituted lactams were synthesized but the stereoselectivity observed was low. The determination of the stereochemical outcome of the reaction proved that the 1,4 chirality transfer was not complete: a Mosher analysis of an appropriate azoninone gave a loss of about 10% of the chiral information. A chiral carbene complex **328** (R^1 = oxazolidinyl) was found to have a negligible influence on the stereoselectivity of the rearrangement. Generally, the present variant of the rearrangement was found to be very sensitive to any sterical hindrance. Additional substituents in any position (eg. $R^2 \neq 1$ H) led to a severe decrease of the yield and the stereoselectivity [Eq. (29)]. Additionally, one example rearranging a 2-vinylpiperidine **330** was given. The corresponding azecinone **331** was formed in about 33% yield, Eq. (30).

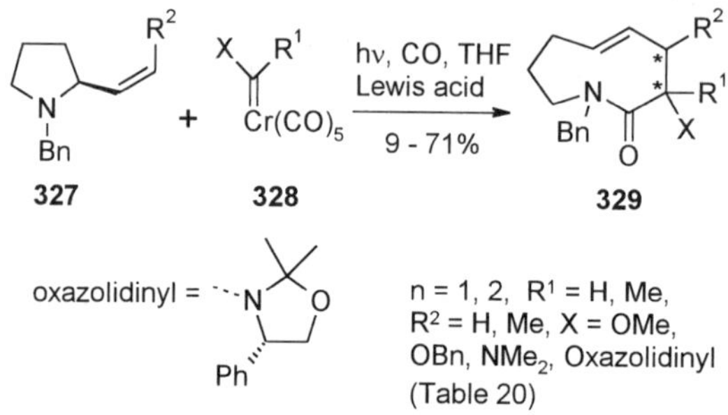

(29)

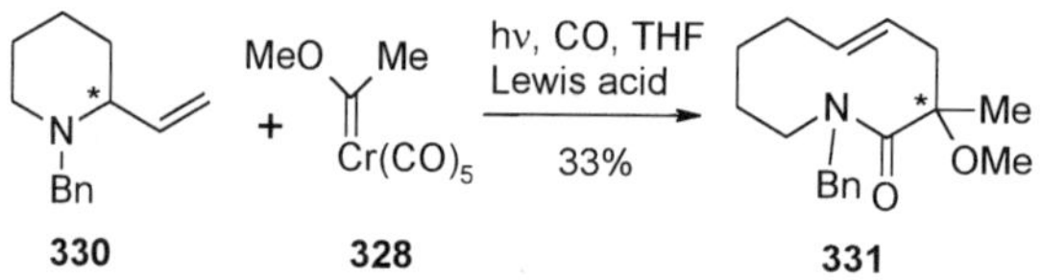

(30)

Some details are outlined in Table 21.

Table 21. Ref. [61]

Entry	n	R^1	X	R^2	Lewis acid	Yield (%)	Ratio **329/331**
1	1	Me	OMe	H	$ZnCl_2$	71	–
2	1	Me	OBn	H	$ZnCl_2$	66	62% de[a]
3	1		-$(CH_2)_3$-O-	H	$ZnCl_2$	15	–
					Me_2AlCl	22	–
4	1	H	NMe_2	H	Me_2AlCl	9	–
5	1	H	oxazolidine[b]	H	$ZnCl_2$	19	74% de[a]
6	1	Me	OMe	Me	$ZnCl_2$	20	60% de[c]
7	1	Me	OBn	Me	$ZnCl_2$	40	33%de[c]
8	2	Me	OMe	H	Me_2AlCl	33	–

[a] Determined via Mosher analysis of a derivative.
[b] Chiral oxazolidine.
[c] Mixture of 3,4 diastereomers.

In analogy to Edstrom's experiments, the nine- and ten- membered ring lactams underwent regio and stereoselective transannular ring contractions to give the corresponding indolizidinones and quinolizidinones, respectively (vide supra Scheme 56) (Table 21) [61].

6 Fragmentations

Although well known for synthesizing medium sized rings, fragmentation reactions had been used only sparsely to generate lactam structures. In principle, a fragmentation should be a useful process. Bicyclic reactants bearing a framework of annulated five and six membered rings with a range of defined stereogenic centers were easily available, the final breaking of the central bond promised a smooth formation of versatile substituted lactams. On the other hand, the formation of a medium-sized ring generated a species suffering from severe transannular repulsive and/or attractive interactions, particularly with regard to eight to eleven-membered constrained systems. The nitrogen atom of the lactam had to be considered as a potential nucleophile interacting with acceptor substituted positions at the opposite face of the ring inducing transannular reactions (vidre supra, Sect. 5.2). Regarding potential nine or ten-membered ring systems **333** and **335**, the existence of the bicyclic form **332** and **334** was favored, respectively (Fig. 5). In contrast, preventing or overcoming such interactions as a mandatory prerequisite for fragmentations caused some difficulties. On investigating the γ-oxoazoninone **338** (n = 1) and a δ-oxo azecinone **338** (n = 2) as the core systems of indolizidine and mitomycin natural products, the metal organyl mediated cyclization of imide **336** via **337** succeeded in the formation of the bicyclic semiaminal structures **338**. The corresponding monocyclic lactam was found as a side product only in the case of the ten-membered ring **338** (n = 2), Eq. (31) [62].

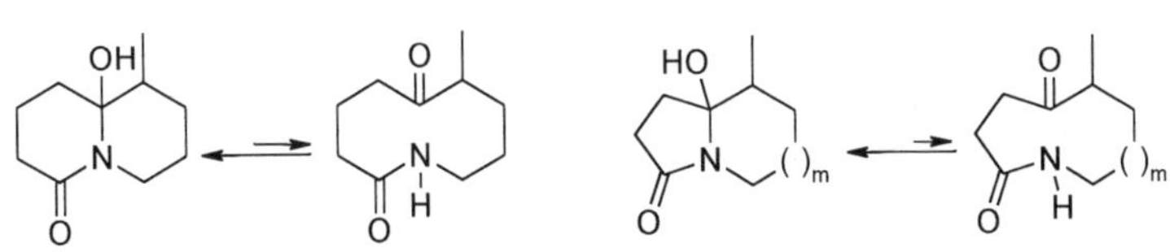

Fig. 5 **332** 5 : 1 **333** **334** >95 : 5 **335**

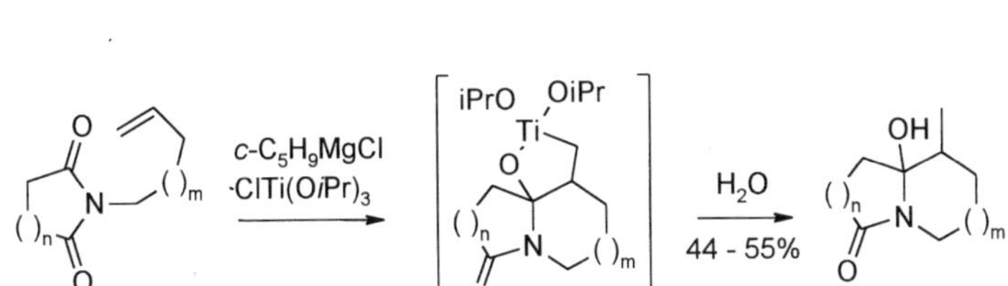

336 **337** **338** (31)

Consequentially, the driving force had to be increased on synthesizing medium-sized ring lactams by means of a fragmentation. Until now, by far the majority of reactions involved the breaking of a small annulated ring (Exception:

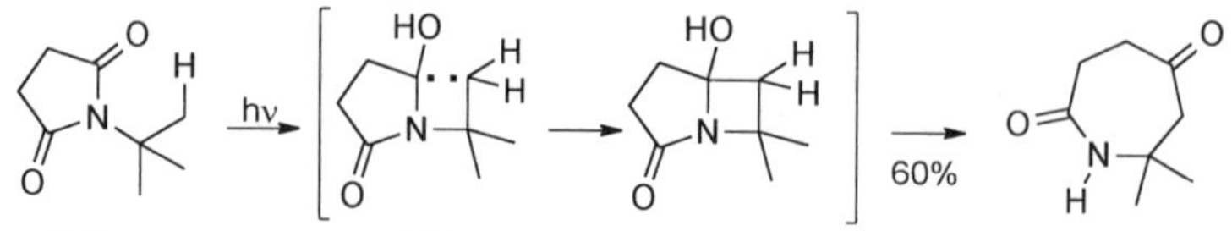

339 340 341 342

Scheme 60

cleavage of DBU [63]). The additional loss of ring constraint was used as the guiding principle to enforce the fragmentation. Prior to the synthesis of the medium sized ring, an efficient cyclization or cycloaddition had to be carried out to prepare the system for the consecutive fragmentation, which were achieved as anionic, radical or pericyclic processes. Azepinones such as **342** were synthesized from **339** via a five-membered ring/four-membered ring annulation through biradical **340** to give the intermediate **341** and a consecutive fragmentation (Scheme 60). The generation of the four-membered ring was carried out as an anionic condensation [64] or some photochemically mediated ring closure [65]. A solid state Norrish type II reaction should serve as an example [65].

Furthermore, the ring enlargement of a range of benzyl azides was described by assuming the intermediate participation of a cyclohexadieno azirine [66].

Unsaturated azocinones were synthesized by 2 + 2 cycloaddition of benzonitriles to phenols and a final radical mediated fragmentation [67].

A useful method to synthesize ten and fourteen-membered ring imides **346** involved an initial condensation of macrocyclic β-ketoestes **343** with alkyl or aryl isocyanates and carbodiimides, respectively, in the presence of a base [68]. After a nucleophilic attack of the enolate on the isocyanate C, the resultant amide N anion **344** induced a ring closure by addition to the keto group. Then, the intermediately formed four-membered ring **345** underwent a fragmentation

Table 22. Ref. [66]

Entry	n	R	Yield (%) **346**	Ref.
1	5	Ph	75	a
2	5	Bu	72	a
3	5	$-(CH_2)_3-Cl$	74	a
4	5	$-(CH_2)_4-Br$	61	a
5	5	$-(CH_2)_2-Cl$	45	a
6	1	allyl	80	b
	5		86	b
7	1	2-(allyloxy)-phenylmethyl	68	c
	5		71	c
8	1	2-bromo-phenylmethyl	80	c
	5		82	c

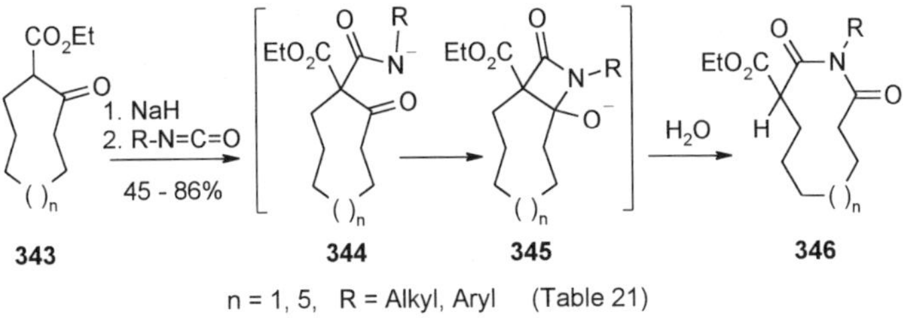

Scheme 61

to give a quite stable β-dicarboxylate anion of **346**. When employing ω-halogen isocyanates, competing additional ring closures were observed. The yield of the imides depended on the sterical and electronical properties of the isocyanates: bulky and electron rich substituents led to disappointing yields. Otherwise, up to 86% of the desired products **346** were isolated (Scheme 61, Table 22) [68].

As an alternative with respect to the Schmidt rearrangements of azidoketones (vide supra, Sect. 5.2) a process of cyclization and a consecutive fragmentation has been developed by Benati [69a] and Kim [69b]: α-Azido-β-ketoesters **347** were treated with Bu_3SnH to induce a regioselective ring expansion to **348** following a radical mechanism (Scheme 62): Initially, an aminyl radical **A** was formed which underwent a 3-exo cyclization with the adjacent keto group. The alkoxy radical **B** effected a regiospecific C–C bond breaking to generate a resonance stabilized C-radical **C** in α-position of the ester, which was finally trapped by a H transfer of the Bu_3SnH to result **348**, the α-amino ketone **349** was found as a side product. Seven- to nine-membered monocyclic (**347** $\rightarrow$ **348**, Scheme 62) or bridged **350** $\rightarrow$ **351** lactams were synthesized in 65 to 81% yield, Eq. (32). In contrast, the generation of bicyclic annulated systems gave the desired ring expansions in varying yields, **352** $\rightarrow$ **353**, Eq. (33), **355** $\rightarrow$ **356**, Eq. (34). A reductive radicalic deazidation [**354** (X = H), **357**] and the azide reduction to give the amines [**349**, **354** (X = NH_2)] were reported as the major competing reactions. Some initial efforts to trap the ring enlarged radical (**C**, Scheme 62) with allyl donors (allyl-$SnBu_3$) failed. (Scheme 62, Eqs. [32–33]) [69a].

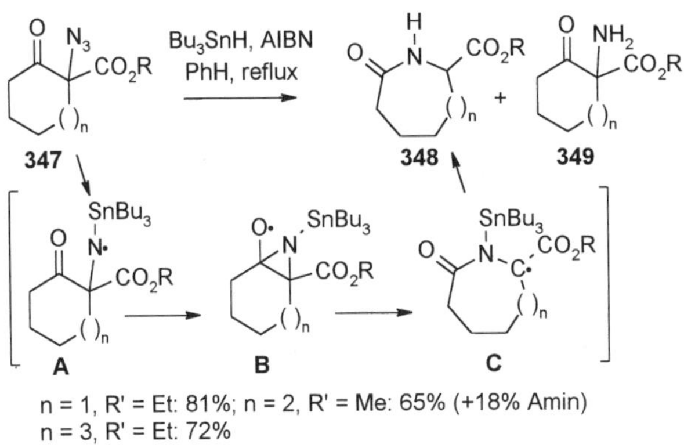

Scheme 62

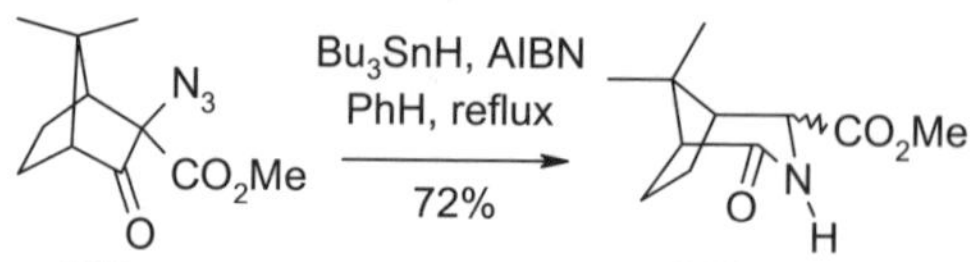

350 351 (32)

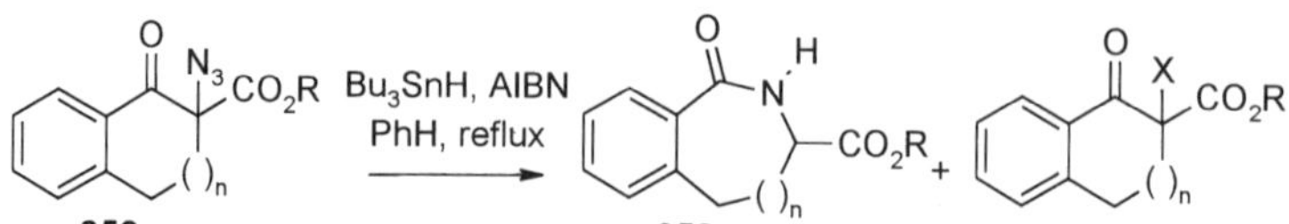

352 353 354

n = 1: 68% (**353**), 27% (**354**, X = H)
n = 2: 28% (**353**), 12% (**354**, X = H), 28% (**354**, X = NH_2) (33)

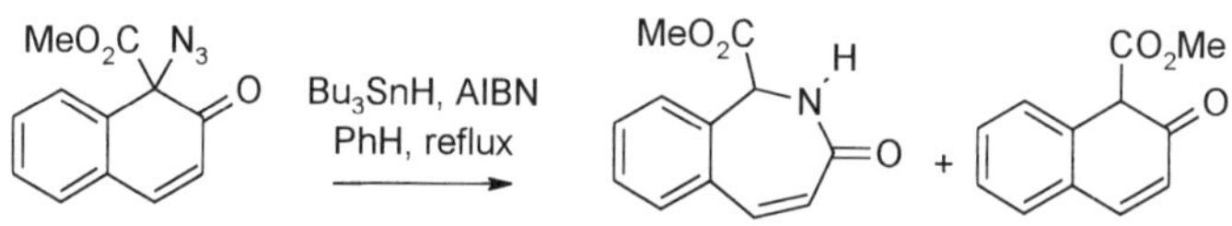

355 15% **356** 55% **357** (34)

Kim described somewhat earlier such a radical path to achieve a ring expansion by more than a single atom [69b]. In analogy to the series described by Aubé and Dumas [50], ω-azidoketones **358** served as starting materials, but the radicalic activation of the azide induced a reaction cascade as shown in Scheme 62 (**A** **C**): in the final step the C–C bond delivering the most stabilized alkyl or allyl radical **C** was broken resulting in the medium-sized lactams **359** regioselectively (Scheme 63): α-(2-Azidoethyl)- (m = 1) and α-(3-azidopropyl)-ketones **358** (m = 2) were expanded by three and four additional atoms to give **359** in 48 to 96% yield, respectively. In several runs, some bicyclic imine **360** was found as a side product. The synthesis of a range of eight to eleven-membered rings has been reported, Scheme 63, Eqs. (35–37), the yields varied from 48 to 96%. In the absence of a radical stabilizing ester group, R = H, Eq. (35), the yield of the lactam **362** decreased (**361** → **362**, R = CO_2Et: 91%, R = H: 75%). As shown in two additional examples, the driving force of the

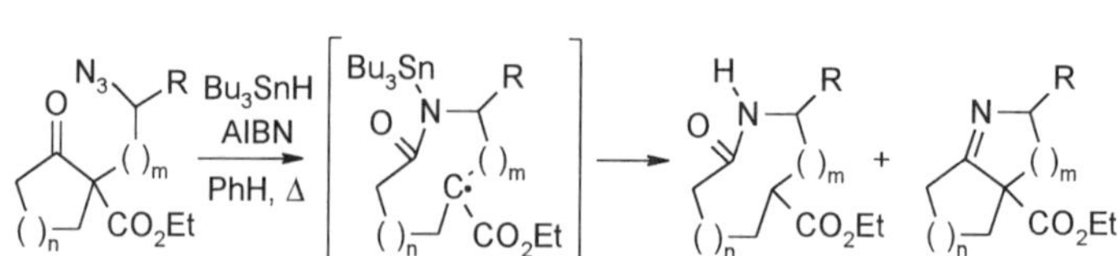

358 C 359 360

n, m = 1, R = H: 93% (**359**); n, m = 1, R = OEt: 83% (**359**);
n = 1, m = 2, R = H: 96% (**359**); n = 2, m = 1, R = H: 68% (**359**), 23% (**360**)
n, m = 2, R = H: 92% (**359**), 5% (**360**); n = 3, m = 2, R = H: 48% (**359**), 44% (**360**)

Scheme 63

fragmentation could be enhanced again by breaking a small four-membered ring: spiro cyclobutanone **363** and cyclobutanone **365** were converted into the corresponding azocinones **364** and **366** in 91 and 96% yield, respectively, Eqs. (36, 37) [69b].

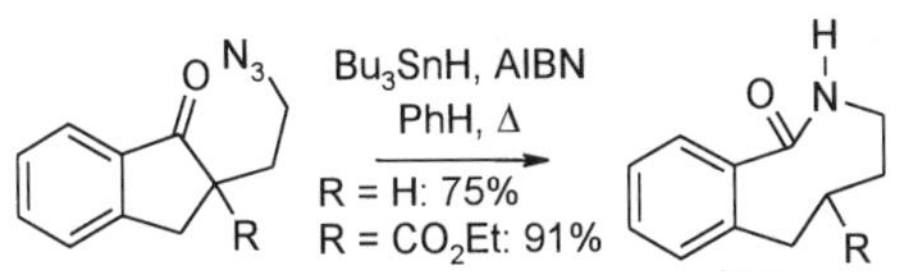

361 **362** (35)

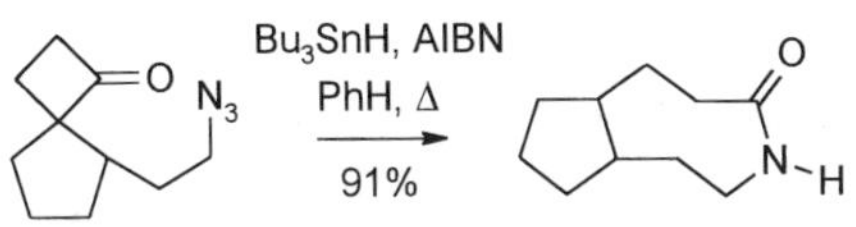

363 **364** (36)

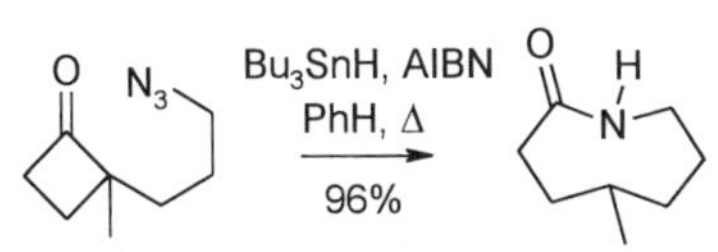

365 **366** (37)

The skeleton of tuberostemonone was characterized by a bicyclic framework of bridged seven and nine-membered ring lactams. With the intention to find an efficient method to generate the azoninone fragment of the core structure, an oxidative ring expansion reaction was developed by Wipf. Firstly, a hydroindole **368** bearing a bridgehead OH function is formed by an oxidative cyclization of the optically active protected tyrosine **367**. Optionally, further stereogenic centers could be introduced. Then, treatment of the bicyclic system **368** with iodobenzene diacetate in the presence of iodine initiated a formal alkoxy radical fragmentation, hypothetically via **369**, alkylradical **370** and iodide **371** to provide the nine-membered ring **372** in about 80% yield. A subsequent oxidation with *m*-CPBA in dichloromethane and BF$_3$-etherate initially led to the peroxide **373**, which gave the corresponding lactam **374** in 83% yield after a final treatment with pyridine (Scheme 64) [70].

In conclusion, the new radical fragmentation path served as an intriguing strategy for synthesizing highly functionalized optically active medium-sized ring lactams as demonstrated generating **377** from **375** in only two steps (Scheme 65). The combination of stereoselective synthesis and defined ring enlargement opens versatile perspectives for investigating the formation of medium-sized rings (Scheme 65) [70].

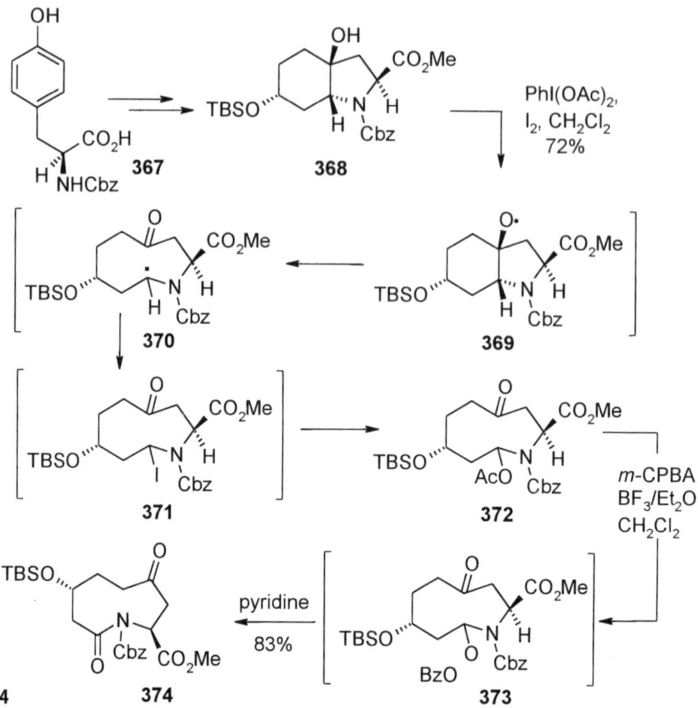

Scheme 64

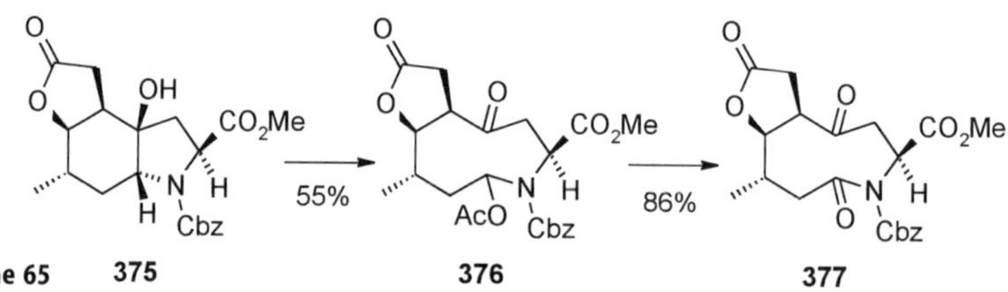

Scheme 65

7 References

1 a) Hegde VR, Patel MG, Gullo VP, Ganguly AK, Sarre O, Puar MS (1990) J Am Chem Soc 112:6403; b) Hegde VR, Patel MG, Gullo VP, Puar MS (1991) J Chem Soc Chem Commun 810; c) Hegde VR, Patel MG, Gullo VP, Horan AC, King AH (1993) J Antibiot 46:1109; d) Ui H, Imoto M, Umezawa K (1995) J Antibiot 48:387; e) Adamczeski M, Quinoa E, Crews P (1990) J Org Chem 55:240; f) D'Auria MV, Giannini C, Minale L, Zampella A, Debitus C, Frostin M (1997) J Nat Prod 60:814; g) Fernández R, Dherbomez M, Letourneux Y, Nabil M, Verbist JF, Biard JF (1999) J Nat Prod 62:678; h) Groweiss A, Newcomer JJ, O'Keefe BR, Blackman A, Boyd MR (1999) J Nat Prod 62:1691; i) Lane SJ, Marshall PS, Upton RJ, Ratledge C, Ewing M (1995) Tetrahedron Lett 36:4129; j) Velten R, Erdelen C, Gehling M, Göhrt A, Gondol D, Lenz J, Lockhoff O, Wachendorff U, Wendisch D (1998) Tetrahedron Lett 39:1737; k) Lin W, Xu R, Wang R, Mak TCR (1991) J Crystallogr Spectrosc Res 21:189; l) Evans PA, Holmes AB (1991) Tetrahedron 47:9131

2. a) Keenan RM, Miller WH, Kwon C, Ali FE, Callahan JF, Calvo RR, Hwang SM, Kopple KD, Peishoff CE, Samanen JM, Wong AS, Yuan CK, Huffman WF (1997) J Med Chem 40:2289; b) Angelucci L, Calvisi P, Catini R, Cosentino U, Cozzolino R, de Witt P, Ghirardi O, Giannessi F, Giuliani A, Guaraldi D, Misiti D, Ramacci MT, Scolastico C, Tinti MO (1993) J Med Chem 36:1511; c) Natchus MG, Cheng M, Wahl CT, Pikul S, Almstead NG (1998) Bioorg Med Chem Lett 8:2077; d) Owens TD, Semple Jedward (1998) Bioorg Med Chem Lett 8:3683; e) Semple JE, Minami NK, Tamura SY, Brunck TK, Nutt RF, Ripka WC (1997) Bioorg Med Chem Lett 7:2421 f) Xu Y, Miller MM (1998) J Org Chem 63:4314
3. a) Mehlfuehrer M, Thirring K, Berner H (1997) J Org Chem 62:4078; b) Belyaev AA (1995) Tetrahedron Lett 36:439
4. Ugi I, Demharter A, Hoerl W, Schmid T (1996) Tetrahedron 52:11657
5. a) Nagasaka T, Kato H, Hayashi H, Shioda M, Hikasa H, Hamaguchi F (1990) Heterocycles 30:561; b) Gavagan JE, Fager SK, Fallon RD, Folsom PW, Herkes FE, Eisenberg A, Hann EC, DiCosimo R (1998) J Org Chem 63:4792; c) Lu W, Sih CJ (1999) Tetrahedron Lett 40: 4965
6. a) Nowak W, Gerlach, H (1993) Liebigs Ann Chem 153; b) Winkler JD, Stelmach JE, Siegel MG, Haddad N, Axten J, Dailey WP (1997) Isr J Chem 37:47; c) Erickson SD, Simon JA, Still WC (1993) J Org Chem 58:1305 d) Newlander KA, Callahan JF, Moore ML, Tomaszek TA, Huffman WF (1993) J Med Chem 36:2321; e) Schmidt B, Lindman S, Tong W, Lindeberg G, Gogoll A, Lai Z, Thörnvall M, Synnergren B, Nilsson A, Welch CJ, Sohtell M, Westerlund C, Nyberg F, Karlén A, Hallberg A (1997) J Med Chem 40:903; f) Wipf P, Kim Y, Goldstein DM (1995) J Am Chem Soc 117:11106; g) Amblard M, Daffix I, Berge G, Calmes M, Dodey P, Pruneau D, Paquet JL, Luccarini JM, Belichard P, Martinez J (1999) J Med Chem 42: 4193
7. Mladenova M, Alami M, Linstrumelle G (1995) Synth Commun 25:1401
8. Mukaiyama T (1979) Angew Chem Int Ed Engl 18:707
9. Rico I, Halvorsen K, Dubrule C, Lattes A (1994) J Org Chem 59:415
10. Meyers AI, Knaus G, Kamata K. Ford ME (1976) J Am Chem Soc 98:567
11. a) Rottmann A, Bartoczek M, Liebscher J (1997) Synthesis 313; b) Bischoff C, Schroeder E (1991) J Prakt Chem 333:181
12. a) Robl JA, Cimarusti MP (1994) Tetrahedron Lett 35:1393; b) Robl JA, Sulsky R, Sieber-McMaster E, Ryono DE, Cimarusti MP, Simpkins LM, Karanewsky DS, Chao S, Asaad MM, Seymour AA, Fox M, Smith PL, Trippodo NC (1999) J Med Chem 42:305; c) Robl JA, Sieber-McMaster E, Sulsky R (1996) Tetrahedron Lett 37:8985; d) Okayama T, Seki S, Ito H, Takeshima T, Hagiwara M, Morikawa T (1995) Chem Pharm Bull 43:1683
13. a) Trost BM, Ceschi MA, König B (1997) Angew Chem Int Ed Engl 36:1486; b) Martín M, Mas G, Urpí F, Vilarrasa J (1999) Angew Chem Int Ed 38:3068, (1999) Angew Chem 111:3274; c) Huang W, Kalivretenos AG (1995) Tetrahedron Lett 36:9113
14. Dutton FE, Lee BH (1998) Tetrahedron Lett 39:5313
15. Inanaga J, Hirata K, Hiroko S, Katsuki T, Yamaguchi M (1979) Bull Chem Soc Jpn 52: 1989
16. a) Derrer S, Feeder N, Teat SJ, Davies JE, Holmes AB (1998) Tetrahedron Lett 39:9309; b) Long DD, Stetz RJE, Nash RJ, Marguess DG, Lloyd JD, Winters AL, Asano N, Fleet GWJ (1999) J Chem Soc Perkin Trans 1:901; c) Herdeis C, Mohareb RM, Neder RB, Schwabenländer F, Telser J (1999) Tetrahedron Asymmetry 10:4521; d) Michaela Hamerníková M, Pakhomova S, Havlícek J, Hana Votavová H, Kefurt K (2000) Carbohydr Res 325:56
17. a) Bartra M, Vilarrasa J (1991) J Org Chem 56:5132; b) Bosch I, Romea P, Urpi F, Vilarrasa J (1993) Tetrahedron Lett 34:4671
18. Hasegawa M, Takabatake T (1991) J Heterocycl Chem 28:1079
19. Quinkert G, Nestler HP, Schumacher B, del Grosso M, Duerner G, Bats JW (1992) Tetrahedron Lett 33:1977
20. MaGee DI, Ramaseshan M, Leach JD, (1995) Can J Chem 73:2111
21. a) Robl JA, Cimarusti MP, Simpkins LM, Weller HN, Pan YY (1994) J Am Chem Soc 116:2348; b) Robl JA, Cimarusti MP, Simpkins LM, Brown B, Ryono DE (1996) J Med Chem 39:494; c) Robl JA (1994) Tetrahedron Lett 35:393

22. Enders D, Groebner R, Raabe G, Runsink J (1996) Synthesis 941
23. a) Fox ME, Holmes AB, Forbes IT, Thompson M (1992) Tetrahedron Lett 33:7421; b) Fox ME, Holmes AB, Forbes IT, Thompson M, Ziller JW (1992) Tetrahedron Lett 33:7425; c) Bourque E, Grenon M, Laliberté S, Deslongchamps P (1999) Synlett 1115
24. MW, Miller ML, Ray PS (1999) Tetrahedron 55:373
25. a) Johnson EP, Chen GP, Fales KR, Lenk BE, Szendroi RJ, Wang JW, Carlson JA (1995) J Org Chem 60:6595; b) Trost BM, Michellys PY, Gerusz VJ (1997) Angew Chem Int Ed Engl 36:1750
26. Zhang Z, Ojima I (1993) J Organomet Chem 454:281
27. a) Gill GB, Pattenden G, Reynolds SJ (1994) J Chem Soc Perkin Trans 1:369; b) Naito T, Honda Y, Miyata O, Ninomiya I (1995) J Chem Soc Perkin Trans 1:19
28. a) Lamas C, Saá C, Castedo L, Domínguez D (1992) Tetrahedron Lett 33:5653; b) Rodriguez G, Castedo L, Domínguez D, Saá C (1998) Tetrahedron Lett 39:6551
29. Lease TG, Shea KJ (1993) J Am Chem Soc 115:2248
30. For reviews see: Schuster M, Blechert S (1997) Angew Chem Int Ed Engl 36:2036; b) Fürstner A (2000) Angew Chem Int Ed 39:3012
31. a) Fink BE, Kym PR, Katzenellenbogen JA (1998) J Am Chem Soc 120:4334; b) Rutjes FPJT, Schoemaker HE (1997) Tetrahedron Lett 38:677
32. a) Houri AF, Xu Z, Cogan DA, Hoveyda AH (1995) J Am Chem Soc 117:2943; b) Xu Z, Johannes CW, Salman SS, Hoveyda AH (1996) J Am Chem Soc 118:10926; c) Xu Z, Johannes CW, Houri AF, La DS, Cogan DA, Hofilena GE, Hoveyda AH (1997) J Am Chem Soc 119:10302
33. a) Borer BC, Deerenberg S, Bieräugel H, Pandit UK (1994) Tetrahedron Lett 35:3191; b) Pandit UK, Borer BC Bieräugel H (1997) Pure & Appl Chem 68:659; c) Martin SF, Liao Y, Wong Y, Rein T (1994) Tetrahedron Lett 35:691; d) Martin SF, Chen HJ, Courtney AK, Liao Y, Pätzel M, Ramser MN, Wagman AS (1996) Tetrahedron 52:7251
34. Goldring WPD, Hodder AS, Weiler L (1998) Tetrahedron Lett 39:4955
35. a) Maarseveen JHv, den Hartog JAJ, Engelen V, Finner E, Kruse CG (1996) Tetrahedron Lett 37:8249; b) Miller SJ, Blackwell HE, Grubbs RH (1996) J Am Chem Soc 118:9606; c) Piscopio AD, Miller JF, Koch K (1997) Tetrahedron Lett 38:7143; d) Piscopio AD, Miller JF, Koch K (1998) Tetrahedron Lett 39:2667
36. Aubert T, Farnier M, Guilard R (1990) Synthesis 149
37. Alonso DA, Falvello LR, Mancheno B, Nájera C, Tomàs M (1996) J Org Chem 61:5004
38. Baker SR, Parsons AF, Wilson M (1998) Tetrahedron Lett 39:2815
39. a) Barluenga J, Tomas M, Rubio E, Lopez-Pelegrin JA, Garcia-Granda S, Pertierra P (1996) J Am Chem Soc 118:695; b) Barluenga J, Tomas M, Ballesteros A, Santamaria J, Carbajo RJ, López-Ortiz F, Garcia-Granda S, Pertierra P (1996) Chem Europ J 2:88; c) Barluenga J, Tomas M, Ballesteros A, Santamaria, J, Suarez-Sobrino A (1997) J Org Chem 62:9229
40. Hansen DW, Peterson KB, Trivedi M, Kramer SW, Webber RK, Tjoeng FS, Moore WM, Jerome GM, Kornmeier CM, Manning PT, Connor JR, Misko TP, Currie MG, Pitzele BS (1998) J Med Chem 41:1361
41. a) Mukaiyama T, Harada T (1991) Chem Lett 1653; b) Kusama H, Yamashita Y, Narasaka K (1995) Bull Chem Soc Jpn 68:373; c) Laurent A, Jacquault P, Di Martino JL, Hamelin J (1995) J Chem .Soc Chem Commun 1101; d) Baldwin JE, Vollmer HE, Lee V (1999) Tetrahedron Lett 40:5401
42. Olah GA, Fung AP (1979) Synthesis 473
43. a) Lyga JW (1996) J Heterocycl Chem 33:1631; b) Mphahlele MJ, Modro TA (1998) J Chem Res Synop 198; c) Wu MH, Jacobsen EN (1997) Tetrahedron Lett 38:1693
44. a) Gribble GW, Silva RA (1996) Tetrahedron Lett 37:2145; b) Sainsbury M, Mahon MF, Williams CS, Naylor A, Scopes DIC (1991) Tetrahedron 47:4195; c) Diedrichs N, Westermann B (1999) Synlett 1127
45. a) Olson GL, Voss ME, Hill DE, Kahn M, Madison VS, Cook CM (1990) J Am Chem Soc 112:323; b) Komatsu N, Simizu S, Sugita T (1992) Synth Commun 22:277
46. a) Toda F, Akai H (1990) J Org Chem 55:4973; b) Toda F, Mori K, Matsuura Y, Akai H (1990) J Chem Soc Chem Commun 1591

47. a) Kalgutkar AS, Crews BC, Marnett LJ (1996) J Med Chem 39:1692; b) Hoffmann RV, Salvador JM (1991) Tetrahedron Lett 32:2429; c) Vedejs E, Sano H (1992) Tetrahedron Lett 33:3261; d) Evans PA, Modi DP (1995) J Org Chem 60:6662
48. a) MacPherson LJ, Bayburt EK, Capparelli MP, Bohacek RS, Clarke FH, Ghai RD, Sakane Y, Berry CJ, Peppard JV, Trapani AJ (1993) J Med Chem 36:3821; b) Stanton JL, Sperbeck DM, Trapani AJ, Cote D, Sakane Y, Berry CJ, Ghai RD (1993) J Med Chem 36:3829; c) Aubé J, Milligan GL, Mossman CJ (1992) J Org Chem 57:1635; d) DeVita RJ, Goulet MT, Wyvratt MJ, Fisher MH, Lo JL, Yang YT, Cheng K, Roy G (1999) Bioorg Med Chem Lett 9:2621
49. a) Gracias V, Milligan GL, Aubé J (1996) J Org Chem 61:10; b) Gracias V, Frank KE, Milligan GL, Aubé J (1997) Tetrahedron 53:16241; c) Forsee JE, Smith BT, Frank KE, Aubé J (1998) Syn Lett 1258 d) Forsee JE, Aubé J (1999) J Org Chem 64:4381
50. a) Milligan GL, Mossman CJ, Aubé J (1995) J Am Chem Soc 117:10449; b) LeDreau MA, Desmaële D, Dumas F, d'Angelo J (1993) J Org Chem 58:2933; c) Mossman CJ, Aubé J (1996) Tetrahedron 52:3403
51. a) Aubé J, Wang Y, Hammond M, Tanol M, Takusagawa F, Vander Velde D (1990) J Am Chem Soc 112:4879; b) Aubé J, Hammond M, Gherardini E, Takusagawa F (1991) J Org Chem 56:499; c) Kitagawa O, Vander Velde D, Dutta D, Morton M, Takusagawa F, Aubé J (1995) J Am Chem Soc 117:5169; d) Wolfe MS, Dutta D, Aubé J (1997) J Org Chem 62:654; e) Suda K, Sashima M, Izutsu M, Hino F (1994) J Chem Soc Chem Commun 949
52. Endo Y, Kataoka K, Haga N, Shudo K (1992) Tetrahedron Lett 33:3339
53. a) Evans PA, Holmes AB, Russell K (1990) Tetrahedron Asymmetry 1:593; b) Evans PA, Holmes AB, Russell K (1992) Tetrahedron Lett 33:6857; c) Evans PA, Holmes AB, Russell K (1994) J Chem Soc Perkin Trans 1:3397; d) Evans PA, Holmes AB, Collins I, Raithby PR, Russell K (1995) J Chem Soc Chem Commun 2325; e) Evans PA, Holmes AB, McGeary RP, Nadin A, Russell K, O'Hanlon PJ, Pearson ND (1996) J Chem Soc Perkin Trans 1:123; f) Evans PA, Collins I, Hamley P, Holmes AB, Raithby PR, Russel K (1992) Tetrahedron Lett 33:6859
54. Müller P, Imogai H (1999) Helv Chim Acta 82:315
55. Lindstroem UM, Somfai P (1997) J Am Chem Soc 119:8385
56. a) Suh YG, Lee JY, Kim SA, Jung JK (1996) Synth Commun 26:1675; b) Suh YG, Kim SA, Jung JK, Shin DY, Min KH, Koo BA, Kim HS (1999) Angew Chem Int Ed 38:3545
57. a) Edstrom ED (1991) J Am Chem Soc 113:6690; b) Edstrom ED (1991) Tetrahedron Lett 32:5709
58. a) Diederich M, Nubbemeyer U (1995) Angew Chem Int Ed Engl 34:1026; b) Diederich M, Nubbemeyer U (1996) Chem Europ J 2:894; c) Sudau A, Nubbemeyer U (1998) Angew Chem Int Ed Engl 37:1140; d) Laabs S, Scherrmann A, Sudau A, Diederich M, Kierig C, Nubbemeyer U (1999) Synlett 25; e) Sudau A, Münch W, Nubbemeyer U, Bats JW (2000) J Org Chem 65:1710
59. a) White DNJ, Bovill MJ (1977) J Chem Soc Perkin Trans II 1610; b) Guella G, Chiasera G, N'Diaye I, Pietra F (1994) Helv Chim Acta 77:1203; c) Deiters A, Mück-Lichtenfeld C, Fröhlich R, Hoppe D (2000) Org Lett 2:2415
60. a) Sudau A, Münch W, Nubbemeyer U, Bats JW (2001) Chem Eur J 7:611; b) Diederich M, Nubbemeyer U (1999) Synthesis 286
61. Deur CJ, Miller MW, Hegedus LS (1996) J Org Chem 61:2871
62. Lee J, Ha, JD, Cha JK (1997) J Am Chem Soc 119:8127
63. Kraft A (1999) J Chem Soc Perkin Trans I 705
64. Sofan MA, Mashaly MM, El-Shamy NY, El-Hossini MS (1997) Pol J Chem 71:196
65. a) Franzisket L, Heesing A (1991) Chem Ber 124:635; b) Sano T, Horiguchi Y, Tsuda Y (1990) Chem Pharm Bull 38:3283; c) Sano T, Enomoto H, Kurebayashi Y, Horiguchi Y, Tsuda Y (1993) Chem Pharm Bull 41:471; d) Fu TY, Scheffer JR, Trotter J (1994) Can J Chem 72:1952; e) Kollenz G, Terpetschnig E, Sterk H, Peters K, Peters EM (1999) Tetrahedron 55:2973
66. a) Lamara K, Smalley RK (1991) Tetrahedron 47:2277; b) Lamara K, Redhouse AD, Smalley RK, Thompson JR (1994) Tetrahedron 50:5515

67. a) Al-Jalal NA (1990) J Heterocycl Chem 27:1323; b) Al-Jalal NA (1995) J Chem Res Synop 256
68. a) Ognyanov VI, Hesse M (1990) Helv Chim Acta 73:272; b) Koch T, Hesse M (1992) Synthesis 931; c) Koch T, Hesse M (1994) Helv Chim Acta 77:819; d) Ivancev W, Ognyanov VI, Hesse M (1996) Tetrahedron 52:4363
69. a) Benati L, Nanni D, Sangiorgi C, Spagnolo P (1999) J Org Chem 64:7836; b) Kim S, Joe GH, Do JY (1993) J Am Chem Soc 115:3328
70. Wipf P, Li W (1999) J Org Chem 64:4576

Author Index Volume 201 – 216

Author Index Vols. 26 – 50 see Vol. 50
Author Index Vols. 51 – 100 see Vol. 100
Author Index Vols. 101 – 150 see Vol. 150
Author Index Vols. 151 – 200 see Vol. 200

The volume numbers are printed in italics

Albert M, see Dax K (2001) *215*:193 – 275
Angyal SJ (2001) The Lobry de Bruyn-Alberda van Ekenstein Transformation and Related Reactions. *215*:1 – 14
Astruc D, Blais J-C, Cloutet E, Djakovitch L, Rigaut S, Ruiz J, Sartor V, Valério C (2000) The First Organometallic Dendrimers: Design and Redox Functions. *210*:229 – 259
Augé J, see Lubineau A (1999) *206*:1 – 39
Baars MWPL, Meijer EW (2000) Host-Guest Chemistry of Dendritic Molecules. *210*:131 – 182
Ballauff M (2001) Structure of Dendrimers in Dilute Solution. *212*:177 – 194
Baltzer L (1999) Functionalization and Properties of Designed Folded Polypeptides. *202*:39 – 76
Bartlett RJ, see Sun J-Q (1999) *203*:121 – 145
Betzemeier B, Knochel P (1999) Perfluorinated Solvents – a Novel Reaction Medium in Organic Chemistry. *206*:61 – 78
Blais J-C, see Astruc D (2000) *210*:229 – 259
Bogár F, see Pipek J (1999) *203*:43 – 61
Brand SC, see Haley MM (1999) *201*:81 – 129
Bray KL (2001) High Pressure Probes of Electronic Structure and Luminescence Properties of Transition Metal and Lanthanide Systems. *213*:1 – 94
Bunz UHF (1999) Carbon-Rich Molecular Objects from Multiply Ethynylated π-Complexes. *201*:131 – 161
Chamberlin AR, see Gilmore MA (1999) *202*:77 – 99
Cloutet E, see Astruc D (2000) *210*:229 – 259
Cooper DL, see Raimondi M (1999) *203*:105 – 120
Cornils B (1999) Modern Solvent Systems in Industrial Homogeneous Catalysis. *206*:133 – 152
Crooks RM, Lemon III BI, Yeung LK, Zhao M (2001) Dendrimer-Encapsulated Metals and Semiconductors: Synthesis, Characterization, and Applications. *212*:81 – 135
Croteau R, see Davis EM (2000) *209*:53 – 95
Curran DP, see Maul JJ (1999) *206*:79 – 105
Davis EM, Croteau R (2000) Cyclization Enzymes in the Biosynthesis of Monoterpenes, Sesquiterpenes and Diterpenes. *209*:53 – 95
Dax K, Albert M (2001) Rearrangements in the Course of Nucleophilic Substitution Reactions. *215*:193 – 275
de la Plata BC, see Ruano JLG (1999) *204*:1 – 126
de Meijere A, Kozhushkov SI (1999) Macrocyclic Structurally Homoconjugated Oligoacetylenes: Acetylene- and Diacetylene-Expanded Cycloalkanes and Rotanes. *201*:1 – 42
de Meijere A, Kozhushkov SI, Khlebnikov AF (2000) Bicyclopropylidene – A Unique Tetrasubstituted Alkene and a Versatile C_6-Building Block. *207*:89 – 147
de Meijere A, Kozhushkov SI, Hadjiaraoglou LP (2000) Alkyl 2-Chloro-2-cyclopropylideneacetates – Remarkably Versatile Building Blocks for Organic Synthesis. *207*:149 – 227
de Raadt A, Fechter MH (2001) Miscellaneous. *215*:327 – 345
Diederich F, Gobbi L (1999) Cyclic and Linear Acetylenic Molecular Scaffolding. *201*:43 – 79

Diederich F, see Smith DK (2000) *210*:183 – 227
Djakovitch L, see Astruc D (2000) *210*:229 – 259
Donges D, see Yersin H (2001) *214*:81 – 186
Dormán G (2000) Photoaffinity Labeling in Biological Signal Transduction. *211*:169 – 225
Drabowicz J, Mikołajczyk M (2000) Selenium at Higher Oxidation States. *208*:143 – 176
Eder B, see Wrodnigg TM (2001) The Amadori and Heyns Rearrangements: Landmarks in the History of Carbohydrate Chemistry or Unrecognized Synthetic Opportunities? *215*: 115 – 175
Famulok M, Jenne A (1999) Catalysis Based on Nucleid Acid Structures. *202*:101 – 131
Fechter MH, see de Raadt A (2001) *215*:327 – 345
Ferrier RJ (2001) Substitution-with-Allylic-Rearrangement Reactions of Glycal Derivatives. *215*:153 – 175
Ferrier RJ (2001) Direct Conversion of 5,6-Unsaturated Hexopyranosyl Compounds to Functionalized Glycohexanones. *215*:277 – 291
Frey H, Schlenk C (2000) Silicon-Based Dendrimers. *210*:69 – 129
Furukawa N, Sato S (1999) New Aspects of Hypervalent Organosulfur Compounds. *205*: 89 – 129
Gamelin DR, Güdel HU (2001) Upconversion Processes in Transition Metal and Rare Earth Metal Systems. *214*:1 – 56
Gilmore MA, Steward LE, Chamberlin AR (1999) Incorporation of Noncoded Amino Acids by In Vitro Protein Biosynthesis. *202*:77 – 99
Glasbeek M (2001) Excited State Spectroscopy and Excited State Dynamics of Rh(III) and Pd(II) Chelates as Studied by Optically Detected Magnetic Resonance Techniques. *213*: 95 – 142
Glass RS (1999) Sulfur Radical Cations. *205*:1 – 87
Gobbi L, see Diederich F (1999) *201*:43 – 129
Güdel HU, see Gamelin DR (2001) *214*:1 – 56
Hackmann-Schlichter N, see Krause W (2000) *210*:261 – 308
Hadjiaraoglou LP, see de Meijere A (2000) *207*:149 – 227
Häusler H, Stütz AE (2001) D-Xylose (D-Glucose) Isomerase and Related Enzymes in Carbohydrate Synthesis. 215:77 – 114
Haley MM, Pak JJ, Brand SC (1999) Macrocyclic Oligo(phenylacetylenes) and Oligo(phenyldiacetylenes). *201*:81 – 129
Hartmann T, Ober D (2000) Biosynthesis and Metabolism of Pyrrolizidine Alkaloids in Plants and Specialized Insect Herbivores. *209*:207 – 243
Hassner A, see Namboothiri INN (2001) *216*: 1 – 49
Hemscheidt T (2000) Tropane and Related Alkaloids. *209*:175 – 206
Hergenrother PJ, Martin SF (2000) Phosphatidylcholine-Preferring Phospholipase C from *B. cereus*. Function, Structure, and Mechanism. *211*:131 – 167
Hermann C, see Kuhlmann J (2000) *211*:61 – 116
Hricovíniová Z, see Petruš L (2001) *215*:15 – 41
Iwaoka M, Tomoda S (2000) Nucleophilic Selenium. *208*:55 – 80
Iwasawa N, Narasaka K (2000) Transition Metal Promated Ring Expansion of Alkynyl- and Propadienylcyclopropanes. *207*:69 – 88
Imperiali B, McDonnell KA, Shogren-Knaak M (1999) Design and Construction of Novel Peptides and Proteins by Tailored Incorparation of Coenzyme Functionality. *202*:1 – 38
Jenne A, see Famulok M (1999) *202*:101 – 131
Kato S, see Murai T (2000) *208*:177 – 199
Khlebnikov AF, see de Meijere A (2000) *207*:89 – 147
Kirtman B (1999) Local Space Approximation Methods for Correlated Electronic Structure Calculations in Large Delocalized Systems that are Locally Perturbed. *203*:147 – 166
Klopper W, Kutzelnigg W, Müller H, Noga J, Vogtner S (1999) Extremal Electron Pairs – Application to Electron Correlation, Especially the R12 Method. *203*:21 – 42
Knochel P, see Betzemeier B (1999) *206*:61 – 78
Kozhushkov SI, see de Meijere A (1999) *201*:1 – 42

Kozhushkov SI, see de Meijere A (2000) *207*:89 – 147
Kozhushkov SI, see de Meijere A (2000) *207*:149 – 227
Krause W, Hackmann-Schlichter N, Maier FK, Müller R (2000) Dendrimers in Diagnostics. *210*:261 – 308
Kuhlmann J, Herrmann C (2000) Biophysical Characterization of the Ras Protein. *211*:61 – 116
Kunkely H, see Vogler A (2001) *213*:143 – 182
Kutzelnigg W, see Klopper W (1999) *203*:21 – 42
Leitner W (1999) Reactions in Supercritical Carbon Dioxide ($scCO_2$). *206*:107 – 132
Lemon III BI, see Crooks RM (2001) *212*:81 – 135
Levitzki A (2000) Protein Tyrosine Kinase Inhibitors as Therapeutic Agents. *211*:1 – 15
Li X, see Paldus J (1999) *203*:1 – 20
Linclau B, see Maul JJ (1999) *206*:79 – 105
Lubineau A, Augé J (1999) Water as Solvent in Organic Synthesis. *206*:1 – 39
Lundt I, Madsen R (2001) Synthetically Useful Base Induced Rearrangements of Aldonolactones. *215*:177 – 191
Loupy A (1999) Solvent-Free Reactions. *206*:153 – 207
Madsen R, see Lundt I (2001) *215*:177 – 191
Maier FK, see Krause W (2000) *210*:261 – 308
March NH (1999) Localization via Density Functionals. *203*:201 – 230
Martin SF, see Hergenrother PJ (2000) *211*:131 – 167
Maul JJ, Ostrowski PJ, Ublacker GA, Linclau B, Curran DP (1999) Benzotrifluoride and Derivates: Useful Solvents for Organic Synthesis and Fluorous Synthesis. *206*:79 – 105
McDonnell KA, see Imperiali B (1999) *202*:1 – 38
Meijer EW, see Baars MWPL (2000) *210*:131 – 182
Metzner P (1999) Thiocarbonyl Compounds as Specific Tools for Organic Synthesis. *204*:127 – 181
Mezey PG (1999) Local Electron Densities and Functional Groups in Quantum Chemistry. *203*:167 – 186
Mikołajczyk M, see Drabowicz J (2000) *208*:143 – 176
Möller M, see Sheiko SS (2001) *212*:137 – 175
Müllen K, see Wiesler U-M (2001) *212*:1 – 40
Müller G (2000) Peptidomimetic SH2 Domain Antagonists for Targeting Signal Transduction. *211*:17 – 59
Müller H, see Klopper W (1999) *203*:21 – 42
Müller R, see Krause W (2000) *210*:261 – 308
Murai T, Kato S (2000) Selenocarbonyls. *208*:177 – 199
Muscat D, van Benthem RATM (2001) Hyperbranched Polyesteramides – New Dendritic Polymers. *212*:41 – 80
Nakayama J, Sugihara Y (1999) Chemistry of Thiophene 1,1-Dioxides. *205*:131 – 195
Namboothiri INN, Hassner A (2001) Stereoselective Intramolecular 1,3-Dipolar Cycloadditions. *216*: 1 – 49
Narasaka K, see Iwasawa N (2000) *207*:69 – 88
Nishibayashi Y, Uemura S (2000) Selenoxide Elimination and [2,3] Sigmatropic Rearrangements. *208*:201 – 233
Nishibayashi Y, Uemura S (2000) Selenium Compounds as Ligands and Catalysts. *208*: 235 – 255
Noga J, see Klopper W (1999) *203*:21 – 42
Nubbemeyer U (2001) Synthesis of Medium-Sized Ring Lactams. *216*: 125 – 196
Nummelin S, Skrifvars M, Rissanen K (2000) Polyester and Ester Functionalized Dendrimers. *210*:1 – 67
Ober D, see Hemscheidt T (2000) *209*:175 – 206
Osanai S (2001) Nickel (II) Catalyzed Rearrangements of Free Sugars. *215*:43 – 76
Ostrowski PJ, see Maul JJ (1999) *206*:79 – 105
Pak JJ, see Haley MM (1999) *201*:81 – 129

Paldus J, Li X (1999) Electron Correlation in Small Molecules: Grafting CI onto CC. *203*: 1–20
Paulmier C, see Ponthieux S (2000) *208*:113–142
Petruš L, Petrušová M, Hricovíniová Z (2001) The Bílik Reaction. *215*:15–41
Petrušová M, see Petruš L (2001) *215*:15–41
Pipek J, Bogár F (1999) Many-Body Perturbation Theory with Localized Orbitals – Kapuy's Approach. *203*:43–61
Ponthieux S, Paulmier C (2000) Selenium-Stabilized Carbanions. *208*:113–142
Raimondi M, Cooper DL (1999) Ab Initio Modern Valence Bond Theory. *203*:105–120
Renaud P (2000) Radical Reactions Using Selenium Precursors. *208*:81–112
Rigaut S, see Astruc D (2000) *210*:229–259
Riley MJ (2001) Geometric and Electronic Information From the Spectroscopy of Six-Coordinate Copper(II) Compounds. *214*:57–80
Rissanen K, see Nummelin S (2000) *210*:1–67
Røeggen I (1999) Extended Geminal Models. *203*:89–103
Ruano JLG, de la Plata BC (1999) Asymmetric [4+2] Cycloadditions Mediated by Sulfoxides. *204*:1–126
Ruiz J, see Astruc D (2000) *210*:229–259
Rychnovsky SD, see Sinz CJ (2001) *216*: 51–92
Salaün J (2000) Cyclopropane Derivates and their Diverse Biological Activities. *207*:1–67
Sanz-Cervera JF, see Williams RM (2000) *209*:97–173
Sartor V, see Astruc D (2000) *210*:229–259
Sato S, see Furukawa N (1999) *205*:89–129
Scherf U (1999) Oligo- and Polyarylenes, Oligo- and Polyarylenevinylenes. *201*:163–222
Schlenk C, see Frey H (2000) *210*:69–129
Sheiko SS, Möller M (2001) Hyperbranched Macromolecules: Soft Particles with Adjustable Shape and Capability to Persistent Motion. *212*:137–175
Shen B (2000) The Biosynthesis of Aromatic Polyketides. *209*:1–51
Shogren-Knaak M, see Imperiali B (1999) *202*:1–38
Sinou D (1999) Metal Catalysis in Water. *206*:41–59
Sinz CJ, Rychnovsky SD (2001) 4-Acetoxy- and 4-Cyano-1,3-dioxanes in Synthesis. *216*: 51–92
Skrifvars M, see Nummelin S (2000) *210*:1–67
Smith DK, Diederich F (2000) Supramolecular Dendrimer Chemistry – A Journey Through the Branched Architecture. *210*:183–227
Steward LE, see Gilmore MA (1999) *202*:77–99
Stocking EM, see Williams RM (2000) *209*:97–173
Stütz AE , see Häusler H (2001) 215:77–114
Sugihara Y, see Nakayama J (1999) *205*:131–195
Sun J-Q, Bartlett RJ (1999) Modern Correlation Theories for Extended, Periodic Systems. *203*:121–145
Sun L, see Crooks RM (2001) *212*:81–135
Surján PR (1999) An Introduction to the Theory of Geminals. *203*:63–88
ten Holte P, see Zwanenburg B (2001) *216*: 93–124
Thiem J, see Werschkun B (2001) *215*:293–325
Thutewohl M, see Waldmann H (2000) *211*:117–130
Tiecco M (2000) Electrophilic Selenium, Selenocyclizations. *208*:7–54
Tomoda S, see Iwaoka M (2000) *208*:55–80
Ublacker GA, see Maul JJ (1999) *206*:79–105
Uemura S, see Nishibayashi Y (2000) *208*:201–233
Uemura S, see Nishibayashi Y (2000) *208*:235–255
Valdemoro C (1999) Electron Correlation and Reduced Density Matrices. *203*:187–200
Valério C, see Astruc D (2000) *210*:229–259
van Benthem RATM, see Muscat D (2001) *212*:41–80
Vogler A, Kunkely H (2001) Luminescent Metal Complexes: Diversity of Excited States. *213*: 143–182

Vogtner S, see Klopper W (1999) *203*:21 – 42
Waldmann H, Thutewohl M (2000) Ras-Farnesyltransferase-Inhibitors as Promising Anti-Tumor Drugs. *211*:117 – 130
Weil T, see Wiesler U-M (2001) *212*:1 – 40
Werschkun B, Thiem J (2001) Claisen Rearrangements in Carbohydrate Chemistry. *215*: 293 – 325
Wiesler U-M, Weil T, Müllen K (2001) Nanosized Polyphenylene Dendrimers. *212*:1 – 40
Williams RM, Stocking EM, Sanz-Cervera JF (2000) Biosynthesis of Prenylated Alkaloids Derived from Tryptophan. *209*:97 – 173
Wirth T (2000) Introduction and General Aspects. *208*:1 – 5
Wrodnigg TM, Eder B (2001) The Amadori and Heyns Rearrangements: Landmarks in the History of Carbohydrate Chemistry or Unrecognized Synthetic Opportunities? *215*: 115 – 175
Yersin H, Donges D (2001) Low-Lying Electronic States and Photophysical Properties of Organometallic Pd(II) and Pt(II) Compounds. Modern Research Trends Presented in Detailed Case Studies. *214*:81 – 186
Yeung LK, see Crooks RM (2001) *212*:81 – 135
Zhao M, see Crooks RM (2001) *212*:81 – 135
Zwanenburg B, ten Holte P (2001) The Synthetic Potential of Three-Membered Ring Aza-Heterocycles. *216*: 93 – 124

Printing (Computer to Film): Saladruck, Berlin
Binding: Stürtz AG, Würzburg